CONTESTING WORLD SOCIOECONOMIC RIGHTS A JUSTICE MOVEMEN

What do equality, dignity and rights mean in a world where eight men own as much wealth as half the world's population? *Contesting World Order? Socioeconomic Rights and Global Justice Movements* examines how global justice movements have engaged the language of socioeconomic rights to contest global institutional structures and rules responsible for contributing to the persistence of severe poverty. Drawing upon perspectives from critical international relations studies and the activities of global justice movements, this book evaluates the 'counter-hegemonic' potential of socioeconomic rights discourse and its capacity to contribute towards an alternative to the prevailing neo-liberal 'common sense' of global governance.

JOE WILLS is a lecturer at the School of Law, University of Leicester, where he conducts research in the fields of human rights, animal rights and legal and political theory. He is one of eight researchers on a British Academy-funded project investigating how the United Kingdom and South Africa compensate private losses resulting from failures to give effect to the special duties human and constitutional rights impose on public authorities. He has also contributed to an Academy of Finland and University of Turku-funded project entitled 'Imagining Post-Neoliberal Regulatory Subjectivities'. He has published in the *Leiden Journal of International Law* and the *Indiana Journal of Global Legal Studies*.

GLOBALIZATION AND HUMAN RIGHTS

Series Editors
Malcolm Langford
César Rodríguez-Garavito

Forthcoming Books in the Series

César Rodríguez-Garavito, *Business and Human Rights: Beyond the End of the Beginning*

Jeremy Perelman, *The Rights-ification of Development: Global Poverty, Human Rights, and Globalization in the Post-Washington Consensus*

Katherine G. Young, *The Future of Economic and Social Rights*

CONTESTING WORLD ORDER? SOCIOECONOMIC RIGHTS AND GLOBAL JUSTICE MOVEMENTS

JOE J. WILLS
University of Leicester

CAMBRIDGE
UNIVERSITY PRESS

University Printing House, Cambridge CB2 8BS, United Kingdom

One Liberty Plaza, 20th Floor, New York, NY 10006, USA

477 Williamstown Road, Port Melbourne, VIC 3207, Australia

314-321, 3rd Floor, Plot 3, Splendor Forum, Jasola District Centre, New Delhi - 110025, India

79 Anson Road, #06-04/06, Singapore 079906

Cambridge University Press is part of the University of Cambridge.

It furthers the University's mission by disseminating knowledge in the pursuit of education, learning and research at the highest international levels of excellence.

www.cambridge.org
Information on this title: www.cambridge.org/9781316628249
DOI: 10.1017/9781316809921

© Joe Wills 2017

This publication is in copyright. Subject to statutory exception
and to the provisions of relevant collective licensing agreements,
no reproduction of any part may take place without the written
permission of Cambridge University Press.

First published 2017
First paperback edition 2018

A catalogue record for this publication is available from the British Library

ISBN 978-1-107-17614-0 Hardback
ISBN 978-1-316-62824-9 Paperback

Cambridge University Press has no responsibility for the persistence or accuracy of URLs for external or third-party internet websites referred to in this publication, and does not guarantee that any content on such websites is, or will remain, accurate or appropriate.

CONTENTS

Acknowledgements   ix
List of Abbreviations   x

Introduction   1
Pathologies of Power   2
Accumulation and Its Dispossessed   5
'New Rights Advocacy'   6
Global Justice Movements   9
A Neo-Gramscian Framework   10
Frameworks, Discourses and Ideologies   11
Structure of the Book   13

1 **Power, Hegemony and World Order**   16
   1.1 Introduction   16
   1.2 Hegemony and Counter-Hegemony   17
       1.2.1 Structure and Superstructure: The Historic Bloc   18
       1.2.2 Political and Civil Society: The Integral State   20
       1.2.3 Securing Hegemony   22
       1.2.4 Counter-Hegemony   24
   1.3 Neo-Liberal Globalisation and Global Hegemony   25
       1.3.1 What Is Neo-Liberalism?   26
       1.3.2 From Intellectual Movement to State Power   27
       1.3.3 From the Nation State to Global Governance:
              The Globalisation of Neo-Liberalism   29
       1.3.4 Global Hegemony and the Global Integral State   31
       1.3.5 The End of Neo-Liberalism? The Aftermath of the Global
              Financial Crisis   40
   1.4 The Prospects for Counter-Hegemonic Global Justice
       Movements   42
       1.4.1 Sites of Resistance   45
       1.4.2 Obstacles to Building Global Counter-Hegemony   46
   1.5 Conclusion   47

2 **Neo-Liberal Globalisation and Socioeconomic Rights: An Overview**   49
  2.1 Introduction   49
  2.2 Human Rights: A Neo-Gramscian Framework   50
    2.2.1 What Are Human Rights?   50
    2.2.2 Human Rights and Social Change   55
  2.3 Socioeconomic Rights and Neo-Liberalism   61
    2.3.1 The Meaning of Socioeconomic Rights   61
    2.3.2 Socioeconomic Rights and Neo-Liberalism: Discursive Tensions   62
  2.4 The Development of Socioeconomic Rights under International Law   67
    2.4.1 'In from the Cold': The Clarification of Socioeconomic Rights Standards   68
    2.4.2 The UN Human Rights Framework and Globalisation   74
  2.5 Critically Evaluating the Counter-Hegemonic Potential of Socioeconomic Rights   79
    2.5.1 Potential Strengths of Socioeconomic Rights Discourse   80
    2.5.2 Possible Limitations of Socioeconomic Rights Discourse   85
  2.6 Conclusion   93

3 **Food Security vs. Food Sovereignty: The Right to Food and Global Hunger**   94
  3.1 Introduction: The Right to Food and World Hunger   94
  3.2 The Political-Institutional Setting: The Neo-Liberal Food Regime   96
    3.2.1 Criticisms of the Neo-Liberal Food Regime   98
  3.3 The Politics of Food: Food Security vs. Food Sovereignty   101
    3.3.1 The Neo-Liberal Ethico-Political Framework: Food Security   102
    3.3.2 The Counter-Hegemonic Ethico-Political Framework: Food Sovereignty   105
  3.4 Discursive Contestation over the Right to Food under International Law   109
    3.4.1 The Legal Basis for the Right to Food   110
    3.4.2 The World Food Summit, Rome 1996   115
    3.4.3 The Draft International Code of Conduct on the Human Right to Adequate Food   118
    3.4.4 General Comment 12   120
    3.4.5 The World Food Summit: Five Years Later   122
    3.4.6 The Voluntary Guidelines on the Right to Food   126
    3.4.7 The Right to Food after the Voluntary Guidelines: The World Food Crisis of 2008   137

        3.4.8 The Right to Food and La Via Campesina's Campaigns against Land Grabs   140
        3.4.9 Limitations of the Right to Food and the Struggle for 'Peasant Rights'   142
    3.5 Food Sovereignty, the Right to Food and Counter-Hegemonic Strategy   144
    3.6 Conclusion   149

4 **Intellectual Property, the Right to Health and the Global Access to Medicines Campaign   151**
    4.1 Introduction   151
    4.2 The Political-Institutional Context: The Global Intellectual Property Regime   154
        4.2.1 Criticisms of the Global IPR Regime in Relation to Access to Medicines: 'A Tragic Double Jeopardy'   157
    4.3 The Politics of Knowledge: Ownership vs. Access   158
        4.3.1 The Neo-Liberal Ethico-Political Framework: Intellectual Property Rights   159
        4.3.2 The Alternative Ethico-Political Framework: Access to Medicines   163
    4.4 The Right to Health and Access to Medicines Campaign   166
        4.4.1 The Right to Health as Oppositional Frame   169
        4.4.2 The Right to Health and the Reform of TRIPS   179
        4.4.3 After Doha: The Right to Health in the 'TRIPS Plus' Era   183
        4.4.4 Beyond TRIPS: The Right to Health and Alternatives to IP   188
    4.5 Conclusion   193

5 **A Commodity or a Right? Evoking the Human Right to Water to Challenge Neo-Liberal Water Governance   196**
    5.1 Introduction   196
    5.2 The Political-Institutional Framework: The Neo-Liberal Water Regime   198
        5.2.1 Criticisms of the Neo-Liberal Water Regime   200
    5.3 Duelling Visions: 'Water as Commodity' vs. 'Water as Commons'   201
        5.3.1 The Neo-Liberal Discursive Framework: Water as a Commodity   201
        5.3.2 The Counter-Hegemonic Discursive Framework: Water as 'Commons'   204
    5.4 The Right to Water   208
        5.4.1 The Right to Water in the Twentieth Century   208
        5.4.2 The Neo-Liberalisation of the Right to Water   210
        5.4.3 The Emergence of the Right to Water as Oppositional Frame   212

5.5 The United Nations and the Right to Water 216
  5.5.1 The United Nations and the Right to Water: Background 217
  5.5.2 The Preliminary Discussions on the General Comment on the Right to Water 217
  5.5.3 General Comment 15 219
  5.5.4 Developments in Relation to the Right to Water since GC15 222
5.6 (Re)Incorporating the Right to Water into the Neo-Liberal Framework 224
  5.6.1 The World Bank 225
  5.6.2 The World Water Council 228
5.7 Case Study: The Right to Water in South Africa 232
  5.7.1 Background to the Free Basic Water Policy 232
  5.7.2 The Free Basic Water Policy 234
  5.7.3 The Free Basic Water Policy Contested: The Mazibuko Ruling 236
5.8 The Right to Water as a Counter-Hegemonic Strategy 239
5.9 Conclusion 245

Conclusion 248

C.1 A Tripartite Model of Counter-Hegemonic Socioeconomic Rights Praxis 252
  C.1.1 Tactical Participation in Inter-governmental Settings 252
  C.1.2 Invoking the Jurisprudence of International Human Rights Bodies 254
  C.1.3 Building Subaltern Counterpublics 256

Bibliography 260
Index 289

ACKNOWLEDGEMENTS

This book arose out of my doctoral thesis and as such I would like to thank my supervisors, Loveday Hodson and Paul O' Connell, for all of their assistance and help throughout the time that it has taken to research and write this thesis. They have inspired me through their own work, assisted in my intellectual development, and provided me with much needed moral encouragement. I am also immensely grateful to William Twinning, Mark Bell, Sally Kyd, José Miola and Chris Clarkson, Margot Salomon, Jill Marshall, François du Bois and Anna Grear for their wonderful support and encouragement. I would also like to thank my friends and family for their incredible help, especially Annie, Frank, Laura, Sammy, Ruth, Lina and Paula.

ABBREVIATIONS

A2K	Access to Knowledge
ABRANDH	Brazilian Action for Nutrition and Human Rights
ALBA	Bolivarian Alliance for the Peoples of Our America
ANC	African National Congress
AoA	Agreement on Agriculture
APF	Anti-Privatization Forum
ARV	Anti-Retroviral
ASEAN	Association of Southeast Asian States
BIT	Bilateral Investment Treaty
BTA	Bilateral Trade Agreement
CAFTA	Central American Free Trade Agreement
CALS	Centre for Applied Legal Studies
CAWP	Coalition against Water Privatization
CEDAW	Convention on the Elimination of All Forms of Discrimination Against Women
CESCR	Committee on Economic, Social and Cultural Rights
CEWG	Consultative Expert Working Group on Research and Development
CFS	Committee on World Food Security
CIPIH	Commission on Intellectual Property Rights and Public Health
COHRE	Centre on Housing Rights and Evictions
CPTech	Consumer Project on Technology
CRC	Convention on the Rights of the Child
CSM	Civil Society Mechanism of the FAO Committee on World Food Security
ECJ	European Court of Justice
EU	European Union
FAO	Food and Agriculture Organization
FBW	Free Basic Water
FIAN	Food First Information and Action Network
FTA	Free Trade Agreement
FTAA	Free Trade Area of the Americas
G8	Group of 8

GATS	General Agreement on Trade in Services
GATT	General Agreement on Tariffs and Trade
GC	UN Global Compact
GMO	Genetically Modified Organism
HAI	Health Action International
ICCPR	International Covenant on Civil and Political Rights
ICESCR	International Covenant on Economic, Social and Cultural Rights
ICSID	International Centre for the Settlement of Investment Disputes
IGO	Inter-governmental Organisation
IGWG	Inter-governmental World Group
IFG	International Forum on Globalisation
IFI	International Financial Institution
IPC	International Planning Committee for Food Sovereignty
IPR	Intellectual Property Rights
IMF	International Monetary Fund
JNSC	Joint North-South Contribution
MERCUSOR	European Free Trade Area Court and the Southern Common Market
MeTA	Medicines Transparency Alliance
MRDT	Medical Research and Development Treaty
MSF	Médecins Sans Frontières
NAFTA	North American Free Trade Agreement
NDG	Neglected Diseases Group
NGO	Non-governmental Organisation
NTHB	Neoliberal Transnational Historical Bloc
OHCHR	Office of the High Commissioner for Human Rights
OP-ICESCR	Optional Protocol to the International Covenant on Economic, Social and Cultural Rights
PhRMA	Pharmaceutical Research and Manufacturers of America
PPM	Prepaid Meter
PRSP	Poverty Reduction Strategy Paper
PWC	Post Washington Consensus
R&D	Research and Development
SAP	Structural Adjustment Program
TAC	Treatment Action Campaign
TAN	Transnational Advocacy Network
TCC	Transnational Capitalist Class
TCP	Peoples' Trade Agreement
TNC	Transnational Corporation
TNI	Transnational Institute
TRIMs	Agreement on Trade-Related Investment Measures
TRIPS	Agreement on Trade-Related Aspects of Intellectual Property Rights

TWN	Third World Network
UDHR	Universal Declaration of Human Rights
UN	United Nations
UNCTC	United Nations Centre on Transnational Corporations
UNCTD	UN Conference on Trade and Development
UNDP	United Nations Development Program
VGs	Voluntary Guidelines to Support the Progressive Realisation of the Right to Adequate Food in the Context of National Food Security
WANAHR	World Alliance for Nutrition and Human Rights
WBI	World Bank Institute
WFS	World Food Summit
WFS:fyl	World Food Summit: Five years later
WHA	World Health Assembly
WHO	World Health Organization
WSF	World Social Forum
WTO	World Trade Organization
WWC	World Water Council
WWF	World Water Forum

INTRODUCTION

'Everyone has the right to a standard of living adequate for the health and well-being of himself and his family, including food, clothing, housing, and medical care.

Everyone is entitled to a social and international order in which the rights and freedoms set forth in this Declaration can be fully realised'.[1]

'In just seventeen years since the end of the Cold War, over 300 million human beings have died prematurely from poverty related causes, with some 18 million more added each year. Much larger numbers of human beings live in conditions of life-threatening poverty ... This catastrophe was and is happening, foreseeably, under a global economic order designed for the benefit of the affluent countries' governments [and] corporations'.[2]

The philosopher Thomas Pogge has identified a contradiction that lies at the heart of contemporary global affairs: international law simultaneously recognises and violates the human rights of the global poor.[3] It recognises the global poor's rights to the degree that rights closely related to poverty elimination – such as rights to health, education, housing, food, water, social security and employment – are enunciated in numerous international treaties. Yet international law also systematically violates these rights by establishing and maintaining institutional structures designed to contribute to the persistence of severe poverty. Pogge argues that core international institutions such as the World Trade Organization (WTO), the International Monetary Fund (IMF) and the World Bank all

---

[1] Universal Declaration of Human Rights 1948 GA Res 217A(III), 10 December 1948, A/810, arts. 22 & 28.

[2] Thomas Pogge, 'Severe Poverty as a Human Rights Violation' in Thomas Pogge (ed.), *Freedom from Poverty as a Human Rights* (Oxford University Press, 2007) 51–52.

[3] Thomas Pogge, 'Recognized and Violated by International Law: The Human Rights of the Global Poor' (2005) 18(4) *Leiden Journal of International Law* 717.

significantly contribute towards 'a collective human rights violation of enormous proportion'.[4]

This book examines how global justice movements have engaged the language of socioeconomic rights to contest global institutional structures and rules responsible for contributing to the persistence of severe poverty. Drawing upon a range of perspectives from critical international studies and critical political economy – most notably the neo-Gramscian perspective – the book will evaluate the 'counter-hegemonic' potential of socioeconomic rights discourse, which is to say, its capacity to contribute towards an alternative to the prevailing neo-liberal 'common sense' of contemporary global governance. This introduction will provide the backdrop against which this issue will be examined.

PATHOLOGIES OF POWER

The fall of the Berlin Wall in 1989 and the collapse of the Soviet Union in 1991 were hailed by many as the beginning of a new era of increased prosperity and respect for human rights.[5] In spite of such optimism, the ensuing ten years were described by the United Nations Development Programme (UNDP) as 'a decade of despair' for many.[6] By the new millennium some fifty-four countries were poorer than they were in 1990.[7] During that same period inequality rose exorbitantly, both between and within countries, with the richest 1 percent of the world's population receiving as much as the poorest 57 percent.[8] Within that decade the number of people living on less than $1 a day increased in the Arab States, Central and Eastern Europe, Sub-Saharan Africa, Latin America and the Caribbean.[9] By 2003 it was estimated that 10 million children were dying of preventable

---

[4] ibid.
[5] For the most infamous expression of this idea see Francis Fukuyama, *The End of History and the Last Man* (Penguin, 1992).
[6] United Nations Development Programme (UNDP), *Summary: Human Development Report 2003: Millennium Development Goals: A Compact among Nations to End Human Poverty* (Oxford University Press, 2003) 2 (hereafter 'UN Human Development Report').
[7] ibid.
[8] UNDP, *Human Development Report 2003: Millennium Development Goals: A Compact among Nations to End Human Poverty* (Oxford University Press, 2003) 39. Such inequality was only the tip of the iceberg. A recent report from Oxfam suggests that the bottom half of the globe's population now owns the same as the world's richest eight people. Oxfam, 'An Economy For the 99%' (16 January 2017) https://www.oxfam.org/sites/www.oxfam.org/files/file_attachments/bp-economy-for-99-percent-160117-en.pdf accessed 30 January 2017.
[9] ibid, 5.

illnesses every year[10] and some 115 million were not attending primary school.[11] On average more than 500,000 women died in pregnancy and childbirth every year.[12]

Further shocking developments unfolded after the turn of the millennium. In 2006 rapid and sharp increases in staple food prices placed basic grains beyond the reach of millions of the World's poor.[13] According to the United Nations (UN) Food and Agriculture Organization (FAO) in 2009 more than one billion became undernourished worldwide, the largest number in recorded history.[14] The on-going economic and financial crisis that began in 2008 has also generated 'bleak prospects for social development'.[15] Global unemployment rose sharply from 178 million people in 2007 to 205 million in 2009, triggering increased levels of vulnerability, particularly in countries without comprehensive social protection.[16] The growing pressure for austerity measures, ostensibly for reasons of fiscal consolidation, is putting pressure on social protection, public health and education programmes, as well as economic recovery measures.[17] The joint IMF and World Bank Global Monitoring 2010 report estimated that an additional 64 million people fell into extreme poverty as a result of the economic crisis alone.[18]

As Paul Farmer has pointed out, the persistence of poverty is neither an 'accident' nor 'random in distribution and effect', but is rather the symptom of 'deeper pathologies of power' that are intimately connected to social conditions that determine who will suffer, and who will benefit, from particular arrangements.[19] It is no secret who has benefited the most from the post-cold-war 'Washington Consensus' of the globalisation era. In a 2014 Report entitled *Working for the Few: Political Capture and Economic*

---

[10] ibid, 8. This is perhaps understandable when one considers that only 10 percent of global spending on medical research and development is directed at the diseases of the poorest 90 percent of the world's people. See ibid, 12.
[11] ibid, 6.   [12] ibid.
[13] Food and Agriculture Organization of the United Nations (FOA), 'More People than Ever Are Victims of Hunger: Background Document' (2009). www.unmalawi.org/news room/press_release/press_release_june_en.pdf accessed 17 January 2012.
[14] ibid.
[15] United Nations Department of Economic and Social Affairs, *The Global Social Crisis: Report on the World Social Situation 2011* (15 June 2011) UN Doc. ST/ESA/334 24.
[16] ibid, iii.   [17] ibid, 6.
[18] World Bank and International Monetary Fund, *Global Monitoring Report 2010: The MDGs after the Crisis* (World Bank, 2010) viii.
[19] Paul Farmer, *Pathologies of Power: Health, Human Rights and the New War on the Poor* (University of California Press, 2005) 7.

Inequality, the anti-poverty global confederation Oxfam International affirmed what many 'anti-globalisation' activists had been arguing for decades: that over the course of the last thirty years the arena of policy making has been progressively colonised by the very wealthy.[20] As a result of 'political and regulatory capture', policies and laws have increasingly been skewed in favour of the rich, entrenching and enhancing their privileged status whilst further marginalising the poor.[21]

This phenomenon is perhaps nowhere more pronounced than in relation to the formulation of international trade rules. To illustrate, in the run up to the negotiations for the proposed European Union-United States trade deal known as the 'Transatlantic Trade and Investment Partnership' (or 'TTIP') in 2012 and early 2013, 92 percent of the 'stakeholders' in consultations with the European Commission's trade department were business lobbyists. By contrast, a mere 4 percent of consultations were with public interest groups.[22] Corporate and financial power has been both the driving force and the principle beneficiary of the neo-liberal economic policies advanced by the WTO, IMF and World Bank since the early 1990s. Policies and laws requiring public spending reduction, the removal of price controls, trade liberalisation, financial deregulation and the privatisation of public utilities and services have facilitated unprecedented capital accumulation on a world scale by removing trade barriers, creating new markets and reducing the taxation and regulation of corporate and financial activity.[23]

At the same time, these policies are implicated in undermining socio-economic rights. Many commenters have documented how trends such as the increased reliance upon the market, the diminution in the role of State provision of social services and the deregulation of financial and labour markets have exposed workers, poor people and vulnerable groups to the vicissitudes of the market and made the objects of their socioeconomic rights less secure.[24] This book takes as its starting point

---

[20] Oxfam, 'Working for the Few' 187 Oxfam Briefing Paper (20 January 2014) www.oxfam.org/sites/www.oxfam.org/files/bp-working-for-few-political-capture-economic-inequality-200114-summ-en.pdf accessed 28 January 2014.

[21] ibid.

[22] Corporate Europe Observatory, 'Who Lobbies Most for TTIP' (*Corporate Europe Observatory* 8 July 2014) http://corporateeurope.org/international-trade/2014/07/who-lobbies-most-ttip accessed 25 June 2015.

[23] Gerard Dumenil and Dominique Levy, 'The Neoliberal (Counter-) Revolution' in Alfredo Saad-Fiho and Deborah Johnson (eds.) *Neoliberalism: A Critical Reader* (Pluto Press, 2005) 13–19.

[24] See discussion in Chapter 2, Section 2.2.3.

the belief that the 'neo-liberal turn' in global governance has engendered widespread, profound and intolerable injustices for the world's poor. It therefore follows Conor Gearty in arguing that human rights must be 'subversive rather than supportive of such a brutal *status quo*' in order to survive as a meaningful language of emancipation.[25]

ACCUMULATION AND ITS DISPOSSESSED

Neo-liberal globalisation has undoubtedly served vested corporate and financial interests but in so doing it has also adversely impacted the lives of millions of the world's poor. Global rules and agreements have in many instances had the effect of eroding social protection and expropriating local communities and primary producers of the means of subsistence. These processes, which the Marxist geographer David Harvey has termed 'accumulation by dispossession'[26] form an integral and on-going dimension of the capitalist mode of production and have intensified in nature and scope in the contemporary era of neo-liberal globalisation.

Accumulation by dispossession is a multifaceted process that has involved, *inter alia*, the commodification and privatisation of land and the forceful expulsion of peasant populations, the conversion of various forms of property rights (common, collective, state, etc.) into exclusively private property rights, the expansion of the domain of intellectual property rights (IPRs) and the suppression of the rights of the commons.[27] These processes are evidenced in, for example, the WTO's agreement on Trade-Related Intellectual Property Rights (TRIPs) requiring extensive protection of IPRs in the areas of medicine and agriculture, and IMF and World Bank loan conditionalities mandating the privatisation of stated owned or commonly held property.

In response to these trends towards the enclosure of the global 'commons' (i.e. those shared spaces and forms of property relations that are not privatised or commodifed) a number of what Karl Polanyi called 'counter-movements' have developed. Such movements, Polanyi argued, emerge spontaneously in response to the chaos and poverty caused by marketisation and seek to restrain the market through political and

---

[25] Conor Gearty, *Can Human Rights Survive?* (Cambridge University Press, 2006) 141.
[26] David Harvey, 'Neoliberalism as Creative Destruction' (2007) 610 *The Annals of the American Academy of Political and Social Science* 21, 34–35.
[27] ibid; David Harvey, *A Brief History of Neoliberalism* (Oxford University Press, 2005) 160.

institutional change.[28] In the era of neo-liberal globalisation such movements have increasingly become *transnational* or even 'global', both in terms of their composition and their aims and objectives.[29] The globalisation of communication systems has enabled the sharing of strategies and the development of alliances between geographically dispersed movements, facilitating the formation of a vast global 'set of networks, initiatives, organisations and movements that fight against the economic, social, and political outcomes of hegemonic globalisation, challenge the conceptions of world development underlying the latter, and propose alternative conceptions'.[30]

'NEW RIGHTS ADVOCACY'

The last three decades have also witnessed 'an increasingly expansive array of international instances' that have generated socioeconomic rights jurisprudence.[31] UN human rights bodies, agencies and special rapporteurs have steadily taken more steps to assist the development of more rigorous awareness, monitoring and implementation of socioeconomic rights.[32] The UN Committee on Economic, Social and Cultural Rights (CESCR) was formed in 1987 to monitor States Parties' compliance with their obligations under the International Covenant on Economic, Social and Cultural Rights (ICESCR) and has subsequently developed an international jurisprudence attempting to clarify the content of these rights. The UN Commission on Human Rights has appointed special rapporteurs on education, food, housing and highest attainable standard of health. UN agencies such as the UNDP, the World Health Organization (WHO) and the FAO have

---

[28] Karl Polanyi, *The Great Transformation: The Political and Economic Origins of Our Time* (Beacon Press, 1944) 130.

[29] Barry K Gills, 'Introduction: Globalization and the Politics of Resistance' in Barry K Gills (ed.), *Globalization and the Politics of Resistance* (Macmillan, 2000) 3–11.

[30] Boaventura de Sousa Santos, 'Beyond Neoliberal Governance: The World Social Forum as Subaltern Politics and Legality' in Boaventura de Sousa Santos and Cesar A Rodriguez-Garivito (eds.), *Law and Globalization from Below: Towards a Cosmopolitan Legality* (Cambridge University Press, 2005) 29.

[31] Philip Alston, 'Foreword' in Malcolm Langford (ed.), *Social Rights Jurisprudence: Emerging Trends in International and Comparative Law* (Cambridge University Press, 2008) 3.

[32] Paul J Nelson and Ellen Dorsey, *New Rights Advocacy: Changing Strategies of Development and Human Rights NGOs* (Georgetown University Press, 2008) 45–46.

also developed rights-based approaches that incorporate concern for socioeconomic rights.³³

The combination of growing forms of transnational resistance to neoliberal globalisation and efforts by international bodies to clarify the normative content of socioeconomic rights has created the material and ideational conditions for a global 'new rights advocacy' based around the protection and promotion of socioeconomic rights. Since the 1960s, international human rights advocacy had been devoted almost exclusively to civil and political rights and such advocacy still remains the predominant focus of Western human rights NGOs.³⁴ However, from the mid-1990s advocacy work around socioeconomic rights has become visible within human rights NGOs which have 'joined in human rights-driven social movements for food, health, education, water and other rights'.³⁵ This movement gained momentum towards the end of the 1990s as the global justice movement began to form and it 'became common to hear human rights language associated with criticism of neoliberal globalization'.³⁶

Over the past two decades, new human rights NGOs have been founded to focus on specific socioeconomic rights, such as the Food First Information and Action Network (FIAN) and the Centre on Housing Rights and Evictions (COHRE).³⁷ National advocacy organisations for socioeconomic rights have also emerged in Nigeria, Ecuador, New Zealand, Canada and many other locations.³⁸ Western Human Rights NGOs such as Amnesty International, Human Rights Watch and Human Rights First, which for decades worked exclusively on civil and political rights, also began to expand their mandates to encompass a range of socioeconomic rights.³⁹ UN-Sponsored conferences have brought diverse NGOs and civil society organisations together on the global stage.⁴⁰ The reaffirmation of the interdependence of all human rights at the Vienna World Conference in 1993 and subsequent world conferences in Copenhagen, Johannesburg and elsewhere provided further opportunities to discuss and support socioeconomic rights.⁴¹ A number of these

---

[33] ibid, 46.   [34] ibid, 13.   [35] ibid, 14.
[36] Andrew Lang, *World Trade Law after Neoliberalism: Re-Imagining the Global Economic Order* (Oxford University Press, 2011) 81.
[37] See www.fian.org/ accessed 21 September 2013; www.cohre.org/ accessed 21 September 2013.
[38] Nelson and Dorsey (n 32) 70–71.   [39] ibid, 31.   [40] ibid.
[41] Daniel PL Chong, *Freedom from Poverty: NGOs and Human Rights Praxis* (University of Pennsylvania Press, 2010) 159.

actors have also converged at other international gatherings such as the annual World Social Forum (WSF) events, where they raised similar concerns in relation to globalisation and socioeconomic rights.[42]

This book examines the role that this 'new rights advocacy' can play in relation to global justice movements that contest the ideologies, institutions and outcomes of neo-liberal transnational governance. As a number of scholars and NGO activists have noted, the language of socioeconomic rights is, in a number of respects, useful for contesting neo-liberalism. It is argued, for example, that socioeconomic rights imply certain forms of wealth and resource distribution, place limits on privatisation and commodification and challenge the logic of unfettered economic rationality.[43]

Nevertheless, human rights have also come under attack from a number of sources, and notably from scholars and activists on the political left. It is argued that human rights, particularly in their legal form, seek only to ensure minimum levels of protection to the most marginalised and fail to address structural factors (political, social, cultural, economic, etc.) that produce and sustain injustice and inequality.[44] Thus, Samuel Moyn has described human rights discourse as 'a powerless companion in the age of neoliberalism', ill-suited and unable to deliver substantive socioeconomic equality.[45]

Other critics have gone further and argued that human rights discourse is not only an ineffective tool for contesting neo-liberalism but itself constitutes part of neo-liberal hegemony.[46] Such critics stress the historic

---

[42] Maria Luisa Mendonca, 'Human Rights: Conference Synthesis on Economic, Social and Cultural Rights' in William F Fisher and Thomas Ponniah (eds.), *Another World Is Possible: Popular Alternatives to Globalization at the World Social Forum* (Fernwood, 2003) 309–316.

[43] See, e.g., Shareen Hertel and Lanse Minkler, 'Economic Rights: The Terrain' in Shareen Hertel and Lanse Minkler (eds.), Economic Rights: Conceptual, Measurement, and Policy Issues (Cambridge University Press, 2007) 1; Oriol Mirosa and Leila M Harris, 'Human Right to Water: Contemporary Challenges and Contours of a Global Debate' (2011) 44(3) *Antipode* 932, 933; Priscilla Claeys, 'From Food Sovereignty to Peasants' Rights: An Overview of Via Campesina's Struggle for New Human Rights' (*La Via Campesina*, 15 May 2013) 2. http://viacampesina.org/downloads/pdf/openbooks/EN-02.pdf accessed 3 January 2014.

[44] Alicia Ely Yamin, 'The Future in the Mirror: Incorporating Strategies for the Defense and Promotion of Economic, Social and Cultural Rights into the Mainstream Human Rights Agenda' (2005) 27 *Human Rights Quarterly* 1200, 1221.

[45] Samuel Moyn, 'A Powerless Companion: Human Rights in the Age of Neoliberalism' (2014) 77(4) *Law and Contemporary Problems* 147.

[46] Naomi Klein, *The Shock Doctrine: The Rise of Disaster Capitalism* (Penguin, 2007) 118–128.

role of economic liberalism in human rights discourse and its emphasis on individual appropriation and exclusive ownership of resources at the expense of equitable redistributive goals.[47] Furthermore, the form of rights discourse – which emphasises the formal equality and abstract freedom of juridical individuals – not only masks social inequalities but also makes those inequalities appear natural and legitimate.[48]

This book adopts neither an uncritical nor a dismissive account of the potential of human rights discourse for challenging neo-liberal globalisation. Rather, following Alan Hunt, it will argue that while socioeconomic rights are neither perfect nor exclusive vehicles for emancipation, they can nevertheless operate as constituents of a strategy of social transformation when 'they become part of an emergent "common sense" and are articulated within social practices'.[49]

GLOBAL JUSTICE MOVEMENTS

This book bases its assessment of socioeconomic rights upon three case studies of global justice movements that have engaged the discourse of socioeconomic rights in their campaigning activity. These movements, discussed below, are *the food sovereignty movement, the access to medicines movement* and *the water justice movement*.

Whilst a number of different case studies could have been chosen, these three were selected on the basis that they share a number of key features that make them appropriate for this enquiry. First, all of these movements can be understood as truly *global* in nature, spanning every continent and involving actors and movements ranging from more formally structured Non-Government Organisations (NGOs) to grassroots collectives.

Second, these movements challenge particular regimes associated with neo-liberal global governance, as well as the values that inform them. In particular, each of these movements seeks to contest the logic of commodification, privatisation or liberalisation in relation to resources that are essential to human dignity and indeed basic survival.

---

[47] John Charvet and Elisa Kaczynska-Nay, *The Liberal Project and Human Rights: The Theory and Practice of a New World Order* (Cambridge University Press, 2008) 11–12.
[48] Karl Marx, 'On the Jewish Question', reproduced in Joseph O'Malley (ed.), *Marx: Early Political Writings* (Cambridge University Press, 1993) 28.
[49] Alan Hunt, 'Rights and Social Movements: Counter-Hegemonic Strategies' (1990) 17(3) *Journal of Law and Society* 309, 325.

Third, these movements engage substantially with the discourse of socioeconomic rights, often drawing on standards recognised under international law.

And finally, all of these movements contain at least significant components that are engaged in *transformative politics*, by which I mean they are concerned not only with winning piecemeal reforms within the existing global political-economic structures, but rather seek to fundamentally transform and transcend those structures. It is in that sense that these movements can be understood as potentially counter-hegemonic.

A NEO-GRAMSCIAN FRAMEWORK

In assessing the counter-hegemonic potential of socioeconomic rights discourse this book will draw theoretical insights from the neo-Gramscian approach to international relations. The neo-Gramscian framework will be more fully explained in Chapter 1, but for now it will suffice to note that, for neo-Gramscians, a hegemonic world order is understood more broadly than the economic or military preponderance of a powerful state or states and instead involves 'a coherent ... fit between a configuration of material power, the prevalent collective image of world order (including certain norms) and a set of institutions which administer the order with a certain semblance of universality'.[50] In other words, hegemony requires ideological legitimation and at least a degree of consent from some sections of the subaltern classes.

For neo-Gramscians, the ideological legitimation of neo-liberal globalisation takes place in the sphere of 'global civil society', understood roughly as the global domain of uncoerced human association and the sets of relational transnational networks that fill that space.[51] Whilst global civil society is a discursive space that helps to reproduce global hegemony, it is also viewed as a platform to contest dominant models of globalisation.[52] Human rights remain ubiquitous within global civil society today.[53] As already noted, there is disagreement over the counter-hegemonic potential

---

[50] Robert W Cox, 'Social Forces, States and World Orders: Beyond International Relations' (1981) 10(2) *Millennium: Journal of International Studies* 126, 139.

[51] Michael Walzer, 'The Civil Society Argument' in Chantal Mouffe (ed.) *Dimensions of Radical Democracy: Pluralism, Citizenship, Community* (Phronesis, 1992) 89.

[52] Lucy Ford, 'Challenging Global Environmental Governance: Social Movement Agency and Global Civil Society' (2003) 3(2) *Global Environmental Politics* 120, 129.

[53] Costas Douzinas, *Human Rights and Empire: The Political Philosophy of Cosmopolitanism* (Routledge, 2007) 33.

of human rights discourse. But even assuming that socioeconomic rights is well-suited as an oppositional vocabulary, a related insight of neo-Gramscian analysis is that hegemony requires the consent of the governed and therefore involves various forms of ideological and material compromises aimed at incorporating aspects of subaltern demands.[54] It is therefore possible that socioeconomic rights discourse could be interpreted or recast in such a way that helps to ideologically stabilise neo-liberal globalisation rather than contest it. The case study chapters will therefore pay attention to the potential for the appropriation of socioeconomic rights discourse by social forces involved in the neo-liberal project.

FRAMEWORKS, DISCOURSES AND IDEOLOGIES

This work is concerned with the effectiveness of socioeconomic rights discourse in contesting neo-liberalism at the global level. At this point I should clarify exactly what I mean by 'effectiveness'. Effectiveness in the context of this book refers to the capacity of socioeconomic rights to be mobilised as part of a 'counter-hegemonic' *narrative* to contest extant neo-liberal norms. As such, this book is not primarily concerned with the effectiveness of socioeconomic rights in achieving material gains, such as reversals of privatisation, wealth redistribution, trade union rights etc.[55] Rather, the focus will largely be on the effectiveness of socioeconomic rights in contesting neo-liberalism at the level of collective frameworks, discourses and ideologies, or what might collectively be referred to as 'the ideational' effectiveness of socioeconomic rights. Given this enquiry's ideational emphasis it will principally be interpretative/hermeneutic rather than 'empirical' in nature.

The book's methodology is entirely literature-based, drawing upon both primary and secondary legal materials as well as scholarship around the issues of neo-liberalism, globalisation, political-economy and international relations. Although exclusively literature-based, this form of research requires analytical engagement with the source material. Such

---

[54] Antonio Gramsci in Quintin Hoare and Geoffrey Nowell Smith (eds. and trans.), *Selections from the Prison Notebooks of Antonio Gramsci* (Lawrence and Wishart, 1971) 58–59.
[55] I say 'primarily' because, as will be spelled out in subsequent chapters, the theoretical framework of this book eschews any neat compartmentalisation of the material and the ideational. Nevertheless, the analysis of the book will be skewed towards the ideational in its focus.

an engagement will take the form of what can be loosely termed 'discourse analysis'. This mode of textual engagement rejects the realist view of language as 'simply a means of reflecting or describing the world' and instead is premised upon the belief that discourse has central importance in 'constructing social life'.[56] As Stewart Hall, following Michel Foucault, put it:

> Discourse ... governs the way that a topic can be meaningfully talked about and reasoned about. It also influences how ideas are put into practice and used to regulate the conduct of others. Just as discourse 'rules in' certain ways of talking about a topic, defining an acceptable and intelligible way to talk, write, or conduct oneself, so also, by definition, it 'rules out' and restricts other ways of talking, of conducting ourselves in relation to the topic or constructing knowledge about it.[57]

Neo-liberal global governance is mobilised around a number of explicit and implicit narrative frames – for example, free trade, market rationality, possessive individualism and so on – that are not merely descriptive of the world, but also prescriptive of how the world ought to be as well as performatively constructive of it. The effectiveness of socioeconomic rights discourse will therefore be measured to the extent that it is able to identify and contest these neo-liberal prescriptions.

This will be assessed in the context of three case studies of global justice movements that have engaged the language of socioeconomic rights. Each chapter will follow the same basic structure. They begin by identifying the *political-institutional context* of the neo-liberal global governance regime in question (agriculture, medicines and water). Following this, what will be termed the *ethico-political (i.e. legitimising) framework* of each political-institutional regime will be considered ('food security', 'IPRs' and 'market environmentalism/ water as commodity', respectively). This legitimising framework will then be contrasted with alternative ethico-political frameworks articulated by various global justice movements ('food sovereignty', 'access to medicines' and 'water as commons', respectively). Whilst these competing discourses are fluid and changing, this book will adopt a Weberian 'ideal type' analysis, whereby a one-sided accentuation of

---

[56] Rosalind Gill, 'Discourse Analysis' in Martin W Bauer and George D Gaskell (eds.), *Qualitative Researching with Text, Image and Sound: A Practical Handbook for Social Research* (Sage, 2000) 172.

[57] Stewart Hall, 'The Work of Representation' in Stewart Hall (ed.), *Representations: Cultural Representations and Signifying Practices* (Sage, 1997) 44.

key features of each discourse will be unified into more stable constructs for comparative analytic purposes.[58]

It is after establishing this broader context of discursive contestation that the role of the respective socioeconomic rights (to food, health and water, respectively) will be considered. Each of the case studies will document the ways in which global justice movements used the language of socioeconomic rights in their struggles, how these articulations are (usually selectively and cautiously) reflected in the statements and jurisprudence issued by UN human rights bodies, and how key states and actors within neo-liberal governance have sought to marginalise, water-down or co-opt these socioeconomic rights discourses.

STRUCTURE OF THE BOOK

Chapter 1 introduces the neo-Gramscian framework for analysing global hegemony and global counter-hegemony within the context of neo-liberal globalisation. The chapter begins by summarising relevant theories of Antonio Gramsci, the Italian Marxist who provided the analytic framework for understanding the operation of hegemony in class societies. Following this, the insights of neo-Gramscians and other critical globalisation scholars are used to analyse the operation of global hegemony within the contemporary neo-liberal period. The chapter concludes by assessing the prospects and challenges faced by efforts to build global justice movements to contest neo-liberal globalisation.

Chapter 2 provides a neo-Gramscian framework for understanding socioeconomic rights and the role they (could) play in social transformation. This chapter rejects both the metaphysical abstraction associated with 'foundationalist' approaches to human rights and the narrow focus of legal positivism. Instead, it will adopt a neo-Gramscian-informed social constructivist framework that understands human rights as neither rooted in a transcendental morality nor merely as a product of sovereign legislation, but rather as social forms which emerge in specific historical contexts. It is argued that human rights are discursively open concepts and hence cannot be definitively categorised as inherently hegemonic or counter-hegemonic. Rather, rights are understood as expressions of the social conflicts and contradictions embedded in social life and therefore are potentially supportive of very different kinds of social visions and

---

[58] Edward A Shills and Henry A Finch (trans. and ed.) *The Methodology of the Social Sciences (1903–17)* (Free Press, 1997) 90.

political projects. The chapter will critically evaluate the potential strengths and weaknesses of socioeconomic rights discourse in contributing to counter-hegemonic praxis.

Chapter 3 is the first case study chapter. It looks at what is termed 'the Food Sovereignty Movement'. This movement developed in opposition to agricultural liberalisation, embodied in the WTO Agreement on Agriculture (AoA), the North American Free Trade Agreement (NAFTA) and IMF and World Bank-imposed structural adjustment programmes. This campaign has prominently featured FIAN, a German-based NGO committed to the right to food, and La Via Campesina, an international peasants movement comprised of small-scale farmers, pastoralists, fisherfolk, indigenous peoples and landless peasants. These organisations, along with a host of other national and international movements, have coalesced around the ideas of food sovereignty and the right to food. This chapter critically evaluates the strengths and weaknesses of invoking the right to food to challenge the global neo-liberal food regime.

Chapter 4 focuses on the 'access to medicines' movement. A global public health movement, spearheaded by HIV/AIDS activists from sub-Saharan Africa, Brazil and elsewhere in alliance with international organisations such as Médecins Sans Frontiers and the Third World Network, has mobilised around constitutionally and internationally enshrined articulations of the right to health to challenge the IPRs of pharmaceutical corporations as protected under the WTO Trade Related Aspects of Intellectual Property Rights (TRIPS) agreement and other regional and bilateral trade agreements. This chapter critically evaluates the extent to which appeals to 'the right to health' have strengthened the ability of states in the Global South to not only make greater use of 'TRIPs flexibilities' (e.g. in relation to compulsory licensing and parallel importing of medicines) but also to push for alternative models of medical research and development that are orientated towards health needs rather than the commercial interests of pharmaceutical corporations.

Chapter 5 examines the Water Justice Movement. For over a decade social movements from around the world have been resisting the privatisation and commercialisation of water, a project pursued by, amongst others, the World Bank, the World Water Forum and Transnational Corporations. A global movement has emerged that has challenged corporate private sector involvement in the supply of water services and has been arguing for – and putting into practice – alternatives aimed at being inclusive, participatory, democratic, equitable and sustainable.

Whilst there is diversity in the water justice movements, they share the belief that water is a common good and therefore must not be treated as a private commodity to be bought, sold or traded for profit. The strengths and pitfalls of this movement engaging the 'right to water' in its advocacy efforts are considered in this chapter.

Finally, this book concludes by reflecting upon these case studies. These studies, while not identical, all document some common features: global justice movements using the language of socioeconomic rights in their struggles; these articulations (usually selectively and cautiously) are reflected in the statements and jurisprudence issued by UN human rights bodies, and; key states and actors within neo-liberal governance seeking to marginalise, water-down or co-opt these socioeconomic rights discourses.

Drawing upon the praxis of the global justice movements discussed in the case studies, this book concludes by arguing that the dangers of co-option identified above can be minimised, or at any rate mitigated, through what will be termed a 'tripartite model of counter-hegemonic rights praxis'. This entails global justice movements tailoring socioeconomic rights tactics to three different planes of global civil society: (1) participation in inter-governmental and other 'official' settings primarily to gain visibility, co-ordinate movement activity and advance incremental discursive shifts in global governance; (2) strategic alliances with UN agencies, human rights bodies and special rapporteurs that are marginalised or peripheral to the neo-liberal global order so as to gain legitimacy, expertise and resources; and (3) connecting socioeconomic rights standards to counter-hegemonic models of governance within 'subaltern counterpublics' (i.e. transnational spaces outside of the domain of global officialdom) in order to guard against the co-option and dilution of oppositional ideologies.

1

POWER, HEGEMONY AND WORLD ORDER

> The "dual" perspective in political action ... can be reduced to two fundamental levels, corresponding to the dual nature of Machiavelli's Centaur – half animal and half human. They are the levels of force and consent, authority and hegemony, violence and civilisation, of the individual moment and the universal moment.[1]

1.1 INTRODUCTION

This book explores the extent to which socioeconomic rights can and have been mobilised by global justice movements to contest dominant narratives of neo-liberal globalisation. In order to do so, the study will rely on a neo-Gramscian framework for analysing the operation of 'hegemony' and 'counter-hegemony' in world order.

This gives rise to a number of questions: How are the concepts of 'hegemony' and 'counter-hegemony' to be understood? Do these terms retain validity as analytic categories today? Against what structures and relations of power are these concepts to be evaluated? Which social forces are, or potentially are, counter-hegemonic? Can hegemony and counter-hegemony really exist on a global level, as opposed to merely at the level of the nation state?

The purpose of this chapter is to seek to provide some answers to these questions. First, it will elucidate the particular meanings that 'hegemony' and 'counter-hegemony' will have in this book. It will do so by locating these terms within the highly original and influential Marxist theoretical corpus of the Italian political theorist Antonio Gramsci. The first substantive part of this chapter will introduce key concepts and ideas from

---

[1] Antonio Gramsci, *Selections from the Prison Notebooks of Antonio Gramsci* (Quintin Hoare and Geoffrey Nowell Smith eds. and trans.) (Lawrence & Wishart, 1971) 169–170 (hereafter 'Gramsci').

Gramsci's political philosophy to help inform the analysis of hegemony and counter-hegemony in the book. The second part of the chapter is concerned with clarifying the meaning and nature of world order in the contemporary era of 'neo-liberal globalisation'. This section will apply Gramsci's theories, as well as the perspectives brought to bear by neo-Gramscian scholars and other critical theorists, to the study of neo-liberal globalisation. In particular, it will examine how hegemony operates at the level of 'global civil society' within the neo-liberal global order. It will also consider whether the current economic crisis marks the end of neo-liberal hegemony. The final part of this chapter will look at the prospects for global counter-hegemony as well as consider the obstacles that could undermine or prevent efforts to build such movements.

1.2 HEGEMONY AND COUNTER-HEGEMONY

Hegemony is a term deployed within the social sciences in a variety of different ways.[2] This book is concerned with the Marxist concept of hegemony as refined by Antonio Gramsci.[3] Gramsci used the term as part of an attempt to provide a systematic analysis of the ways in which ruling classes succeeded in establishing 'moral and intellectual leadership'[4] over non-dominant ('subaltern') social classes.[5] Hegemony conveys the idea

---

[2] For example, in the field of International Relations 'hegemony' is used to both describe the domination of an international order by a powerful state through its preponderance in economic and military capacities and also a situation where the dominance of a powerful state is supplemented by consensual mechanisms of global governance. Contrast Robert Gilpin, *War and Change in World Politics* (Cambridge University Press, 1981) and Robert W Cox, 'Gramsci, Hegemony and International Relations' in Stephen Gill (ed.), *Gramsci, Historical Materialism and International Relations* (Cambridge University Press, 1993) 49. Within Marxism the term is used 'in two diametrically opposed senses: first to mean domination, as in "hegemonism"; and secondly, to mean leadership, implying some notion of consent'. Anne Showstack Sassoon, 'Hegemony' in Tom Bottomore (ed.), *A Dictionary of Marxist Thought* (2nd edn., Blackwell, 2006) 229–230. There is also a division between the way in which Marxists and Post-Marxists use the term. Whereas Marxists understand hegemony as ultimately deriving from social classes, Post-Marxists reject the notion of a single overarching hegemony rooted in class and instead posit the existence of 'multiple hegemonies' based on plural forms of domination (race, gender, disability etc.). Contrast Gramsci (n 1) with Ernesto Laclau and Chantal Mouffe, *Hegemony and Socialist Strategy: Towards a Radical Democratic Politics* (Verso, 1985).
[3] '[Hegemony's] full development as a Marxist concept can be attributed to Gramsci. Most commentators agree that hegemony is the key concept in Gramsci's Prison Notebooks and his most important contribution to Marxist theory'. Sassoon (n 2) 230.
[4] Gramsci (n 1) 57. [5] ibid, 52.

that the ruling capitalist class not only control the means of production and the repressive apparatuses of the state but also the ideological apparatuses that are used to create a general consensus that the *status quo* is legitimate and beneficial.[6] This section will seek to elaborate on Gramsci's conception of hegemony as well as related concepts in his theoretical corpus that are relevant to this thesis.

1.2.1 STRUCTURE AND SUPERSTRUCTURE: THE HISTORIC BLOC

In order to be able to make sense of the significance of the concept of hegemony, it is important to begin by locating it within Gramsci's contribution to the development of historical materialism in the Marxist tradition. Historical materialism is a method which seeks to account for historical change in human societies on the basis of the nexus between economic, technological, and more broadly, material developments on the one hand and more 'cultural' phenomena such as law, politics, philosophy, ideology or religion on the other.[7] It was an analytical method developed by Karl Marx in opposition to idealist accounts of historical change that stressed the primacy of thought and ideas in the development of human society.[8] A central concern of the historical materialist approach is to account for the relationship between the most basic and fundamental level of economic existence (the 'structure') in society and the ideas, beliefs and institutions that make up the rest of basic human culture (the 'superstructure').[9] The most reductionist forms of Marxism (what might be termed 'crude materialism') espouse a deterministic causality from the 'material' structure to the 'ideal' superstructure, with

---

[6] ibid, 12.
[7] Karl Marx and Frederick Engels, *The German Ideology Part One* (Lawrence & Wishart, 1970) 47 ('The production of ideas, of concepts, of consciousness, is at first directly interwoven with the material activity and the material intercourse of men ... The same applies to mental production as expressed in the language of politics, laws, morality, religion, metaphysics etc'.).
[8] ibid, 39–42.
[9] The structure-superstructure metaphor is summarised by Marx as follows: 'In the social production of their existence, men inevitably enter in to definite relations, which are independent of their will, namely [the] relations of production appropriate to a given stage in the development of their material forces of production. The totality of these relations of production constitutes the economic structure of society, the real foundation, on which arises a legal and political superstructure, and to which correspond definite forms of social consciousness'. Karl Marx, *A Contribution to the Critique of Political Economy* (International Publishers Company, 1970) 20.

the latter being reduced to mere epiphenomena produced by, and reflecting, the deeper logic of the former.[10]

Gramsci sought to charter an alternate course, one that would pave the way for 'an integral and original philosophy which opens up a new phase of history ... [that] goes beyond both traditional idealism and traditional materialism ... while retaining their vital elements'.[11] To do so Gramsci did not reject the structure-superstructure metaphor but, in place of accounts that stressed a rigid hierarchical division between these two components, Gramsci advanced the concept of the 'historic bloc' as embodying the 'reciprocity between structure and superstructure'.[12] As Gramsci puts it:

> Material forces are the content and ideologies are the form, though the distinction between form and content has purely 'didactic' value, since the material forces would be inconceivable historically without form and the ideologies would be individual fancies without the material forces.[13]

In other words, the superstructural actions involved in the production of law, art, policies and so on, are not regarded as epiphenomenal, but rather as being as operative and 'real' as economic phenomena. After all, ideologies '"organise" human masses, and create the terrain on which men move [and] acquire consciousness of their position'.[14]

Whilst Gramsci accepted Marx's analysis of the dynamics and structure of capitalism he rejected the more mechanistic and economistic readings of Marx within the socialist tradition. The 'immanent unity of material and ideational conditions'[15] that is at the core of Gramsci's social ontology implies an internal relationship of theory and practice 'such that progressive social change does not automatically follow in train behind economic developments, but must instead be produced by historically situated social agents whose actions are enabled and constrained by their social self-understandings'.[16] For Gramsci, history 'is at once freedom and necessity' in that whilst individuals are constrained

---

[10] For Gramsci's critique of crude materialist tendencies in Marxism see Gramsci (n 1) 419–472.
[11] Gramsci (n 1) 435.   [12] ibid, 366.   [13] ibid, 377.   [14] ibid, 376–377.
[15] A Claire Cutler, 'Towards a Radical Political Economy Critique of Transnational Economic Law' in Susan Marks (ed.) *International Law on the Left: Re-Examining Marx Legacies* (Cambridge University Press, 2008) 216.
[16] Mark Rupert, 'Reading Gramsci in an Era of Globalising Capitalism' in Andreas Bieler and Adam David Morton (eds.) *Images of Gramsci: Connections and Contentions in Political Theory and International Relations* (Routledge, 2006) 93.

by structural forces they also have the capacity to transcend them.[17] It is the ability of people to realise the desirability of changing society and their capacity to do so that enables the passage from 'necessity to freedom': 'Structure ceases to be an external force which crushes man, assimilates him to itself and makes him passive; and is transformed into a means of freedom, an instrument to create a new ethico-political form and a source of new initiatives'.[18] Thus social and historical change is bound up with the ability of social classes to both contest and sustain dominant ideologies that shape people's perceptions about what is possible. The dialectical relationship between theory and practice is captured in Gramsci's work by the term praxis,[19] a term that will be frequently deployed throughout this thesis to capture the 'reflexivity of "theory" and "practice" – the way the two co-construct each other'.[20]

1.2.2 POLITICAL AND CIVIL SOCIETY: THE INTEGRAL STATE

Prior to Gramsci, Marxist analysis of state power was primarily concerned with the coercive measures used to suppress the intractable conflicts that arose in class-divided society.[21] Gramsci's expanded conception of class rule identified two spheres in which the bourgeois exerted its domination and leadership as a ruling class: (1) civil society, which is 'the ensemble of organisms commonly called "private"' and (2) political society, in other words 'the State'.[22] It is through political society – that is, the army, the courts, the criminal justice system and so on – that the ruling group exercises domination and force to subordinate or eliminate

---

[17] Gramsci, cited in Andreas Bieler and Adam David Morton, 'The Gordian Knot of Agency–Structure in International Relations: A Neo-Gramscian Perspective' (2001) 7 (1) *European Journal of International Relations* 5, 19.

[18] Gramsci (n 1) 367.

[19] Gramsci developed his writings on hegemony under conditions of imprisonment by Mussolini's fascist regime. In order to evade the prison censor he often resorted to obfuscation. Throughout his work he frequently uses 'the philosophy of praxis' as a code term for Marxism. Nevertheless, the term is indicative of the centrality of the idea of praxis in Gramsci's Marxism. See Peter Thomas, *The Gramscian Moment: Philosophy, Hegemony and Marxism* (Brill, 2009) 105–108.

[20] Martti Koskenniemi, 'A Response' (2006) 7(12) *German Law Journal* 1103, 1103.

[21] In *The State and Revolution* Vladimir Lenin approvingly cites Friedrich Engels' definition of the State as consisting 'not merely of armed men but also of material adjuncts, prisons, and institutions of coercion of all kinds'. See Vladimir Lenin, *The State and Revolution* (Penguin, 1992) 8.

[22] Gramsci (n 1) 12.

hostile forces, that is, those who will not actively or passively consent to the dominant group's rule.[23] In contrast, it is through civil society – for example, the schools, churches and trade unions – that the dominant class exercises intellectual and moral leadership to win and maintain the consent of the masses.[24] It is important to stress here that 'consent' is not necessarily understood as formal or explicit acceptance, but rather as a sort of 'saturation of the consciousness' involving naturalising the *status quo* as constituting the boundaries of what is possible or desirable.[25] According to Gramsci the strongest states are not those that operate through naked force and violence, but those that maintain themselves primarily though hegemony.[26]

It should also be noted that the distinctions that Gramsci draws between political and civil society are presented as methodological rather than 'organic' distinctions[27] and 'in concrete reality, civil society and state are one and the same'.[28] Because the executive, legislative and judicial functions of the state are in fact constrained by the hegemony a dominant class exercises in civil society, a wider conception of the state that includes the underpinnings of the political structure of civil society is required.[29] Rather than identifying the state with governmental coercive power, Gramsci advanced a broader conception which includes 'the state proper and civil society ... the entire complex of practical and theoretical activities with which the ruling class not only justifies and maintains its domination, but manages to win the active consent of those whom it rules'.[30] In Gramsci's analysis, the hegemony of the dominant class in effect bridges the conventional categories of state and civil society, which although retaining analytical validity, have 'ceased to correspond to separable entities in reality'.[31]

Gramsci identified the political–civil society complex as the 'integral state'.[32] The idea of the integral state is based on the rejection of any artificial separation of 'the ethical-political aspect of politics' from the aspects of 'force and economics'.[33] Every instance of hegemony in the private sphere is backed by physical force on some level, and every act of physical force is also a symbolic performance and a hegemonic statement

---

[23] ibid.   [24] ibid, 258.
[25] Raymond Williams, 'Base and Superstructure in Marxist Cultural Theory' (1973) 87 *New Left Review* 3, 8.
[26] Gramsci (n 1) 238.   [27] ibid, 160.   [28] ibid.   [29] Cox (n 2) 51.
[30] Gramsci (n 1) 244.   [31] Cox (n 2) 51.   [32] Thomas (n 19) 180.
[33] Gramsci (n 1) 207 (editors' comment).

about the legitimacy of the state.[34] As Louis Althusser put it 'the Army and Police also function through ideology both to ensure their own cohesion and reproduction, and in the "values" they propound externally'[35] whilst schools and churches use 'suitable methods of punishment, expulsion, selection, etc., to "discipline" not only their shepherds, but also their flocks'.[36]

1.2.3 SECURING HEGEMONY

Hegemony is the process through which dominant classes are able to acquire political, economic and ideological legitimacy and the consent of the governed.[37] Gramsci suggested that there are two different paths to achieving hegemony. The first involves a carefully co-ordinated and well-defined plan orchestrated by members of the ruling class through formal regimes and specific rules.[38] The second way to achieve hegemony, which Gramsci referred to as 'liberal' or 'molecular', involves the unco-ordinated actions of individuals initially acting independently of one another, but whom become 'spontaneously' 'naturally condensed' due to their shared class interests.[39] The informal nature of these relationships means that it will often appear as if there has been no collusion in the formation of certain ideas, world views and policies, making the appearance of genuine 'consensus' more plausible.[40]

In order to achieve hegemony, a ruling class must present its own sectional interests as the common interests of all and articulate values and norms in such a way that they take on a universal appeal.[41] Such universal norms are promoted by 'organic intellectuals': the thinking and organising element of a particular fundamental social class.[42] The defining feature of these intellectuals is not so much their profession, but rather their function in directing the ideas and aspirations of the class to which they organically belong. As such, they can include journalists, teachers, lawyers, entrepreneurs and so forth in addition to the 'traditional' intellectuals.[43] Furthermore, in order to achieve alliances with the subaltern classes, a ruling class may also make economic concessions and sacrifices

---

[34] Douglas Litowitz, 'Gramsci, Hegemony and the Law' (2000) *Brigham Young University Law Review* 515, 527.
[35] Louis Althusser, 'Ideology and Ideological State Apparatuses' in Louis Althusser, *Lenin and Philosophy and other Essays* (Monthly Review Press, 1972) 137.
[36] ibid.   [37] Gramsci (n 1) 12.   [38] ibid, 60.   [39] ibid.   [40] ibid.   [41] ibid, 181.
[42] ibid, 5.   [43] ibid, 6.

which take account of the interests and tendencies of the subaltern social groups.[44] This can result in a process that Gramsci termed 'passive revolution': elite-led reform directed towards preventing and disabling efforts at transformation by subaltern movements. Within such 'compromises' the interests of the dominant group prevail and therefore must not touch upon 'the decisive nucleus of economic activity' of the ruling class.[45] On the other hand these concessions must go beyond the narrowly corporate economic interests of the ruling class and offer real material benefits and incentives to the subaltern classes.[46] When successful, the incorporative dimensions of hegemony result in what Gramsci termed *trasformismo*: the process whereby leaders and potential leaders of subordinate groups are co-opted into the dominant hegemonic project in an effort to neuter and forestall challenges to the rule of the dominant class.[47]

Whilst the forgoing analysis might suggest that hegemony is a uniform phenomenon, Joseph Femia suggests that it is possible to identify three different levels of hegemony in Gramsci's work.[48] The paradigmatic example of hegemony is what Femia calls *integral hegemony*.[49] This entails mass affiliation, unqualified commitment and a strong, organic relationship between rulers and ruled. The immediate situation in revolutionary France was for Gramsci an exemplifier of such integral hegemony. The second level of hegemony is *decadent hegemony*.[50] This is when bourgeois economic dominance starts to decay and becomes increasingly difficult to represent as furthering everybody's interests. Conflict lurks beneath the surface of society, and instability characterises the relations between the dominant and non-dominant classes. Finally, the third and lowest form of hegemony is *minimal hegemony*.[51] This type of hegemony 'rests on the ideological unity of the economic, political and intellectual elites'[52] along with 'aversion to any intervention of the popular masses in State life'.[53] In such circumstances hegemonic activity is limited, confined primarily to the upper and middle classes and therefore becomes 'merely an aspect of the function of domination'.[54] Femia's tripartite reading of Gramsci's concepts of hegemony and consent reveals that they are not uniform concepts, but vary in degree and intensity according to various socio-political contingencies.

---

[44] ibid, 161.   [45] ibid.   [46] ibid, 182.   [47] ibid, 58–59.
[48] Joseph V Femia, *Gramsci's Political Thought: Hegemony, Consciousness, and the Revolutionary Process* (Clarendon Press, 1981) 46.
[49] ibid.   [50] ibid, 47.   [51] ibid.   [52] ibid.   [53] ibid (quoting Gramsci).
[54] ibid, 48.

1.2.4 COUNTER-HEGEMONY

The need for hegemony to continuously produce and reproduce the consent of the governed and incorporate the subaltern classes into the hegemonic project means that it should be understood as an incomplete and highly unstable *process*, susceptible to contestation by *counter-hegemonic* forces. For Gramsci the historic unity of the ruling classes lies in the integral state. By contrast, the subaltern classes are 'by definition . . . not unified and cannot unite until they are able to become a "State". Their history is therefore, intertwined with that of civil society'.[55] Gramsci argued that in a society with highly developed ruling class hegemony, direct attacks upon the state (what he termed 'wars of movement') would be ineffective strategies because the institutions of civil society would be able to reassert bourgeois dominance.[56]

As an alternative strategy Gramsci realised the importance of what he called *war of position*:[57] a process which 'slowly builds up the strength of the social foundations of a new state' by 'creating alternative institutions and alternative intellectual resources within existing society'.[58] The aim here is to capture the agencies of civil society – the trade unions, the arts, the mass media and so forth – in order to advance a proletarian counter-hegemony. Counter-hegemony thus involves a conscious strategy of destabilising the consensus – what Gramsci termed 'common sense' – upon which the ruling world view rests.[59] Common sense, although inadequate in itself to be able to transcend bourgeois domination, is nonetheless the starting point for a counter-hegemonic strategy.[60] Common sense is not understood as 'monolithic or univocal' but rather as 'a syncretic historical residue, fragmentary and contradictory, open to multiple interpretations and potentially supportive of very different kinds of social visions and political projects'.[61] Because of this, in the construction of counter-hegemony 'it is not a question of introducing from scratch a scientific form of thought into everyone's individual life, but of making "critical" an already existing activity'.[62] Counter-hegemonic

---

[55] Gramsci (n 1) 52.   [56] ibid, 239.   [57] ibid, 238–239.   [58] Cox (n 2) 53.
[59] Gramsci (n 1) 134, 197, 195, 229, 413.
[60] As Thomas Nemeth put it 'Clearly since we are to proceed to coherence from incoherence, to systemization from confusion, to reflection and self-consciousness from naivety, we must start from that philosophy and attitude which is so much a part of everyday life'. Thomas Nemeth, *Gramsci's Philosophy: A Critical Study* (Harvester Press, 1981) 75.
[61] Rupert (n 16) 488.   [62] Gramsci (n 1) 330–331.

strategy therefore entails 'the "reworking" or "refashioning" of elements which are constitutive of the dominant hegemony'.[63]

It should be recalled that hegemony itself is not simply a dominant ideology that shuts out all alternative visions and political projects. One of the important aspects of a hegemonic project is the way that it absorbs counter-hegemonic ideas through at least partially addressing some of the concerns and demands of the subaltern classes. The task of a counter-hegemonic strategy must therefore be to 'tease out' and supplement aspects of the hegemonic apparatus that embody the 'good sense' of the subaltern classes and refashion them into an ideologically mature alternative to the extant hegemony.[64] However, as Cox points out, this gives rise to complex strategic implications: 'It means actively building a counter-hegemony within an established hegemony while resisting the pressures and temptations to relapse into pursuit of incremental gains for subaltern groups within the framework of bourgeois hegemony'.[65] Counter-hegemonic strategy is therefore a highly delicate balancing act between co-option and subversion.

This brief exploration of the concept of hegemony reveals an idea of a complex and multifaceted process that cannot be reduced to mere manipulation or propaganda. Hegemony is not ideology in the narrow sense, but rather the construction of an entire world view or 'common sense'. Rather than being a one-way or top-down process, it is a process of (asymmetric) dialogue between the ruling group and the subaltern classes. The consent of the masses will always be conditional and subject to constant renegotiation. Counter-hegemonic strategies involve the contestation of the legitimacy of the world view of the dominant hegemonic bloc by unifying pre-existing oppositional tendencies within the subaltern classes into a coherent and systematic alternative world view that can win broad-based support across the subaltern classes.

1.3 NEO-LIBERAL GLOBALISATION AND GLOBAL HEGEMONY

This section applies Gramsci's theories, as well as theories of contemporary scholars, to an analysis of the emergence of neo-liberal globalisation as a hegemonic project over the last thirty years. It will begin with a brief account of how neo-liberalism will be understood in this thesis. It will

---

[63] Alan Hunt, 'Rights and Social Movements: Counter-Hegemonic Strategies' (1990) 17(3) *Journal of Law and Society* 311, 313.
[64] Gramsci (n 1) 326–327.   [65] Cox (n 2) 53.

then trace neo-liberalism's development from an intellectual movement to a form of state power and its movement from a national to a transnational phenomenon. Next, this section will consider if the aftermath of the 2007 global financial crisis constitutes a crisis for neo-liberal hegemony or even whether it is arguable that we now live in a 'post-neo-liberal world'.

1.3.1 WHAT IS NEO-LIBERALISM?

For the last twenty years the term 'neo-liberalism' has become ubiquitous in the social sciences. It has been argued that '[w]e live in an age of neo-liberalism'[66] and that it is 'the common sense way many of us interpret, live in, and understand the world'.[67] Increased usage of the term does not, however, signify widespread consensus about its meaning. On the contrary, the term neo-liberalism has been subject to much critical contestation with regard to its origins, meaning and trajectories. Wendy Larner has argued that neo-liberalism should be understood as comprising three intertwined dimensions: a set of policies, an ideology and a form of 'governmentality'.[68] Others have suggested the term 'neo-liberalism' can be invoked to describe anything from an overarching economic, or even cultural, system to particular attitudes or inclinations towards entrepreneurship, competition, responsibility and self-improvement.[69]

Whilst this author believes that all of the above descriptions contain valid insights into the meaning of neo-liberalism, this thesis will follow David Harvey in suggesting that neo-liberalism should be understood fundamentally as 'a project to achieve the restoration of class power'[70] advanced to 'reflect the interests of private property owners, businesses, multinational corporations and financial capital'.[71] Since the 1980s neo-liberalism can be identified as 'a particular form of class rule and state power that intensifies competitive imperatives for both firms and workers, increases dependence on the market in daily life and reinforces

---

[66] Alfredo Saad-Fiho and Deborah Johnson, 'Introduction' in Alfredo Saad-Fiho and Deborah Johnson, *Neoliberalism: A Critical Reader* (Pluto Press, 2005) 1.
[67] David Harvey, *A Brief History of Neoliberalism* (Oxford University Press, 2005) 3.
[68] Wendy Larner, 'Neo-Liberalism: Policy, Ideology, Governmentality' (2000) 63 *Studies in Political Economy* 5.
[69] Andrew Kipnis, 'Neoliberalism Reified: Suzhi Discourse and Tropes of Neoliberalism in the People's Republic of China' (2007) 13 *Journal of the Royal Anthropological Institute* 383, 383.
[70] Harvey (n 67) 16.   [71] ibid, 7.

the dominant hierarchies of the world market, with the U.S. at its apex'.[72] Gramsci observed that prior to the creation of new forms of state 'there can, and indeed must, be hegemonic activity even before the rise to power, and that should not count only on ... material force'[73] but also on 'intellectual, moral and political hegemony'.[74] This section will therefore provide a brief genealogy of neo-liberalism's transformation from a counter-hegemonic intellectual movement into 'a hegemonic concept that is seeping into and co-opting the whole spectrum of political life'.[75]

1.3.2 FROM INTELLECTUAL MOVEMENT TO STATE POWER

The intellectual origins of neo-liberalism have been traced to the ideals of strong individualism and property rights advanced by the eighteenth-century French physiocrats,[76] the critique of mercantilism and the defence of free markets found in Adam Smith's *Wealth of Nations*,[77] and the nineteenth-century *laissez-faire* economics of the 'Manchester School'.[78] After the Second World War classical liberal economics fell from grace within the capitalist order as a result of the catastrophic consequences of the Wall Street Crash and the Great Depression in the 1930s.[79] New state forms and international relations were engendered to ensure domestic peace and international stability within the framework of a new 'class compromise between capital and labour'.[80] This period in world politics has been termed by some as the era of 'embedded liberalism'[81], characterised by the sharing of productivity gains through rises in wages, active fiscal and monitory policies aimed at the constant

---

[72] Greg Albo, Sam Gindin and Leo Panitch, *In and Out of the Crisis: The Global Financial Meltdown and Left Alternatives* (PM Press, 2010) 28.
[73] Gramsci, 59.   [74] ibid, 58.
[75] Philip Cerny, 'Embedding Neoliberalism: The Evolution of a Hegemonic Paradigm' (2008) 2(1) *The Journal of International Trade and Diplomacy* 1, 3.
[76] Bo Strath, 'The Liberal Dilemma: The Economic and the Social, and the Need for a European Contextualization of a Concept with Universal Pretensions' in Ben Jackson and Marc Stears (eds.) *Liberalism as Ideology: Essays in Honour of Michael Freeden* (Oxford University Press, 2012) 101.
[77] Simon Clarke, 'The Neoliberal Theory of Society' in Alfredo Saad-Fiho and Deborah Johnson (eds.) (n 66) 50.
[78] Thomas Palley, 'From Keynesianism to Neoliberalism: Shifting Paradigms in Economics' in Saad-Fiho and Johnson (eds.) ibid, 20.
[79] Harvey (n 67) 9–10.   [80] ibid, 10.
[81] John Gerard Ruggie, 'International Regimes, Transactions, and Change: Embedded Liberalism in the Postwar Economic Order' (1982) 36(2) *International Regimes* 379–415.

progression of demand, extensive welfare state provisions and state-regulated financial systems.[82]

Neo-liberalism can be understood as a movement that emerged in opposition to post-war embedded liberalism and instead agitated for the resurrection and reinterpretation of the values and economic perspectives associated with classical liberalism such as individualism, efficiency, competition and minimal state intervention in the economy.[83] Initially, a small group of academics, economists, historians and lawyers formed the Mont Pelerin Society in 1947 to co-ordinate and formulate policies as well as agitate against Marxist theories of centralised planning and Keynesian policies of state intervention in the economy. Whilst these ideas were able to secure influence in a small number of intellectual nerve centres, such as in Freiberg and Chicago, they remained on the margins of policy considerations for the first two decades after the Second World War.[84] The Keynesian solution to the perceived threats of economic collapse, disorder and communism held sway over the elites of the western world.[85]

This situation began to change in the 1970s when a 'structural' crisis within the Keynesian world economic order facilitated the shift from neo-liberalism as an intellectual movement into a project for the reconfiguration of socioeconomic and state power.[86] Faced with a 'crisis of capital accumulation' business interests and state bureaucrats turned to the political economy of neo-liberal thinkers like Friedrich Von Hayek, Milton Friedman and James Buchanan. The macroeconomic crisis conditions of the 1970s provided the basis for a critique of Keynesian financial regulation, union power, corporatist planning, state ownership and 'overregulated' labour markets.[87] In place of these policies neo-liberals advanced arguments for sharply cutting back government spending, deregulating

---

[82] Robert Boyer, *The Future of Economic Growth: As Old Becomes New* (Edward Elgar, 2004) 61–68.

[83] James H Mittelman, *Wither Globalization?* (Routledge, 2004) 5.

[84] Jodi Dean, *Democracy and Other Neoliberal Fantasies: Communicative Capitalism and Left Politics* (Duke University Press, 2009) 52.

[85] Describing the dominant form of state that emerged in the aftermath of the Second World War in Western Europe, Japan and North America, Harvey argues 'What all these various state formations had in common was an acceptance that the state should focus on full employment, economic growth, and the economic welfare of its citizens, and that state power should be freely deployed, alongside of or, if necessary, intervening in or even substituting for market practices to achieve these ends.' Harvey (n 67) 8.

[86] Jamie Peck and Adam Tickell, 'Neoliberalizing Space' (2002) 34(3) *Antipode* 380, 388.

[87] ibid.

labour and financial markets, privatising public services and opening national economies to free trade and multinational capital investments. In short, these policies aimed at 'rolling back' the state and 'rolling out' the 'free market'.[88] Agitation for the erosion of policies of extensive welfare and economic planning in favour of financial deregulation and privatisation began to take hold as the neo-liberal viewpoint became 'bureaucratized and institutionalized' with different degrees of intensity across a range of public and private spheres.[89]

1.3.3 FROM THE NATION STATE TO GLOBAL GOVERNANCE: THE GLOBALISATION OF NEO-LIBERALISM

Although neo-liberalism was originally 'a nation-state-level phenomenon' it soon developed in nature alongside 'structurally transformative transnational and globalising developments'.[90] From the mid-1970s onwards, three paradigmatic shifts in the global economy paved the way for the phenomenon that has been termed 'neo-liberal globalisation'.[91] First, financial deregulation and computerisation phased out most significant geographic barriers in relation to financial activities, meaning that capital became more geographically mobile and less embedded within a national state context. Second, Transnational Corporations (TNCs) consolidated their global productivity extensively, making them even more influential and powerful economic agents at both the national and supranational levels. Third, the transnational structures regulating the global economy transformed in both nature and size and were thus able to exercise extraordinary leverage in the implementation of neo-liberal reforms.[92]

These shifts have all significantly strengthened the *structural* power of *global* capital to limit the range of public policy options available to national states.[93] Stephen Gill argues that these shifts have provided the

---

[88] Robert Pollin, *Contours of Descent* (Verso, 2005) 173; Rick Rowden, *The Deadly Ideas of Neoliberalism: How the IMF Has Undermined Public Health and the Fight against AIDS* (Zed, 2009) 66.
[89] Stephen Gill, 'Globalisation, Market Civilisation, and Disciplinary Neoliberalism' (1995) 24 *Millennium: Journal of International Studies* 399, 412.
[90] Cerny (n 75) 2.   [91] Colin Leys, *Market-Driven Politics* (Verso, 2001) 13.
[92] ibid, 14–21.
[93] Stephen Gill and David Law, 'Global Hegemony and the Structural Power of Capital' in Stephen Gill (ed.), *Gramsci, Historical Materialism and International Relations* (Cambridge University Press, 1993) 93–124.

basis for what he terms 'disciplinary neo-liberalism': the socioeconomic project of transnational capital to 'expand the scope and increase the power of market-based structures and forces so that governments and other economic agents are disciplined by market mechanisms'.[94]

The interrelated economic, ideological and technological shifts described above have rapidly increased the global mobility of capital and enabled the development of stateless forms of capital accumulation.[95] The emergence of neo-liberalism as a global phenomenon has been analysed by a school of thought within international relations that has been identified as 'neo-Gramscianism'.[96] Neo-Gramscians take Gramsci's analysis of power relations at the nation state level and apply it to the study of World Order.[97] For Robert Cox, the pioneer of neo-Gramscian analysis, the state is the basic entity of international relations and therefore where the hegemonies of social classes are built.[98] A world hegemony is thus in its beginnings an outward expansion of the internal hegemony established by nationally based ruling classes.[99] However, a growing body of literature has since identified the rise of a transnational capitalist class (TCC) – comprising the owners and managers of TNCs and private financial institutions – that has become increasingly autonomous from specific national state formations.[100] It is argued that under globalisation a new class fractionation has evolved between national and transnational fractions of classes.[101]

---

[94] Stephen Gill and A Claire Cutler (eds.), *New Constitutionalism and World Order* (Cambridge University Press, 2014) 315. For a detailed analysis see Alasdair Roberts, *The Logic of Discipline: Global Capitalism and the Architecture of Government* (Oxford University Press, 2010).

[95] Angus Cameron and Ronen Palan, 'The Imagined Economy: Mapping Transformations in the Contemporary State' (1999) 28(2) *Millennium* 267–288.

[96] John Bayles and Steve Smith, *The Globalization of World Politics* (Oxford University Press, 2001) 235–239.

[97] Cox (n 2) 59–62.   [98] ibid, 61.   [99] ibid.

[100] See, for example, Martin Shaw, *Theory of the Global State: Globality as an Unfinished Business* (Cambridge University Press, 2000); William I Robinson and Jerry Harris, 'Towards a Global Ruling Class? Globalization and the Transnational Capitalist Class' (2000) *Science & Society* 64 (1) 11–54; Leslie Sklair, *The Transnational Capitalist Class* (Blackwell, 2001); William Robinson, 'Social Theory and Globalisation: The Rise of the Transnational State' (2001) 30 *Theory and Society* 157–200; William I Robinson, *Global Capitalism and the Crisis of Humanity* (Cambridge University Press, 2014).

[101] William I Robinson, 'Gramsci and Globalisation: From Nation-State to Transnational Hegemony' in Andreas Bieler and Adam David Morton (eds.), *Images of Gramsci: Connections and Contentions in Political Theory and International Relations* (Routledge, 2006) 170.

1.3.4 GLOBAL HEGEMONY AND THE GLOBAL INTEGRAL STATE

How do neo-Gramscians understand the concept of hegemony? The term is used and understood in a variety of ways within the discipline of international relations (IR). In the predominant neo-realist approach, hegemony is conventionally employed to signify the dominance of one state over several others.[102] The 'hegemon' has the capacity to exercise substantial control over the international system by virtue of some advantage that it possesses such as its geographic location, natural resources, economic power, military capacity, technological innovation and so on.[103] Neo-realist proponents of 'hegemonic stability theory' (HST) believe that the existence of a hegemon is necessary to provide a stable framework for an open and liberal word market to function.[104]

HST has been critiqued by those who adopt a 'neo-liberal' approach to IR. Neo-liberals argue that whilst international regimes are the product of a hegemonic power, they are not reducible to that hegemonic power. Even after the decline of a hegemon an international system can continue to function through its institutions, which exert an independent causal effect on world politics.[105] Thus, on the neo-liberal account, hegemony can survive beyond the hegemon that established it in the first place.[106]

Neo-Gramscian scholars, drawing on Gramsci's analytic framework, have argued that the conception of hegemony in IR must move beyond both state-centric and institutional-based analysis to an understanding of the world system as a 'totality'.[107] Hegemony at the international level is not merely amongst states, but comprises a social, economic and political structure wherein world hegemony 'is expressed in universal norms, institutions and mechanisms which lay down general rules of behaviour for states and those who support the dominant mode of production'.[108] Neo-Gramscianism thus locates the subject of hegemony in neither a

---

[102] For a classic statement of this perspective see generally, Robert Gilpin (n 2).
[103] Hans Morgenthau and Kenneth W Thompson, *Politics among Nations: The Struggle for Power and Peace* (McGraw-Hill, 2005) 124–164.
[104] Charles Kindleberger, *The World in Depression, 1929–39* (University of California Press, 1973) 288–306.
[105] Stephen Krasner, 'Structural Causes and Regime Consequences: Regimes as Intervening Variables' (1982) 36(2) *International Organizations* 185.
[106] Robert Keohane, *After Hegemony: Cooperation and Discord in the World Political Economy* (Princeton University Press, 1984).
[107] Stephen Gill, 'Epistemology, Ontology and the "Italian School"' in Stephen Gill (ed.), *Gramsci, Historical Materialism and International Relations* (n 2) 41–42.
[108] Cox (n 2) 61–62.

powerful state nor in the institutional architecture of a world order, but rather in international social forces that control the means of production.

Neo-Gramscianism also builds upon Gramsci's analysis of power as comprising elements of both coercion and consent, represented in political society and civil society, respectively, and together comprising an 'integral state'. While some have questioned whether there is a global civil society that is organically related to a global political society (or global state),[109] a number of studies reveal that while global governmental and civil societal institutions 'may be rudimentary, they are succeeding rather well in facilitating the transnational expansion and consolidation of capitalism and a transnational class'.[110] The following sections will therefore explore the nature of global political society and global civil society in the context of neo-liberal global political-economic order.

1.3.4.1 GLOBAL POLITICAL SOCIETY

Following Gramsci's conception, global political society can be understood as the predominantly coercive apparatus of neo-liberal globalisation.[111] Contemporary global political society is constituted through a network of international institutions and powerful states whose function is to enforce the interests of transnational capital and powerful states in the international system.[112] Importantly, a number of sovereign economic decision-making powers have been removed from nation states and placed in the hands of international economic institutions (IFIs) that possess effective judicial and economic enforcement powers.[113] The socioeconomic project of disciplinary neo-liberalism requires the adoption of a political-juridical form that Stephen Gill terms 'new constitutionalism'.[114] The political-juridical project aims to entrench the power of capital through a series of pre-commitment mechanisms such as multilateral agreements and structural adjustment programmes. These mechanisms 'serve to constitute the limits of political possibility and to

---

[109] Randall Germain and Michael Kenny, 'Engaging Gramsci: International Relations Theory and the New Gramscians' (1998) 24(3) *Review of International Studies* 3–21.
[110] A Claire Cutler, 'Gramsci, Law, and the Culture of Global Capitalism' (2005) 8(4) *Critical Review of International Social and Political Philosophy* 527, 536–537.
[111] See Section 1.2.
[112] BS Chimni, 'International Institutions Today: An Imperial State in the Making' (2004) 15 *European Journal of International Law* 1.
[113] ibid.
[114] Stephen Gill, 'New Constitutionalism, Democracy and Global Political Economy' (1998) 10(1) *Pacifica Review* 23.

inspire the confidence of investors by increasing the role and scope of market values and disciplines'.[115]

There are two components central to the coercive dimension of new constitutionalism. The first is the reconfiguration of state apparatuses to insinuate capital and economic decision-making from popular democracy.[116] In order to do this 'neoliberals have to put strong limits on democratic governance, relying instead upon undemocratic and unaccountable institutions (such as the Federal Reserve or the IMF) to make key decisions'.[117] The second component is the adoption of measures to construct and extend liberal capitalist markets.[118] The neo-liberal state form enables capital to exploit new forms of wealth through what David Harvey calls 'accumulation by dispossession'.[119] This is a multifaceted process that has involved, *inter alia*, the commodification and privatisation of land and the forceful expulsion of peasant populations, the conversion of various forms of property rights (common, collective, state, etc.) into exclusively private property rights, the expansion of the domain of intellectual property rights (IPRs) and the suppression of the rights of the commons.[120]

It is possible to identify three fundamental nodes within global political society that are integral to establishing new constitutionalism: multilateral financial institutions such as the IMF and World Bank; international trade agreements and organisations such as the World Trade Organization (WTO); and powerful states such as those 'Group of 8' (G8) countries that gather routinely to co-ordinate polices and also wield disproportionate influence within the aforementioned international institutions. Since the 1980s the IMF and World Bank have become 'missionary institutions' for neo-liberal reform.[121] The lending practices in these IFIs increasingly became attached to 'structural adjustment' policies. Beginning in the early 1980s these 'structural adjustment programs' involved macroeconomic stabilisation policies such as public spending reduction, the removal of price controls and the devaluation of currency. By the 1990s, these conditionalities had evolved into broader packages of policies including 'trade liberalization and foreign exchange restrictions, deregulation of the economy, privatisation and the other elements of what became known

---

[115] Stephen Gill (ed.), *Global Crisis and the Crises of Global Leadership* (Cambridge University Press, 2012) 257.
[116] ibid.   [117] Harvey (n 67) 69.   [118] Gill, 'New Constitutionalism' (n 114) 26.
[119] David Harvey, 'Neoliberalism as Creative Destruction' (2007) 610 *The Annals of the American Academy of Political and Social Science* 21, 34–35.
[120] ibid; Harvey (n 67) 160.
[121] Joseph Stiglitz, *Globalisation and Its Discontents* (Penguin, 2002) 13.

as the "Washington consensus"'.[122] The IMF and World Bank therefore have much economic leverage to act as enforcers of the Washington Consensus, particularly over aid-dependent, low income countries. The World Bank also operates a trade court called the International Centre for the Settlement of Investment Disputes (ICSID) which facilitates legal disputes between international investors.[123]

International trade agreements and bodies also form a key component of the global political society. The most important of these is the WTO. The WTO, formed in 1995 as the successor to the General Agreement on Trade and Tariffs (GATT), is formally committed to the globalisation and liberalisation of trade and has become a key institution in which the economic sovereignty of nation states has been seceded.[124] Through agreements such as the agreement on Trade-Related Intellectual Property Rights (TRIPs), the agreement on Trade-Related Investment Measures (TRIMs) and the General Agreement on Trade in Services (GATS) the WTO has required states to, *inter alia*, provide extensive protection of IPRs in the areas of medicine and agriculture, allow for the marketisation of services such as health and education and open up domestic markets to foreign investments and competition.[125] These agreements are judicially enforced through the panel and Appellate Body of the WTO.[126] In addition to the WTO there are a number of other regional trade agreements that are enforced through economic courts such as the Association of Southeast Asian States (ASEAN), the European Court of Justice (ECJ), the European Free Trade Area Court and the Southern Common Market (MERCUSOR).[127]

Powerful states within the global order, such as those that comprise the G8 countries or the 'permanent five' members of the United Nations (UN) Security Council, are also an important component of global political society. These states are able to utilise their military, economic, diplomatic, cultural and political hegemony to shape global economic and governance policy.[128] Such states exert enormous influence on countries in the Global

---

[122] Rowden (n 88) 66.
[123] See https://icsid.worldbank.org/ICSID/Index.jsp accessed 28 April 2016.
[124] Chimni (n 112) 7.   [125] ibid, 7–9.   [126] ibid, 9.
[127] Karen J Alter, 'The Multiple Roles of International Courts and Tribunals: Enforcement, Dispute Settlement, Constitutional and Administrative Review' (2012) Faculty Working Papers, Paper 212, 3. http://scholarlycommons.law.northwestern.edu/cgi/viewcontent.cgi?article=1211&context=facultyworkingpapers accessed 16 August 2013.
[128] Robert Gilpin, *Understanding the Global Political Economy* (Princeton University Press, 2001) 15.

South through multilateral institutions. For example, the IMF and World Bank have capital-weighted decision-making processes that give the northern states the dominant say in these institutions. The undemocratic financial share-based structures of the IMF and World Bank mean that the lending patterns of both institutions are 'influenced by the commercial and financial interests of the US and to a lesser extent the E.U'.[129] while the WTO is dominated by the United States, Europe and Japan.[130] Outside of multilateral institutions northern states employ an array of military, economic and diplomatic techniques to secure policies favourable to northern-based capital interests in third-party states. In particular, bilateral trade agreements (BTAs) and bilateral investment treaties (BITs) are now a principle means through which social services are commodified, investor access to investment opportunities is guaranteed, public services are privatised and economic sovereignty is diminished.[131] The relevant institutional subset of global political society that will be explored in each of the case study chapters will be referred to as the 'political-institutional context'.

1.3.4.2 GLOBAL CIVIL SOCIETY

Alongside the globalisation of the social relations of production ('disciplinary neo-liberalism') and the globalisation of the politico-juridical form ('new constitutionalism'), global civil society is the global domain of uncoerced human association and the set of relational transnational networks that fill that space.[132] Global civil society is therefore composed of those non-government organisations (NGOs), social movements, think tanks, political parties, forums, media organisations and other forms of human association that extend beyond the domain of national civil society due to their organisational form or because their participants increasingly see their interests in global rather than national terms. Global civil society is the principle domain in which global hegemony is constructed, reproduced and potentially contested. It is the domain where the values of neo-liberal globalisation are legitimated and potentially counter-hegemonic challenges are defeated or

---

[129] Ricardo Faini and Enzo Grill, 'Who Runs the IFIs?' (2004) Centre for Economic and Policy Research, Discussion Paper no. 4666, 21 www.dagliano.unimi.it/media/WP2004_191.pdf accessed 16 August 2013.
[130] Stiglitz (n 121) 225.
[131] James Thuo Gathii, 'The Neoliberal Turn in Regional Trade Agreements' (2011) 86 *Washington Law Review* 421, 421.
[132] Michael Walzer, 'The Civil Society Argument' in Chantal Mouffe (ed.) *Dimensions of Radical Democracy: Pluralism, Citizenship, Community* (Phronesis, 1992) 89.

co-opted.[133] Across the domains of global political society and global civil society it is possible to identify a *neo-liberal transnational historical bloc* (NTHB) that is wider than the strict class base of the TCC. Such a bloc also encompasses the cadre, managers and technicians of the agencies of global political society discussed above and an array of organic intellectuals operating elite policy-planning institutions, TNCs and global governance organisations.[134]

The NTHB aims to promote consensus around the creation of 'market civilisation': a social order in which the allocation of goods, services, everyday life and culture is mediated by capitalist market mechanisms, market forces and market values.[135] Private organisations like the International Chamber of Commerce, the World Business Council for Sustainable Development and the World Economic Forum legitimate the norms of neo-liberal globalisation through co-ordination strategies, political advocacy and lobbying directed at governments and multilateral organisations.[136] In addition to these private organisations, international institutions like the World Bank and IMF also constitute part of global civil society when exerting non-coercive functions of persuasion and legitimation. The World Bank, for example, has a highly influential research arm called the World Bank Institute (WBI) which promotes the organisation's values and programmes through education and outreach.[137] International institutions therefore help to produce and reproduce hegemony by legitimating the norms of world order, co-opting elites from peripheral states and absorbing counter-hegemonic ideas.[138] As Benjamin Kohl notes, the international financial institutions (IFIs) have been able to persuade 'even progressive governments from the Worker's Party in Brazil to the United Marxist Leninist Party in Nepal and the African National Congress in South Africa that, as Margaret Thatcher insisted, there is no alternative to the market'.[139] Within the case

---

[133] Stephen Gill, *American Hegemony and the Trilateral Commission* (Cambridge University Press, 1990) 52; William K Carroll, 'Crisis, Movements, Counter-Hegemony: In Search of the New' (2010) 2(2) *Interface* 168.

[134] William K Carroll, 'Hegemony and Counter-Hegemony in a Global Field' (2007) 1(1) *Studies in Social Justice* 36, 36 and 38.

[135] Gill, 'Globalisation, Market Civilisation, and Disciplinary Neoliberalism' (n 114) 399.

[136] Carroll, 'Hegemony and Counter-Hegemony in a Global Field' (n 134) 40–52.

[137] The World Bank, 'About WBI' http://wbi.worldbank.org/wbi/about accessed 16 August 2013.

[138] Cox (n 2) 62–64.

[139] Benjamin Kohl, 'Challenges to Neoliberal Hegemony in Bolivia' (2006) 38(2) *Antipode* 305, 307.

study chapters these 'non-coercive' aspects of neo-liberal globalisation will be referred to as the 'legitimising (ethico-political) framework' of the particular global governance regime in discussion.

The UN has also become an important forum for the promotion of neo-liberal governance over the last thirty years. In the 1970s the UN was mandated to regulate and monitor the activities of TNCs through the United Nations Centre on Transnational Corporations (UNCTC). In the 1980s, as global neo-liberal discourse was being consolidated, the UN's approach to TNCs gradually retreated from one of *regulation* to *facilitation*, with agencies like the UN Conference on Trade and Development (UNCTD) helping to secure TNC access to the Global South.[140] Throughout the 1990s US pressure led to the curtailment of efforts to monitor corporate practices and hold corporations accountable for their actions under international law.[141] Under the leadership of Kofi Anan, an MIT business school graduate sympathetic to the neo-liberal agenda, a number of business partnerships were cultivated with the UN, the most notable of which is the UN 'Global Compact' (GC) forged in 2000. The GC is a non-binding voluntary corporate initiative in which member companies were encouraged to learn from other members' best practices.[142] It is widely regarded as both a retreat from earlier attempts to regulate TNCs to ensure compliance with human rights standards and as a strategic attempt – by leading factions within the TCC – to provide legitimacy to neo-liberal global corporate governance.[143] This is not to suggest, however, that the UN is a homogeneously hegemonic institution, and indeed there may be agencies and individuals within it that foster and assist counter-hegemonic discourse.[144]

---

[140] Polaris Institute, 'The Corporate Stranglehold over the United Nations: How Big Business Already Wields Significant Power over the UN Water Agenda' (*Polaris Institute*, October 2009) www.polarisinstitute.org/files/UNreport.pdf accessed 17 August 2013.
[141] Ellen Paine, 'The Road to the Global Compact: Corporate Power and the Battle over Global Public Policy at the United Nations' (*Global Policy Forum*, October 2000). http://dspace.cigilibrary.org/jspui/bitstream/123456789/17581/1/The%20Road%20to%20the%20Global%20Compact.pdf?1) accessed 17 August 2013.
[142] ibid.
[143] Susanne Soederberg, 'Taming Corporations or Buttressing Market-Led Development? A Critical Assessment of the Global Compact' (2007) 4(4) *Globalizations* 500–511.
[144] Importantly, for the context of this study, Upendra Baxi notes that 'the United Nations human rights treaty bodies, the Office of the High Commissioner for Human Rights and the human rights council ... have *often*, although not *always*, provided the space for counter-hegemonic contestation'. Upendra Baxi, 'Adjudicatory Leadership in a Hyper-Globalizing World' in Stephen Gill (ed.), (n 15).

It is important to recall at this stage that hegemony is not simply an attempt to impose a top-down, unified and coherent theory onto a passive populace, but rather an on-going process that requires and presumes the consent of the subordinate classes via an array of concessionary processes. To achieve hegemony, neo-liberal governance must be adaptive to critiques that emanate from 'counter-movements' that spring up in response to the dislocating effects of radical free market policies.[145] Jamie Peck argues that at the heart of the neo-liberalisation process lies a contradictory dialectic of 'market/order' whereby 'the rising costs of deregulatory overreach, public austerity, market failure, and social abandonment typically force neoliberals to engage with a range of unsavoury challenges of intervention, amelioration, and reregulation'.[146] This is not simply out of a concern to advance short-term economic interests, but also a broader attempt to ideologically stabilise the neo-liberal order. In line with such analysis, it is generally recognised that neo-liberal globalisation has gone through two specific phases; 'the first as the shock-therapy associated with Reagan and Thatcher, Latin America, and the Soviet bloc, and the second with the social market, Third-Wayism and the post-Washington consensus'.[147] At the level of World Order, the shift from the Washington Consensus to the Post Washington Consensus (PWC) could be interpreted as an attempt to facilitate the expansion of a hegemonic neo-liberal world order through incorporating *aspects* of the critique of the neo-liberal model into the governance framework of neo-liberal globalisation itself.[148]

One noteworthy shift associated with the PWC is the emergence of 'global governance'.[149] Global Governance can be understood as a multi-layered process of social co-ordination bringing together state actors, businesses, NGOs and other civil society organisations in the global arena for the purposes of rule-making, political co-ordination and problem solving.[150]

---

[145] Thomas Pogge and Karl Polanyi, *The Great Transformation: The Political and Economic Origins of Our Time* (Beacon Press, 1944) 130.

[146] Jamie Peck, *Constructions of Neoliberal Reason* (Oxford University Press, 2010) 23–24.

[147] Ben Fine and Dimitris Milonakis, '"Useless but True": Economic Crisis and the Peculiarities of Economic Science' (2011) 19(2) *Historical Materialism* 3, 6.

[148] Mark Rupert, 'The New World Order: Passive Revolution or Transformative Process?' in Louise Amoore (ed.), *The Global Resistance Reader* (Routledge, 2005) 194–208.

[149] See generally, Ulrich Brand, 'Order and Regulation: Global Governance as a Hegemonic Discourse of International Politics?' (2005) 12(1) *Review of International Political Economy* 12(1) 155–176.

[150] David Held and Anthony McGrew, *Governing Globalization: Power, Authority and Global Governance* (Wiley, 2002) 8.

As Gill notes, a key theme that runs through much of the policies promoted by global governance is that free trade, free competition and free exchange are essential for economic efficiency, social welfare and social progress.[151] Nevertheless, global governance has also emerged as a response to a growing perception that neo-liberalism has not succeeded in consolidating a viable and sustainable mode of regulation capable of protecting the environment and addressing the concerns of the more marginalised sections of the world's population. Global Governance 'recognises the need to solve "world problems" such as economic instability, poverty and ecological destruction, cooperatively and in dialogue, by bringing together not only state actors but NGOs and private enterprises from civil society' on the global stage.[152]

A key illustration of the PWC global governance approach is the shift in the lending practices of the IMF and World Bank from 'structural adjustment' to 'poverty reduction'. In 1999, the IMF and World Bank reformulated their much-criticised SAPs as 'Poverty Reduction Strategy Papers' (PRSPs). In response to criticisms that previous structural adjustment programmes were top-down in nature and failed to adequately integrate pro-poor measures into their strategies, the more recent PRSP model purports to recognise the importance of *national ownership, participation* and *poverty reduction* and emphasises the need for 'broad based participation by civil society'.[153] In accordance with the PRSP approach, the World Bank maintains that 'complementary policies – particularly the provision of an effective social safety net – are ... necessary to minimize adjustment costs and to help make trade reform work for the poor'.[154]

However, poverty reduction is still primarily achieved through economic growth and this requires macroeconomic stability, privatisation and liberalisation.[155] In the new PRSP strategy, the provision of social

---

[151] Gill, 'Globalisation, Market Civilisation, and Disciplinary Neoliberalism' (n 114) 406.
[152] William K Carroll, 'Hegemony, Counter-hegemony, Anti-hegemony' (2006) 2(2) *Socialist Studies* 9, 17.
[153] International Monetary Fund (IMF), 'Poverty Reduction Strategy Papers: A Fact Sheet' www.imf.org/external/np/exr/facts/prsp.htm.
[154] Jeni Klugman et al. (eds.), *A Sourcebook for Poverty Reduction Strategies, Volume II: Macroeconomic and Sectoral Approaches* (World Bank, 2002) 33.
[155] For critique of the PRSP process see: Frances Stewart and Michael Wang, 'Poverty Reduction Strategy Papers within the Human Rights Perspective' in Philip Alston and Mary Robinson (eds.), *Human Rights and Development: Towards a Mutual Reinforcement* (Oxford University Press, 2005) 456–457 (noting that key categories of participants such as parliamentarians, trade unions, women and marginalised groups have been excluded from the PRSP process); Arne Ruckert, 'Towards an Inclusive-Neoliberal

safety nets and other complimentary measures therefore become 'wedded in a marriage of convenience' with traditional neo-liberal economic policy prescriptions.[156] Thus these forms of concessions satisfy Gramsci's formula for ruling class hegemony in the sense that they do not touch upon the essential nucleus of neo-liberal economic relations.[157]

These shifts can be explained by reference to *trasformismo*. While the PRSP strategy and other PWC policies may have failed to create a strong hegemonic world order, particularly in the aftermath of the on-going financial crisis, they may have succeeded in co-opting and forestalling certain manifestations of popular political mobilisation and thereby disabling potentially transformative, self-empowering social movements.[158]

1.3.5 THE END OF NEO-LIBERALISM? THE AFTERMATH OF THE GLOBAL FINANCIAL CRISIS

At the turn of the century Perry Anderson described neo-liberalism as 'the most successful ideology in world history'.[159] However, neo-liberalism has undergone a series of crises in the last ten years, and following the 2007 financial meltdown there has been an unprecedented public debate concerning the relevance, credibility and durability of neo-liberalism as an economic, political and social order.[160] The financial crash, widely attributed to the failure of governments to effectively regulate the financial sector, has undoubtedly dealt a heavy blow to the free market credo that is integral to neo-liberalism's ideological self-representation.[161] Many mainstream commentators have joined radical opponents of neo-liberalism in identifying the unsustainability of the

---

Regime of Development: From the Washington to the Post-Washington Consensus' (2006) 39(1) *Labour, Capital and Society* 35–67; Arne Ruckert, 'Producing Neoliberal Hegemony? A Neo-Gramscian Analysis of the Poverty Reduction Strategy Paper (PRSP) in Nicaragua' (2007) 79 *Studies in Political Economy* 91, 103; Geske Dijkstra, 'The PRSP Approach and the Illusion of Improved Aid Effectiveness: Lessons from Bolivia, Honduras and Nicaragua' (2011) 29(1) *Development Policy Review* 111.

[156] Andrea Cornwall and Karen Brock, 'Beyond Buzzwords: "Poverty Reduction", "Participation" and "Empowerment" in Development Policy' (2005) United Nations Research Institute for Social Development, Program Paper No.10, 8.

[157] Gramsci (n 1) 161. [158] Ruckert, 'Producing Neoliberal Hegemony?' (n 151) 41.

[159] Perry Anderson, 'Renewals' (2000) 1 *New Left Review* 1, 13.

[160] Stephen Gill, 'Introduction: Global Crisis and the Crises of Global Leadership' in Stephen Gill (ed.) (n 115) 4–7; David M Kotz, *The Rise and Fall of Neoliberal Capitalism* (Harvard University Press, 2015).

[161] ibid, 5.

current economic order.¹⁶² Indeed, the impact of the crisis even led the associate editor and chief economics commentator at the Financial Times to declare that 'the world of the last three decades has gone'.¹⁶³

Despite the challenge seemingly posed to the legitimacy of neo-liberalism by the current economic crisis, assumptions that neo-liberalism is dead or that we have now moved to a 'post-neoliberal world' are premature.¹⁶⁴ Indeed, it is clear that the overwhelming response to economic recession by national governments and global governance organisations has been to impose austerity, cut social protection and further privatise and commodify pensions, health and education.¹⁶⁵ In other words, the structural power of disciplinary neo-liberalism combined with the absence of credible alternatives has meant that the economic recession 'has been used by many Western governments as a means of further entrenching the neoliberal model'.¹⁶⁶ The IMF, EU and European Central Bank's joint promotion and enforcement of austerity and privatisation in Greece, Italy, Spain, Portugal and Ireland in response to the economic crisis in the Eurozone demonstrates the continued existence of neo-liberal policy in global governance.¹⁶⁷

Whilst widespread state intervention in the economy in the form of bail outs, quantitative easing, liquidity pumping and stimulus spending appear to contradict the *laissez faire* ethic that underpins much neo-liberal rhetoric, in reality neo-liberal governance has always been characterised not so much by 'deregulation' but rather by new forms of regulation designed to advance the economic interests of the TCC.¹⁶⁸ Measures associated with neo-liberal globalisation such as large state subsidies to the agricultural, pharmaceutical and textile industries (predominantly in the northern states), the introduction of anti-trade union legislation, the creation of new commercial urban spaces shaped by market imperatives and the establishment of tax havens for domestic

---

¹⁶² ibid, 4.
¹⁶³ Martin Wolf, 'Seeds of Its Own Destruction' *Financial Times* (London, 8 March 2009).
¹⁶⁴ Manuel B Aalbers, 'Neoliberalism Is Dead ... Long Live Neoliberalism!' (2013) 37(3) *International Journal of Urban and Regional Research* 1083–1090.
¹⁶⁵ Robin Blackburn, 'Crisis 2.0' (2011) 72 *New Left Review* 33, 34.
¹⁶⁶ Stuart Hall, Doreen Massey and Michael Rustin, 'After Neoliberalism: Analysing the Present' in Stuart Hall, Doreen Massey and Michael Rustin (eds.) *After Neoliberalism? The Kilburn Manifesto* (Surroundings, 2013) 4 available at http://lwbooks.co.uk/journals/soundings/pdfs/manifestoframingstatement.pdf accessed 8 August 2013.
¹⁶⁷ William I Robinson, 'The Global Capital Leviathan' (2011) 165 *Radical Philosophy* 2, 5.
¹⁶⁸ Ronaldo Munck, 'Neoliberalism and Politics, and the Politics of Neoliberalism' in Saad-Fiho and Johnson (eds.) (n 66) 60.

and foreign corporations demonstrate that neo-liberal governance is better understood as a deeply proactive model of governance despite neo-liberalism's self-identification as anti-interventionist. In that sense, whilst the scale of the recent bank bailouts is unprecedented in quantitative terms, qualitatively they do not represent a fundamental rupture with the logic of neo-liberal governance. If neo-liberalism is understood as a class project for the restoration of the optimal conditions for capital accumulation, then the responses governments have taken in the current crisis represent a continuation of this project rather than a departure from it.[169]

1.4 THE PROSPECTS FOR COUNTER-HEGEMONIC GLOBAL JUSTICE MOVEMENTS

Whilst neo-liberalism has held sway over elite economic and policy circles for the last thirty years, as a hegemonic project it faces multiple challenges. The basic mechanisms of transnational neo-liberalism – accumulation by dispossession, market liberalisation, and (increasingly) limited ameliorative measures – are not capable of offsetting the harmful impact of unfettered global capitalism on workers and communities around the world.[170] A necessary condition for the achievement of hegemony by a ruling class is the ability to transcend its own economic interests by a more universal vision. The narrow self-interest demonstrated by international finance capital in the current financial crisis illustrates the incremental thinning of neo-liberalism's hegemonic bloc.[171] Currently, global neo-liberalism can therefore be argued to constitute a *minimal hegemony* in the sense that whilst there is relative ideological unity of the economic, political and intellectual elites, the masses are for the most part excluded from governance.[172]

The failure of the TCC to establish strong global leadership has strengthened possibilities for subaltern social forces to contest neo-liberal globalisation: 'In response to neoliberalism's dynamic of accumulation by dispossession, multifarious movements and campaigns have arisen to

---

[169] On this, see generally, Colin Couch, *The Strange Non-Death of Neoliberalism* (Polity, 2011).
[170] Carroll, 'Hegemony and Counter-Hegemony in the Global Field' (n 134) 37–38.
[171] William K Carroll, 'Crisis, Movements, Counter-Hegemony: In Search of the New' (2010) 2(2) *Interface* 168, 173.
[172] See Femia's typology of hegemony discussed in Section 2.3.

protect and reclaim the commons from privatization and commodification'.[173] Since the early 1990s, a range of subaltern groups opposed to neo-liberal globalisation have emerged and begun to mount concerted struggles against the predations of global capital accumulation.[174] Globalisation has not been limited to rapid advances in capital mobility; it has also involved the globalisation of communication systems and sharing of strategies between geographically dispersed social movements and NGOs. Thus, the same technologies and developments that have enabled the creation of the TCC have also enabled the creation of alternative networks for transnational subaltern movements to contest neo-liberal governance. Initially reactive and protective, various indigenous campaigns against privatisation have opened up 'a political dynamic of social action across the whole spectrum of civil society'.[175]

For the purposes of this thesis, two questions need to be asked about such resistance: first, at what point does it become counter-hegemonic, and second when can it be understood to be transnational or global in nature? With regard to the first question, Robert Cox, following Gramsci, identifies three moments of collective political consciousness in the struggles of the subaltern strata to formulate alternatives to the hegemony of the TCC. The first is the most primitive 'economic-corporate' phase, in which a subaltern group recognises their immediate interests and a sense of opposition to the dominant powers. The second level is 'solidarity or class consciousness', which involves, at the level of a social movement, 'a condition of aroused and motivated collective consciousness'. The final stage is 'counter-hegemony', in which the interests of the subaltern classes are elaborated as an alternative ideology 'expressed in universal terms'.[176] Resistance to neo-liberal globalisation becomes counter-hegemonic at the point at which it transcends various forms of defensive particularism and becomes part of an alternative ethico-political world viewpoint.

With regard to the second question, Richard Falk has argued that, alongside an elitist 'globalisation from above', there has developed an alternative 'globalization from below' made up of the countervailing forces of labour, social movements and progressive states in the Global

---

[173] Carroll, 'Crisis' (n 171) 179.
[174] Barry K Gills, 'Introduction: Globalization and the Politics of Resistance' in Barry K Gills (ed.), *Globalization and the Politics of Resistance* (Macmillan, 2000) 3–11.
[175] Carroll, 'Crisis' (n 171) 197.
[176] Robert Cox, 'The Way Ahead: Toward a New Ontology of World Order' in Catherine Eschle and Brice Maiguashea (eds.), *Critical Theory and World Politics* (Boulder, 2000) 58.

South.[177] Globalisation from below therefore describes the various forms of local, regional, national and transnational resistance to neo-liberal globalisation.[178] Because the organs of global political society remain in the hands of the NTHB it is within the domain of global civil society that an alternative counter-hegemony is constructed: 'Just as hegemony has been increasingly organized on a transnational basis ... counter-hegemony has also taken on transnational features that go beyond the classic organization of left parties into internationals'.[179] Nancy Fraser has identified the formation of 'subaltern counterpublics' as central to counter-hegemonic strategy.[180] These consist of 'parallel discursive areas where members of the subordinated social groups invent and circulate counter discourses, which in turn permit them to formulate oppositional interpretations of their identities, interests and needs'.[181] Counter-hegemonic war of position entails the building of these counterpublics in opposition to neo-liberal hegemony as the basis for an alternative form of globalisation. Transnational movements require the self-consciousness, organisational capacity and ideological maturity to become the basis of a counter-hegemony that can be consolidated concurrently across several countries.[182]

It is important to note at this stage that a global war of position should not be understood as a direct prelude to global war of movement in which subaltern forces seize control of the existing international institutions. As Cox points out, even if these institutions could be taken over by subaltern forces this would achieve nothing because such institutions are inadequately connected with any popular political base.[183] Instead, a global war of position should be understood as enhancing the enabling conditions for *national (and possibly regional) based* forms of social transformation. These nationally based forms of transformation in turn provide the necessary preconditions for transformations within the nature and structure of world order.[184] This book will not examine

---

[177] Richard Falk, 'Resisting "Globalization-from-Above" through "Globalization-from-Below"' in Barry K Gills (ed.) (n 174) 46.
[178] ibid, 48.   [179] Carroll, 'Crisis' (n 171) 177.
[180] Nancy Fraser, 'Rethinking the Public Sphere: A Contribution to the Critique of Actually Existing Democracy' in Francis Barker, Peter Hulme, Margaret Iversen (eds.), *Post Modernism and the Re-Reading of Modernity* (Manchester University Press, 1992) 84.
[181] ibid.
[182] Adam David Morton, 'Mexico, Neoliberal Restructuring and the EZLN: A Neo-Gramscian Analysis' in Barry K Gills (ed.) (n 171) 272.
[183] Cox (n 2) 64.   [184] ibid, 65.

in any detail any instances of national-based social transformations based on socioeconomic rights praxis that have arisen from, or in conjunction with, global counter-hegemonic praxis involving socioeconomic rights. The focus instead is limited to socioeconomic rights within the context of global wars of position as *prefigurative* to broader counter-hegemonic transformations at both the national and global levels.

1.4.1 SITES OF RESISTANCE

Perhaps the exemplar of the alter-globalisation movement would be the annual World Social Forum (WSF) gatherings and its regional variants. These events bring together a vast global 'set of networks, initiatives, organisations and movements that fight against the economic, social, and political outcomes of hegemonic globalisation, challenge the conceptions of world development underlying the latter, and propose alternative conceptions'.[185] In addition to (as well as subsumed under the banner of) the WSF are what Margret Keck and Kathryn Sikkink call 'transnational advocacy networks' (TANs).[186] These consist of 'a set of relevant organizations working internationally with shared values, a common discourse, and dense exchanges of information'.[187] Their aim is not only to influence outcomes, but to shift the terms of the discourse through (re)negotiating meanings, norms and frames.[188] In response to neoliberalism's dynamic of accumulation by dispossession, a number of TANs have emerged to reclaim and protect the commons from privatisation and commodification in the areas of food, health, education, water and housing.[189]

---

[185] Boaventura de Sousa Santos, 'Beyond Neoliberal Governance: The World Social Forum as Subaltern Politics and Legality' in Boaventura de Sousa Santos and Cesar A Rodriguez-Garavito (eds.) *Law and Globalization from Below: Towards a Cosmopolitan Legality* (Cambridge University Press, 2005) 29.
[186] Margret Keck and Kathryn Sikkink, 'Transnational Advocacy Networks in the Movement Society' in Meyer DS and Tarrow S (eds.) *The Social Movement Society* (Rowman and Littlefield, 1998).
[187] Margret Keck and Kathryn Sikkink, *Activists beyond Borders* (Cornell University Press, 1998) 46.
[188] Josee Johnston and Gordon Laxer, 'Solidarity in the Age of Globalization: Lessons from the anti-MAI and Zapatista Struggles' (2003) 32 *Theory and Society* 39, 47–48.
[189] Such movements, and their engagement with socioeconomic rights discourse, will be explored in more detail in the following four chapters.

1.4.2 OBSTACLES TO BUILDING GLOBAL COUNTER-HEGEMONY

Whilst these examples indicate that global civil society is a domain for the contestation of neo-liberalism, it is important to also bear in mind that it is also 'a discursive space, which helps to reproduce global hegemony'.[190] Social movements must therefore be aware that 'they are positioned within the hegemonic constellation, and ... that there are structural and discursive forces at play, of which the very framework of global civil society is itself a part, and which social movements themselves may actually be actively reproducing, rather than challenging'.[191] Global civil society is 'neither a unified agent nor a collection of politically progressive groups but a field in which interests and identities take shape vis-à-vis each other'.[192] The disparities of power that exist in this domain can profoundly shape the ways in which oppositional currents express themselves and can even lead to the reproduction of the dominant values that are being contested. This is particularly the case in relation to north–south axes within transnational activism. As Ronen Shamir argues

> transnational coalitions are typically based on the ability of indigenous groups, grassroots movements, and activists in impoverished countries to establish discursive and organizational ties with relatively resourceful experts and often with highly professionalized NGOs ... we must also worry about the way the perceived grievances of oppressed, marginal, and exploited populations ... are transformed into a meaningful political and legal voice by relatively affluent and secured career-situated experts who often speak the language of and deploy the instruments of hegemonic rational organizational and managerial systems characteristic of contemporary capitalism.[193]

Western NGOs involved in oppositional movements can be particularly susceptible to co-option given that their funding often comes from elite governments, and wealthy private donors.[194] Furthermore, in order to

---

[190] Lucy Ford, 'Challenging Global Environmental Governance: Social Movement Agency and Global Civil Society' (2003) 3(2) *Global Environmental Politics* 120, 129.
[191] ibid.
[192] Carroll, 'Hegemony and Counter-Hegemony in the Global Field' (n 134) 39.
[193] Ronen Shamir, 'Corporate Social Responsibility: A Case of Hegemony and Counter-Hegemony' in Boaventura De Sousa Santos and Cesar A Rodriguez-Garavito (eds.) (n 185) 113.
[194] Julie Mertus, 'Doing Democracy "Differently": The Transformative Potential of Human Rights NGOs in Transnational Civil Society' (1999) 15 *Third World Legal Studies* 205, 219.

participate in various UN forums it is often necessary that they satisfy certain predetermined criteria. NGOs can thus very easily become bureaucratised and divorced from the social movements they are supposed to represent.[195] As such, Louise Amoore and Paul Langley argue that global civil society 'simultaneously holds out the potential for resistance, while it closes down, excludes, controls and disciplines'.[196] A neo-Gramscian analysis of world order therefore argues that global civil society constitutes a terrain on which neo-liberal global hegemony and governance can be contested whilst warning against assumptions that it is either an inherently progressive social force or an agent of emancipation in itself. It also draws our attention to the fact that movements ostensibly committed to contesting hegemonic neo-liberalism can end up inadvertently discursively reproducing it.

1.5 CONCLUSION

Drawing on a neo-Gramscian framework this chapter has argued that neo-liberalism should be understood as a globalising political project pursued by a TCC in order to achieve the optimal conditions for transnational capital accumulation. The TCC has been successful in progressively disembedding market relations at the national level and subsequently using an array of transnational mechanisms to 'discipline' national states to reduce barriers to the international mobility of capital. The network of multilateral institutions, trade regimes and powerful states that help enforce this disciplinary regime has been identified as 'global political society' as it is primarily characterised by its economic and military coercive capacity. Alongside global political society is a global civil society where a transnational neo-liberal historical bloc (TNLB), comprising the TCC and allied social forces, promotes the values of 'market civilisation' and other neo-liberal norms.

Nevertheless, the hegemonic basis for neo-liberal globalisation has been incrementally weakened by a series of 'crises' since neo-liberalism's global ascendency, including the dismal economic position of the former Soviet bloc countries after free market 'shock therapy', the 1997 Asian financial meltdown and the present global financial crisis.[197] The rise of

---

[195] ibid.
[196] Louise Amoore and Paul Langley, 'Ambiguities of Global Civil Society' (2004) 30 *Review of International Studies* 89, 100.
[197] For an account of the first two of these see Stiglitz (n 121) 89–165.

global governance and the Post Washington Consensus indicate efforts by the TNLB to stabilise neo-liberal hegemony but such efforts have only been partially successful and have been undermined, particularly by the decline in living standards experienced by the poor as a result of the austerity measures in response to the current financial crisis. It is argued here that neo-liberal globalisation is underscored by a *minimum hegemony* that is characterised by the relative ideological unity of global elites without popular participation in global governance.

The various crises in neo-liberal globalisation have generated the possibility for subaltern forces to pursue an alternative *globalisation from below.* In response to neo-liberalism's drive to *accumulation by dispossession,* an array of international movements and campaigns have emerged to reclaim and defend the commons from privatisation and commodification. To the extent that these movements are capable of unifying a broad alliance of subaltern forces across national borders and articulating an alternative 'ethico-political' conception of world order they can be understood as global counter-hegemonic movements.

2

NEO-LIBERAL GLOBALISATION AND SOCIOECONOMIC RIGHTS: AN OVERVIEW

'The Sheer ideological promiscuity and slipperiness of rights talk precludes a definitive classification that human rights are inherently "dominant" or "hegemonic" and essential to a US-led neo-liberal political-economic project, or that they are a universal charter for the liberation of the weak and dispossessed.'[1]

2.1 INTRODUCTION

In the previous chapter it was argued that neo-liberalism should be understood as a global hegemonic project advanced by an emergent transnational capitalist class (TCC). Whilst this project remains dominant at the global level, it has been opposed by an array of transnational resistance movements committed to the defence of the commons against neo-liberalism's drive towards accumulation by dispossession.[2] The purpose of this chapter is to explore the possibilities for such movements to mobilise praxis around the discourse of socioeconomic rights.

The first substantive part of this chapter will provide a neo-Gramscian framework for understanding human rights and the role they (could) play in social transformation. Here it is argued that human rights are socially constructed in particular social and historical settings and that as a discursive category they are open-ended enough to both bolster neo-liberal hegemony as well as challenge it. Following this, the relationship of socioeconomic rights discourse to neo-liberal doctrine will be critically assessed. It will be argued that there are a number of material and discursive tensions that characterise this relationship concerning

---

[1] Richard Wilson, 'Afterword to "Anthropology and Human Rights in a New Key": The Social Life of Human Rights' (2006) 108(1) *American Anthropologist* 77, 78.
[2] Stephen Gill, 'Toward a Postmodern Prince? The Battle in Seattle as a Moment in the New Politics of Globalisation' (2000) 29(1) *Millennium: Journal of International Studies* 131; Stephen Gill 'Constitutionalizing Inequality and the Clash of Globalizations' (2002) 4(2) *International Studies Review* 47.

different understandings of freedom, citizenship, the market and the legitimate functions of the state. It is argued that these tensions provide the basis for socioeconomic rights to constitute a counter-hegemonic discourse. Following this there will a discussion of the normative content of socioeconomic rights under international law, as developed by UN human rights bodies, particularly the Committee on Economic Social and Cultural Rights (CESCR). This section will also discuss the cautious stance taken by the UN human rights bodies, again primarily the CESCR, in relation to the negative impact of globalisation on socioeconomic rights. Finally, the chapter will conclude by assessing the relative strengths and weaknesses of socioeconomic rights as a counter-hegemonic discourse contesting neo-liberal globalisation.

2.2 HUMAN RIGHTS: A NEO-GRAMSCIAN FRAMEWORK

2.2.1 WHAT ARE HUMAN RIGHTS?

Within traditional human rights scholarship it is possible to identify three broad perspectives on the question of *what human rights are*. The first perspective can be termed *foundationalism*. Foundationalist approaches have been expressed in a variety of ways, but what all such approaches share in common is the belief that human rights can be discovered in timeless and universal values such as 'the will of God, the natural order of things, or in philosophical argument'.[3] Such approaches can be understood as philosophic attempts to ground or justify the idea of human rights in some sort of normative 'essence' or ethical 'foundation'.[4]

Whilst foundationalist approaches can produce interesting arguments for universal human rights, they are unconvincing in terms of their

---

[3] Philip Harvey, 'Aspirational Law' (2004) 52 *Buffalo Law Review* 701, 715.

[4] The French Declaration of the Rights of the Man and the Citizen 1789, preamble ('The National Assembly recognises and proclaims, in the presence and under the auspices of the Supreme Being, the following rights of man and citizen') http://avalon.law.yale.edu/18th_century/rightsof.asp accessed 19 September 2013; *The Declaration of Independence 1776* ('We hold these truths to be self-evident, that all men are created equal, that they are endowed by their Creator with certain unalienable Rights, that among these are Life, Liberty and the pursuit of Happiness') www.archives.gov/exhibits/charters/declaration_transcript.html accessed 19 September 2013. For modern secular foundationalist approaches see David Little, 'The Nature and Basis of Human Rights' in Gene Outka and John P Reeder, *Prospects for a Common Morality* (Princeton University Press, 1993) 73; Alan Gewirth, *Reason and Morality* (University of Chicago Press, 1978) 13; Arthur Dyck, *Rethinking Rights and Responsibilities: The Moral Bonds of Community* (Georgetown University Press, 1994) 10.

explanation of what human rights *are*. By basing human rights on supposedly timeless, universal truths, foundationalism can obscure the particular historical origins of human rights and the role they play in different social and political contexts.[5] It is extremely difficult, if not impossible, to find a cross-cultural or trans-historical consensus or scientific basis on the nature or meaning of concepts as contested as human need, dignity or flourishing.[6] Such approaches do not pay sufficient attention to the ways in which these supposedly foundational ideas are socially constructed, reconstructed and deconstructed within different social settings over different periods.[7] It is for these reasons that Gramsci rejected such approaches, arguing that attempts to understand the law 'will have to be freed from every residue of transcendentalism and from every absolute; in practice, from every moralistic fanaticism'.[8]

The second approach to human rights is legal positivism. Legal positivist approaches do not attempt to discern and articulate an idea of human rights transcending the 'empirical reality' of existing legal systems.[9] For positivists, legal systems are understood as the ensemble of formal texts, institutions and proceedings that ostensibly separate such systems from other non-legal processes.[10] Human rights exist to the extent that they are defined, expressed and implemented as positive law within the context of such a system. This requires that they are made legally obligatory, defined clearly and precisely, and are monitored, interpreted and enforced by a neutral body.[11] Positivism asserts 'the strict separation of the law in force,

---

[5] Neil Stammers, 'Social Movements and the Social Construction of Human Rights' (1999) 21(4) *Human Rights Quarterly* 980, 990–991.
[6] Alasdair MacIntyre, *After Virtue: A Study of Moral Theory* (Duckworth, 1981) 66–67; Jack Donnelly, *Universal Human Rights in Theory and Practice* (2nd edn., Cornell University Press, 2003) 18–20.
[7] Hannah Arendt, *The Origins of Totalitarianism* (Schocken, 1958) 296–297; MacIntyre, ibid, 65.
[8] Antonio Gramsci, *Selections from the Prison Notebooks of Antonio Gramsci* (Quintin Hoare and Geoffrey Nowell Smith eds. and trans.) (Lawrence & Wishart, 1971) (hereafter Gramsci) 246.
[9] Jeremy Bentham, 'Anarchical Fallacies' in Jeremy Waldron (ed.), *Nonsense upon Stilts: Bentham, Burke and Marx on the Rights of Man* (Methuen, 1987) 53; Jerome J Shestack, 'The Philosophic Foundations of Human Rights' (1998) 20(2) *Journal of Human Rights* 201, 209.
[10] Bruno Simma and Andreas L Paulus, 'The Responsibility of Individuals for Human Rights Abuses in Internal Conflicts: A Positivist View' (1999) 93(2) *American Journal of International Law* 302, 304.
[11] Judith Goldstein, 'Introduction: Legalization and World Politics' (2000) 54(3) *International Organization* 385, 387.

as derived from formal sources that are part of a unified system of law from non-legal factors such as natural reason, moral principles and political ideologies'.[12] For some legal positivists, the absence of coercive enforcement mechanisms in the framework of international law means that human rights are only 'real' rights when they are justiciable (i.e. enforceable in a court of law) at the domestic or regional level.[13] In the absence of such enforcement mechanisms, expressions of human rights are regarded either as 'moral rights' or as quasi-judicial 'soft law' rather than paradigmatic legal rights.

Unlike natural law approaches, legal positivism acknowledges the role of human agency in defining the meaning of human rights. Human rights are seen as the product of human activities such as standard setting, monitoring, reporting and enforcement. In that sense legal positivism avoids some of the problems of ahistorical metaphysical abstraction associated with the natural law approaches.

Nevertheless, legal positivist approaches have their own set of shortcomings. A number of critics of legal positivism have challenged the premise that it is possible to analyse legal decision-making without an awareness of the influence of political, economic, cultural and other factors on the process.[14] Such critics argue that 'the legal order is far less autonomous, far less self-regulating and self-sufficient, than often portrayed'.[15] The functioning of law depends as much on a complex set of social relations as it does on the formal provisions of a legal system. Critical legal scholars have gone further and argued that law is not only influenced by politics but in fact *is* a form of politics, masquerading as a separate process.[16] The outcomes of legal disputes are as much the consequence of the cultural assumptions and individual biases of legal actors and the political pressures exerted upon them as the legal texts themselves. There is no 'bright line' delineating legal from non-legal

---

[12] Simma and Paulus (n 10) 409.
[13] Kenneth W Abbott, 'The Concept of Legalization' (2000) 54(3) *International Organization* 401, 409.
[14] Jonathon Turley, 'The Hitchhikes Guide to CLS: Unger, and Deep Thought' (1987) 81 *Northwestern University Law Review* 593, 599–600 (commenting on legal realism).
[15] Philip Selznick, 'Law in Context Revisited' (2003) 30 *Journal of Law and Society* 177, 178.
[16] Roberto Manabeira Unger, 'The Critical Legal Studies Movement' (1983) 96(3) *Harvard Law Review* 561; David Kennedy 'A New Stream of International Law Scholarship' in Robert J Beck, Anthony Clark Arend and Robert D. Vander Lugt (eds.), *International Rules: Approaches from International Law and International Relations* (Oxford University Press, 1996) 236; Martti Koskenniemi, *From Apology to Utopia: The Structure of International Legal Argument* (Cambridge University Press, 2005) 606–607.

processes for anti-positivist critics, and no clear distinction between 'legal rights' and 'moral rights'.

A third approach to human rights is social constructivism. In opposition to foundationalism and legal positivism, social constructivist approaches do not seek to understand human rights as either rooted in a transcendental morality or merely as a product of sovereign legislation, but rather as particular social forms which emerge in specific historical contexts.[17] To say that human rights are socially constructed 'is to say that ideas and practices in respect of human rights are created, re-created and instantiated by human actors in particular socio-historical settings and conditions'.[18] As such, human rights 'are neither self-generating nor self-enforcing, but rather summarise, make concrete, and depend for any protective effectiveness they may possess, on the nature of wider sets of social relations and developments within them'.[19] Social constructivism attributes an important role to formal national and international legal systems in generating understandings of human rights, but also ascribes importance to the non-state, non-institutionalised and pre-legal forms that human rights articulations take.[20] Non-state actors such as NGOs, social movements, advocacy networks and the entire complex of civil society are not understood as merely lobbying for the adoption or enforcement of pre-existing human rights norms, but rather as actively engaging in 'jurisgenerative' activity, that is, of generating meanings and altering perceptions about *what those human rights norms are* through framing and other discursive strategies.[21]

The social constructivist theories of human rights are the most congruent with the Gramscian account of hegemony. Although Gramsci did not write about the law in any great detail, the scattered observations contained within the *Prison Notebooks* provide a useful starting point to reconcile a social constructivist account of human rights with an understanding of the role that they might play in the context of hegemonic and counter-hegemonic struggles. Gramsci shared the anti-positivist perspective that the legal sphere is not a clearly demarcated zone of human activity. Law, Gramsci argued, is 'wider than purely State and governmental activity and also includes the activity involved in directing civil

---

[17] Tony Evans, *Human Rights in the Global Political Economy: Critical Perspectives* (Lynne Reinner 2011) 44.
[18] Stammers (n 5) 981.   [19] Alan Woodiwiss, *Human Rights* (Routledge, 2005) 3–4.
[20] Stammers (n 5) 992.
[21] Seyla Benhabib, *Dignity in Adversity: Human Rights in Troubled Times* (Polity, 2011).

society, in those zones which the technicians of law call legally neutral – that is, in morality and in custom generally'.[22] It operates where '"coercion" is not a State affair but is effected by public opinion, moral climate etc'.[23] Gramsci's broad conception of the law is further indicated in his discussion of the meaning of a 'legislator'. Gramsci argued that the legislator must be identified with the 'politician' and since 'all men are "political beings", all are also "legislators"':[24]

> Every man, in as much as he is active, i.e. living, contributes to modifying the social environment in which he develops (to modifying certain of its characteristics or to preserving others); in other words, he tends to establish "norms", rules of living and of behaviour. One's circle of activity may be greater or smaller, one's awareness of one's own action and aims may be greater or smaller; furthermore, the representative power may be greater or smaller, and will be put into practice to a greater or lesser extent in its normative, systematic expression by the "represented".[25]

Thus for Gramsci the law is understood not so much as a reified set of norms or institutional practices but rather as a *social relation* which can find expression across an array of public and private settings. Law is conceptualised broadly as the types of activities – both coercive and persuasive – that are capable of generating norm complying, observing or replicating behaviour in others. Whilst the greatest legislative capacity lies with state personnel – who have the repressive and ideological apparatuses of the state at their disposal – Gramsci also notes that the leaders of private organisations 'have coercive sanctions at their disposal too, ranging even up to the death penalty'.[26] Here we may recall Althusser's observation that schools and churches exercise discipline through methods of punishment, expulsion and selection.[27] However, for Gramsci law is most effective and hegemonic when it is capable of generating 'the "spontaneous" consent of the masses who must "live" those directives' thus not requiring recourse to punishment.[28] In that sense an array of organic intellectuals who lack coercive mechanisms – such as journalists, business gurus, life style coaches, public relations experts, trade union activists and so forth – can also be understood to be 'organic legislators' capable of modifying the habits, will and convictions of others.[29]

---

[22] Gramsci (n 8) 195.   [23] ibid, 196.   [24] ibid.   [25] ibid, 265–266.   [26] ibid, 266.
[27] Louis Althusser, 'Ideology and Ideological State Apparatuses' in Louis Althusser, *Lenin and Philosophy and Other Essays* (Monthly Review Press, 1972).
[28] Gramsci (n 8) 266.   [29] ibid.

Applying Gramsci's analysis of the law to human rights, it can be argued that whilst states (responsible for drafting and implementing human rights treaties) and official human rights institutions (responsible for interpreting and ensuring compliance with these treaties) may have the most influence in terms of generating meanings around the nature of human rights, an array of non-state actors and 'non-official' agents such as academics, think tanks, advocacy networks, NGOs, social movements and so on also have human rights norm-generating capacities. This critical sociological perspective is captured in a recent work by Aramline, Glasberg and Purkayastha to describe what they call 'the human rights enterprise':

> The human rights enterprise includes both legal, statist approaches to defining and achieving human rights through agreements amongst duty-bearing states, social movement approaches that manifest as social struggles over power, resources, and political voice. The human rights enterprise offers a way to conceptualise human rights as a terrain of social struggle, rather than a static, contingent legal construct.[30]

As will be discussed in the next section, the diverse array of potential sources of human rights normativity opens up the possibilities for competing conceptions of human rights and indeed opposing hegemonic and counter-hegemonic articulations.

2.2.2 HUMAN RIGHTS AND SOCIAL CHANGE

What role do human rights play in affecting social change? In international relations and legal compliance literature it is possible to identify two main perspectives: realist and constructivist. Realist perspectives are based on *rational choice theory*: the belief that actors (predominantly states) are exclusively motivated by the desire to increase their power within the international order to the maximum extent possible.[31] The creation of international standards for human rights, and compliance with such standards, is therefore driven solely in pursuit of these interests. Thus human rights and other norms only emerge when they serve the interests of the most powerful actors.[32] In such accounts, human

---

[30] William T Armaline, Davita Silfen Glasberg and Bandana Purkayastha, *The Human Rights Enterprise* (Polity, 2015) 14.
[31] Oona A Hathaway, 'Do Human Rights Treaties Make a Difference?' (2002) 111(8) *Yale Law Journal* 1935, 1944.
[32] ibid, 1946.

rights have no causative impact in affecting social change and are only the epiphenomenal expression of distributions of power and interests. By contrast, constructivist theorists argue that human rights and other norms have a direct causal impact upon state behaviour. Whilst realist approaches assume that interests are fixed and predetermined, social constructivists argue 'that ideas and communicative processes define... and ... influence understandings of interests, preferences and political decisions'.[33] Human rights norms therefore help to 'shape an actor's identity, interests, and behaviour and change states understandings of their national interests'.[34]

Both realist and constructivist approaches have strengths and weaknesses from the vantage point of a neo-Gramscian perspective. Neo-Gramscian approaches to international relations accept the realist perspective that the creation, promotion and reconstruction of international norms is directly bound up with the material interests of the actors expressing those norms.[35] However, neo-Gramscian accounts also accept the constructivist viewpoint that norms and ideas are not merely the epiphenomenal product of some deeper truth or interest but rather are 'an objective and operative material reality'.[36] For Gramsci, law is used both 'repressively' for the exercise of direct domination by the ruling class and 'consensually' to 'raise the great mass of the population to a particular cultural and moral level ... which corresponds to the needs of the productive forces of development'.[37] The general absence of coercive enforcement mechanisms associated with international human right law means that the discourse should be largely understood as *ethico-political*, that is to say, involved in the promotion, preservation and contestation of values, norms and frames in the domain of global civil society.

---

[33] Thomas Risse and Kathryn Sikkink, 'The Socialization of International Human Rights Norms into Domestic Practices' in Thomas Risse, Stephen C Ropp and Kathryn Sikkink (eds.), *The Power of Human Rights: International Norms and Domestic Change* (Cambridge University Press, 1999) 7.
[34] Lisa Forman, '"Rights" and "Wrongs": What Utility for the Right to Health in Reforming Trade Rules on Medicines?' (2008) 10(2) *Health and Human Rights* 37, 42.
[35] As Robert Cox argues 'World hegemony ... is expressed in universal norms, institutions, and mechanisms which lay down general rules of behaviour for states and those forces of civil society that act across national boundaries – rules which support the dominant mode of production' Robert W Cox, 'Gramsci, Hegemony and International Relations', in Stephen Gill (ed.), *Gramsci, Historical Materialism and International Relations* (Cambridge University Press, 1993) 62.
[36] Gramsci (n 7) 196.   [37] ibid, 258.

Where a neo-Gramscian account would differ from dominant realist and constructivist approaches is in its identification of social class and the relations of production as the fundamental bases of interest in the construction of norms, as opposed to the nation state.[38] Whilst neo-Gramscian accounts would not dismiss the importance of the nation state as a unit of interest, it is the social relations of production that provide the underlying basis for hegemonic and ideological activity. To the extent that the social relations of production and the social forces they engender have been transnationalised, transnational social forces are also interest-holding agents that engage in norm promotion and contestation, independently, or at least with relative autonomy, from nation state interests.[39] This applies to both the transnational neo-liberal hegemonic bloc (TNHB) that advance the hegemonic project of neo-liberal globalisation on the one hand and the subaltern social forces of global civil society that are committed to alternative models of globalisation on the other.[40]

It should be noted at this point that a number of critical scholars have either dismissed or expressed scepticism about the ability of human rights discourse to critically interrogate extant relations and structures of power in the world.[41] Such arguments suggest that, despite the claims of universality, human rights are in fact tied to Western liberalism and the rise of capitalism and, as such, obscure and recode the various hierarchies around class, gender, ethnicity, and so on, embedded within the liberal capitalist order rather than subverting or challenging them.[42]

---

[38] Cox (n 35) 62. [39] ibid, 61.
[40] Evans, *Human Rights in the Global Political Economy* (n 17) 18.
[41] See, for example, Stewart Scheingold, *The Politics of Rights* (Yale University Press, 1974); Mark Tushnet, 'An Essay on Rights' (1984) 62 *Texas Law Review* 1363; Peter Gabel, 'The Phenomenology of Rights – Consciousness and the Pact of the Withdrawn Selves' (1984) 62 *Texas Law Review* 1563; Alan Freeman, 'Racism, Rights and the Quest for Equality of Opportunity' (1988) 23 *Harvard Civil Rights and Civil Liberties Law Review* 295; Mary Ann Glendon, *Rights Talk: The Impoverishment of Political Discourse* (Free Press, 1991); Wendy Brown, *States of Injury: Power and Freedom in Late Modernity* (Princeton University Press, 1995); Duncan Kennedy 'The Critique of Rights in Critical Legal Studies' in Wendy Brown and Janet Halley (eds.), *Left Legalism/Left Critique* (Duke University Press, 2002) 178–228; David Kennedy, *The Dark Side of Virtue: Reassessing International Humanitarianism* (Princeton University Press, 2004) 3–37; Samuel Moyn, 'A Powerless Companion: Human Rights in the Age of Neoliberalism' (2014) 77(4) *Law and Contemporary Problems* 147.
[42] Tushnet (n 41) 1363; Costas Douzinas, *The End of Human Rights* (Hart 2000) 1; Fiona Robinson 'Human Rights Discourse and Global Civil Society: Contesting Globalisation' Prepared for the 2002 Annual Meeting of the International Studies Association, New

It is also argued that human rights discourse is too narrow and legalistic to be used to challenge the systematic and material bases for social deprivation that are governed by the structural logic and organisation of the global political economy.[43] Such arguments are concerned that rights discourse channels oppositional movements into technical legal disputes around peripheral questions and diverts attention away from the task of building political movements aimed at supporting radical social and political transformations.[44]

Whilst these critiques of human rights discourse are illuminating in so far as they highlight some shortcomings in particular articulations of the discourse and uncover the ways in which human rights have historically been appropriated in the service of dominant powerful interests, it must be open to question whether human rights are simply reducible to these shortcomings or are preordained to exclusively serve the interests of the powerful and the *status quo*. In response to the critics of rights, a number of theorists and activists have advanced compelling arguments that there is still a place for human rights discourse in meaningful social transformation.[45] Patricia Williams makes the striking observation that the problems noted by critics of human rights are not the result of something inherent within the discourse itself but rather reflect the 'constricted referential universe' that it operates within.[46] Similarly, Alan Hunt argues that 'no discursive symbol has a necessary political content' and that the co-option of rights discourse by the powerful is 'only one of the practical manifestations of the social consequences of the real world of hegemony'.[47] Discursive struggle takes place not simply between human rights and alternative discourses but also internally within human rights discourse

---

Orleans, Louisiana, March, 2002 available at http://isanet.ccit.arizona.edu/noarchive/robinson.html accessed 30 September 2011; Makau Mutua, 'Standard Setting in Human Rights: Critique and Prognosis' (2007) 29(3) *Human Rights Quarterly* 547.

[43] David Kennedy (n 41) 11–13; Naomi Klein, *The Shock Doctrine: The Rise of Disaster Capitalism* (Penguin, 2007) 118–128.

[44] ibid.

[45] See, for example, Patricia Williams, 'Alchemical Notes: Reconstructing Ideals from Deconstructed Rights' (1987) 22 *Harvard Civil Rights-Civil Liberties Law Review* 401; Alan Hunt 'Rights and Social Movements: Counter-Hegemonic Strategies' (1990) 17(3) *Journal of Law and Society* 309; Boaventura de Sousa Santos and Cesar A Rodriguez-Garavito (eds.) *Law and Globalization from Below: Towards Cosmopolitan Legality* (Cambridge University Press, 2005); Sally Engle Merry, *Human Rights and Gender Violence: Translating International Law into Local Justice* (Chicago University Press, 2005); Mark Goodale and Sally Engle Merry (eds.), *The Practice of Human Rights: Tracking Law between the Global and the Local* (Cambridge University Press, 2007).

[46] Williams (n 45) 431.   [47] Hunt (n 45) 324.

itself. Recall from the previous chapter Gramsci's observation that counter-hegemonic praxis entails 'making "critical" an already existing activity'.[48]

It follows from such an understanding that, in the context of hegemonic contestations at the global level, human rights can be understood as expressions of 'the social conflicts and contradictions embedded in social life and ... one of the many forms in which these struggles get played out in ways that reflect, albeit in complex and mediated fashions, the prevailing balance of forces'.[49] Thus there are conceptions of human rights that advance the project of neo-liberal globalisation by emphasising 'the freedom of individual action, non-interference in the private sphere of economics and the right to own and dispose of property'.[50] There are also conceptions that challenge the neo-liberal consensus. Boaventura de Sousa Santos and Cesar A. Rodriguez-Garavito note that there are multifarious grassroots struggles for the collective rights to the commons that 'seek to articulate new notions of rights that go beyond the liberal ideal of individual autonomy, and incorporate solidaristic understandings of entitlements grounded in alternative forms of legal knowledge'.[51] As such, Katie Nash has argued:

> Critics argue that human rights are depoliticising, individualising: they are enabling the world to be made secure for neo-liberal global elites rather than ending the suffering of ordinary people. But human rights are not only used in justifications of military adventures on the part of the US led 'coalition of the willing', and nor does using the language of human rights necessarily involve submission to neo-liberalism. Human rights also represent a language within which a variety of claims of justice are articulated against imperialism and neo-liberalism.[52]

It is arguable that such radical divergences in interpretations of human rights discourse have been there since the beginning. Whilst a proto-capitalist conception of natural rights and the Rights of Man can be found in the writings of early theoreticians of rights, such as Thomas

---

[48] Gramsci (n 7) 330–331.
[49] Amy Bartholomew and Alan Hunt, 'What's Wrong with Rights?' (1991) 9 *Law and Inequality* 1, 13.
[50] Tony Evans, 'The Human Right to Health?' (2002) 201 23(2) *Third World Quarterly* 197, 201.
[51] Boaventura de Sousa Santos and Cesar A Rodriguez-Garavito, 'Law, Politics and the Subaltern in Counter-Hegemonic Globalization' in Sousa Santos and Rodriguez-Garavito (eds.) (n 45) 16.
[52] Katie Nash, *The Political Sociology of Human Rights* (Cambridge University Press, 2015) 3.

Hobbes and John Locke,[53] there were also prior social movements that emerged during the English Civil War that articulated an alternative vision of universal rights based around the defence of the commons and opposition to emerging forms of capitalist enclosure.[54] This included movements such as the Diggers that argued for the 'equal right' of the 'poor, oppressed people of the world' to the common treasures of the earth.[55] Similar radical elements emerged during the French and American Revolutions and in the context of the workers movements during the industrial revolution.[56] There have therefore always been two opposing traditions in the history of human rights. As Alan Woodiwiss argues

> For the 'major tradition' within rights discourse that the social contract theorists initiated, the rights related to property and contract represented ... the means (that is owning things and making agreements) by which the essential elements of humanity's supposed primordial liberty were preserved despite the recognition for social order. By contrast, for the 'minor tradition'... of the Diggers and Ranters ... humanity's original position was governed by the principle of reciprocity rather than that of liberty. The result was that the establishment of the same rights of property and contract as were celebrated by the major tradition was seen by the minor tradition as a severe challenge to freedom in the form of the danger that reciprocity might be replaced by selfishness as the core value.[57]

The discursive contestation around the meaning of human rights can be viewed as one of the many forms that hegemonic struggles take in challenging the ethico-political foundations of world order. It has been argued that one manifestation of counter-hegemonic contestation of neo-liberal globalisation is the increasing articulation of demands for the realisation of socioeconomic rights.[58] The next section will introduce the idea of socioeconomic rights and then consider the discursive basis for its utilisation as a counter-hegemonic discourse.

---

[53] CB Macpherson, *The Political Theory of Possessive Individualism: Hobbes to Locke* (Oxford University Press, 1964).

[54] Micheline R Ishay, *The History of Human Rights: From Ancient Times to the Globalization Era* (University of California Press, 2004) 92–93.

[55] Gerrard Winstanley, 'A Declaration from the Poor Oppressed People of England' (first printed 1649) www.bilderberg.org/land/poor.htm accessed 20 September 2013.

[56] Neil Stammers, *Human Rights and Social Movements* (Pluto, 2009) 55–63; Ishay (n 54) 135–145.

[57] Woodiwiss (n 19) 140.

[58] Andrew Lang, *World Trade Law after Neoliberalism: Re-Imagining the Global Economic Order* (Oxford University Press, 2011) 83.

2.3 SOCIOECONOMIC RIGHTS AND NEO-LIBERALISM

2.3.1 THE MEANING OF SOCIOECONOMIC RIGHTS

For the purpose of this book, socioeconomic rights are understood as those rights concerned with the material bases of human wellbeing.[59] Their primary normative function is to secure a basic quality of life for individuals and communities through guaranteeing access to material goods and services such as food, water, shelter, education, healthcare and housing. Such rights are frequently labelled 'second generation' rights and are identified as 'deriving from the growth of socialist ideals in the late nineteenth and early twentieth centuries and the rise of the labour movement in Europe'.[60] They are contrasted with the 'first generation' of 'civil and political rights' associated with the Age of Enlightenment and the English, American and French revolutions in the seventeenth and eighteenth centuries, and the 'third generation' of 'solidarity rights', such as the rights to development and self-determination, identified with the anti-colonial struggles of the 20th century.[61] Other accounts of socioeconomic rights stress their diffuse historic origins and sources of legitimation, ranging from individuals as diverse as Thomas Paine, Karl Marx, Immanuel Kant, John Maynard Keynes, Franklin Roosevelt and John Rawls, through to the injunctions of various religions and revolutionary documents such as the Mexican and Soviet Constitutions of 1917.[62]

Socioeconomic rights have been variously described as economic, social and cultural rights (ESC rights), social rights, economic rights and social welfare rights.[63] For this thesis the term socioeconomic rights is preferred because it conveys '"the inextricable link between the economic and social policy fields" and the law and therefore brings to the fore the relevance of socio-economic rights to the social and economic

---

[59] Mark Tushnet, 'Civil Rights and Social Rights: The Future of the Reconstruction Amendments' (1992) 25 *Loyola of Los Angeles Law Review* 1207, 1207.
[60] Matthew Craven, *The International Covenant on Economic, Social and Cultural Rights: A Perspective on Its Development* (Clarendon Press, 1995) 8.
[61] Karel Vasak 'Pour une troisième génération des droits de l'homme' in Christophe Swinarski (ed.), *Studies and Essays on International Humanitarian Law and Red Cross Principles in Honour of Jean Pictet* (Martinus Nijhoff, 1984) 837.
[62] Henry Steiner, Philip Alston and Ryan Goodman, *International Human Rights Law in Context: Law, Politics, Morals* (3rd edn., Oxford University Press, 2007) 269–270.
[63] For an account of the different classifications of socioeconomic rights see Katherine G Young, 'The Minimum Core of Economic and Social Rights: A Concept in Search of Content' (2008) 33 *Yale Journal of International Law* 113, 118 (fn19).

status of rights claimants, and the implications that the assertion of such rights has for power relations within societies.'[64] Social and economic rights are often grouped with cultural rights. However, due to their somewhat different legal nature from social and economic rights and their distinct historical development, cultural rights will not form part of the consideration of this thesis.[65]

The three rights that are examined in detail in the three case study chapters – the rights to food, medicines and water – have been identified as a subset of socioeconomic rights termed 'subsistence rights' or 'basic rights' in the sense that they are demands that a person's material survival is socially guaranteed and that the realisation of such rights is 'essential to all other rights'.[66] Nevertheless, the term socioeconomic rights is preferred in this thesis because although these rights are indeed essential for survival, it may be too reductive to limit understandings of them to protecting mere subsistence as opposed to broader conceptions of human dignity, flourishing or wellbeing.[67]

2.3.2 SOCIOECONOMIC RIGHTS AND NEO-LIBERALISM: DISCURSIVE TENSIONS

It is widely recognised in the human rights literature that neo-liberalism as a doctrine is hostile to socioeconomic rights at a normative level.[68] Historically, the neo-liberal rejection of socioeconomic rights has been based upon a minimal account of the state and government. Based on a conception of 'negative freedom', neo-liberal doctrine limits human rights to traditional civil and political rights aimed at protecting

---

[64] Paul O'Connell, *Vindicating Socioeconomic Rights: International Standards and Comparative Experiences* (Routledge, 2011) 15 (quoting Ellie Palmer).

[65] See Henry Steiner and Philip Alston, *International Human Rights in Context: Law, Politics, Morals* (2nd edn., Oxford University Press, 2000) 248; Ishay (n 54) 10–13.

[66] Daniel PL Chong, *Freedom from Poverty: NGOs and Human Rights Praxis* (University of Pennsylvania Press, 2010) 6; Henry Shue, *Basic Rights: Subsistence, Affluence and U.S. Foreign Policy* (2nd edn., Princeton University Press, 1996) 19.

[67] Philip Alston, 'Human Rights and Basic Needs: A Critical Assessment' (1979) 12 *Human Rights Journal* 19, 55–56.

[68] See, for example, Philip Alston, 'Resisting the Merger and Acquisition of Human Rights by Trade Law: A Reply to Petersmann' (2002) 13(4) *European Journal of International Law* 815, 826–827; Marius Pieterse, 'Beyond the Welfare State: Globalization of the Neo-Liberal Culture and the Constitutional Protection of Social and Economic Rights in South Africa' (2003) 14 *Stellenbosch Law Review* 3, 14; Raymond Plant, *The Neo-Liberal State* (Oxford University Press, 2010) 116; Paul O'Connell, 'On Reconciling Irreconcilables: Neo-Liberal Globalisation and Human Rights' 7(3) *Human Rights Law Review* (2007) 483, 537.

individuals from the coercive actions of others.⁶⁹ In accordance with this view, to conceive of guaranteed access to a material good or service as a 'right' is fundamentally misconceived because such a right requires a correlative duty on another person to provide that good or service.⁷⁰ Whereas neo-liberals regard rights as primarily consisting of duties of forbearance ('negative obligations'), socioeconomic rights seem to entail positive obligations on duty-bearing agents that compel them to act in ways that are normatively incompatible with the neo-liberal conception of freedom as negative liberty.⁷¹ The institutional arrangements required to realise socioeconomic rights involve coercive acts – such as taxation or wealth distribution – that interfere with individual freedom.⁷² The most notable interference is with the individual's right to private property, which is one of the central rights of a free society for neo-liberals.⁷³

A second aspect of neo-liberalism's historic opposition to socio-economic rights is the belief that such rights constitute unacceptable interference with the 'spontaneous order' of the free market.⁷⁴ In the neo-liberal account, markets are not only an intrinsic expression of freedom but also have instrumental value as vehicles for welfare maximisation, information co-ordination and the guarantee of broader political

---

[69] FA Hayek, *The Constitution of Liberty* (Routledge & Kegan Paul, 1960) 16–17.
[70] FA Hayek, *Law, Legislation and Liberty Volume 2: The Mirage of Social Justice* (Routledge and Kegan Paul, 1976) 102–103.
[71] ibid 103; Robert Nozick, *Anarchy, State and Utopia* (first published 1974, Blackwell, 2001) 238. Nozick is usually classified as a libertarian rather than a neo-liberal. Nevertheless, as Raymond Plant notes, Nozick's theories have been influential in the development of neo-liberalism. Plant (n 68) 96.
[72] ibid 169; Erich Weede, 'Human Rights, Limited Government, and Capitalism' (2008) 28 (1) *Cato Journal* 35, 47 ('Since positive rights or entitlements need funding, the attempt to provide positive rights requires an infringement of negative rights, especially of the right to enjoy the fruits of one's labor').
[73] Hayek, *Constitution of Liberty* (n 69) 140; Milton and Rose Friedman, *Free to Choose: A Personal Statement* (Secker & Warburg, 1980) 67; James M Buchanan, *Property as a Guarantor of Liberty* (Edward Elgar, 1993) 59. Murray Rothbard goes so far as to argue that 'not only are there no human rights, which are also property rights, but the former rights lose their absoluteness and clarity and become fuzzy and vulnerable when property rights are not used as the standard' Murray Rothbard, *The Ethics of Liberty* (New York University Press, 1998) 113. Also see generally, David Kelley, *A Life of One's Own: Individual Rights and the Welfare State* (Cato Institute, 1998).
[74] Hayek, *Law, Legislation and Liberty* (n 70) 103, 107–132; Cass R Sunstein, 'Against Positive Rights' (1993) 2 *East European Constitutional Review* 35 (arguing against constitutionalised socioeconomic rights on the basis that they compel governments to interfere with free markets); Weede (n 72) 40.

freedom.[75] Whereas the state is regarded as bureaucratic, unresponsive and inefficient, markets are held to be flexible, responsive and self-correcting.[76] The superiority of the market stems from its ability to 'spontaneously' co-ordinate the dispersed, separate and partial knowledge of individuals through the price mechanism and the laws of supply and demand.[77] Markets are threatened by central authority interventions that seek to achieve particular outcomes because such interventions distort the information-coordinating role of markets.[78] Furthermore, as spheres of un-coerced space, markets provide a vital counter-weight to the coercive power of the state.[79] The more that the state is able to interfere in the private sphere of markets, the more powerful, bureaucratic and 'totalitarian' it becomes, diminishing the political and civil freedoms of its citizens in the process.[80]

Socioeconomic rights are at least in part concerned with achieving particular outcomes for certain individuals and therefore favour the distribution of resources according to normative criteria such as human dignity or need.[81] In order to achieve this, a central authority would have to determine how and on what basis goods and services should be distributed. The effect of such interference would be to both 'distort' the information-coordinating role of markets and increase the 'totalitarian' control of the government over its citizens.[82] Hence neo-liberals argue for a strict separation of the political sphere of the state, which has the responsibility of upholding fundamental civil and political rights, and the economic sphere of the market, which should be left to its own mechanisms to determine social and economic entitlement.[83] Particular levels of education, healthcare, social security and so forth are not regarded as legal or moral entitlements, but rather as commodities to be acquired through the market.[84]

---

[75] Friedman, *Free to Choose* (n 70) 9–38; Hayek, *Constitution of Liberty* (n 69) 120; Milton Friedman, *Capitalism and Freedom* (University of Chicago Press, 1962) 10.

[76] Friedman, *Free to Choose*; ibid, 9–70.

[77] ibid, 13–24; Hayek, *Constitution of Liberty* (n 69) 120.   [78] ibid, (n 69) 128–129.

[79] ibid, 139–142.

[80] FA Hayek, *The Road to Serfdom* (first published 1944, ARK, 1986) 66–75.

[81] The Universal Declaration of Human Rights states that everyone is entitled, 'as a member of society', to the realisation of 'the economic, social and cultural rights indispensable for his dignity and the free development of his personality'. Universal Declaration of Human Rights (UDHR) (adopted 10 December 1948 UNGA Res 217 A(III) article 22.

[82] Friedman, *Free to Choose* (n 70) 17; Hayek, *The Road to Serfdom* (n 80) 75.

[83] Tony Evans, *The Politics of Human Rights: A Global Perspective* (Pluto Press, 2005) 79–80.

[84] Hayek, *Law, Legislation and Liberty* (n 70) 106.

The neo-liberal opposition to socioeconomic rights should not be taken as a mere matter of intellectual abstraction. Many advocates of socioeconomic rights maintain that the policy prescriptions and trends associated with neo-liberal governance undermine the material conditions for the realisation of such rights.[85] It is argued that trends such as the increased reliance upon the market, the diminution in the role of State provision of social services and the deregulation of financial and labour markets have exposed workers, poor people and vulnerable groups to the vicissitudes of the market in ways that make the objects of their socioeconomic rights less secure.[86] Such policy trends are currently being intensified in the context of the on-going global economic crisis, thereby undermining the social environment required for the realisation of socioeconomic rights.[87]

Structural Adjustment Programs (SAPS) imposed by the International Monitory Fund (IMF) and World Bank have been particular targets for socioeconomic rights advocates.[88] In their global, comparative analysis of 131 developing countries between 1981 and 2003, M. Rodwan Abouhard and David Cingranelli found that the adoption of SAPs 'causes governments to lessen respect for the economic and social rights of their citizens, including the rights to decent jobs, education, health care, and

---

[85] Pieterse (n 68) 15–19; O'Connell, 'On Reconciling Irreconcilables' (n 68) 507.
[86] See, e.g., CESCR 'Globalization and Economic, Social and Cultural Rights' (May 1998) UN Doc. E/1999/22-E/C.12/1998/26; Economic and Social Council, 'The right to food, Report by the Special Rapporteur on the right to food, Mr. Jean Ziegler' (10 January 2002) UN Doc. E/CN.4/2002/58 para.110.
[87] United Nations Department of Economic and Social Affairs, The Global Social Crisis: Report on the World Social Situation 2011 (15 June 2011) UN Doc. ST/ESA/334 iii ('The global economic downturn has had wide-ranging negative social outcomes for individuals, families, communities and societies, and its impact on social progress in areas such as education and health will only become fully evident over time'; OHCHR, 'Report of the United Nations High Commissioner for Human Rights: Austerity Measures and Economic, Social and Cultural Rights' (7 May 2013) UN Doc. E/2013/82, para.70 ('Many States have responded to the recent global financial crisis with austerity measures that ... have resulted in the denial or infringement of economic, social and cultural rights').
[88] A 2002 study from the Structural Adjustment Participatory Review Initiative Network (SAPRIN) found that the IMF and World Bank SAPs had increased economic inequality, resulting 'in less secure employment, lower wages, fewer benefits and an erosion of workers' rights and bargaining power', and increased poverty through 'privatization programs, the application of user fees, budget cuts and other adjustment measures that have reduced the role of the state in providing or guaranteeing affordable access to essential quality services'. SAPRIN, 'The Policy Roots of Economic Crisis and Poverty: A Multi-Country Participatory Assessment of Structural Adjustment' (1st edn., April 2002) 174 accessed 22 September 2013.

housing'.[89] The rules governing the World Trade Organization (WTO) have also come under intense scrutiny from socioeconomic rights advocates. A report published in 2000 by the UN Sub-Commission on the Promotion and Protection of Human Rights argued that the structure of the WTO and the trade rules are 'grossly unfair and ... serves only to promote dominant corporate interests' whilst undermining the progressive realisation of socioeconomic rights.[90] This has led some advocates of socioeconomic rights to argue that violations of socioeconomic rights are embedded within the structures of the global political economy[91] and unless such structures are challenged 'the mandate of "progressive realisation" of social, economic and cultural rights' will remain nothing more than 'an on-going cruel hoax'.[92]

Given both the rejection of socioeconomic rights in orthodox neo-liberal doctrine and the apparent material incommensurability of neo-liberal policy prescriptions with the promotion and protection of socioeconomic rights, it can be argued that the evocation of such rights could form part of a counter-hegemonic challenge to the socioeconomic basis and ethico-political framework of neo-liberal globalisation.[93] There are at least three reasons for this. First, in response to neo-liberalism's dynamic of accumulation by dispossession, which involves the suppression of the rights of the commons through privatisation and commodification, a socioeconomic rights perspective holds that those goods and services that are vital for human flourishing and dignity – or indeed basic survival – should be made available to everybody on the basis of human need rather than ability to pay.[94]

---

[89] M Rodwan Adouharb and David Cingranelli, *Human Rights and Structural Adjustment* (Cambridge University Press, 2007) 4–5.

[90] Sub-Commission on the Promotion and Protection of Human Rights, 'The Realization of Economic, Social and Cultural Rights: Globalization and Its Impact on the Fully Enjoyment of Human Rights' (15 June 2000) UN Doc. E/CN.4/Sub.2/2000/13. See also Gillian Moon, 'Trading in Good Faith? Importing States' Economic Human Rights Obligations into the WTO's Doha Round Negotiations' (2013) 13(2) *Human Rights Law Review* 245.

[91] Asbjørn Eide, 'The Importance of Economic and Social Rights in an Age of Economic Globalization' in Wenche Barch Eide and Uwe Kracht (eds.), *Food and Human Rights in Development Volume 1: Legal and Institutional Dimensions in Selected Topics* (Intersentia, 2005) 3; O'Connell, 'On Reconciling Irreconcilables' (n 68) 508.

[92] Upendra Baxi, *The Future of Human Rights* (Oxford University Press, 2006) 248.

[93] Marius Pieterse, 'Eating Socioeconomic Rights: The Usefulness of Rights Talk in Alleviating Social Hardship Revisited' (2007) 29 *Human Rights Quarterly* 796, 803 ('many of the weaknesses of liberal rights discourse ... may be countered by guaranteeing socioeconomic rights alongside civil and political rights').

[94] See, for example, Committee on Economic, Social and Cultural Rights (hereafter 'CESCR'), 'General Comment No. 15: The Right to Water' (20 January 2003) UN Doc.

Second, whereas neo-liberalism normatively prescribes a 'minimal state' limited to upholding negative rights and the rule of law, socioeconomic rights discourse conceives of the State as the duty-holding entity tasked with ensuring the progressive realisation of universal access to healthcare, education, social security and so forth for its citizenry. In place of the 'minimal state', socioeconomic rights discourse invites the possibility of a 'social state' that plays a more pro-active role in distributing resources and regulating markets to ensure the material wellbeing and dignity of its population at large.[95]

Finally, while neo-liberal discourse has regarded poverty and material deprivation as 'problems' to be addressed through technical solutions related to securing the macroeconomic conditions for economic growth, socioeconomic rights approaches raise the notion that certain forms of deprivation constitute 'violations' that give rise to binding obligations on States to take concrete steps towards ameliorating and reversing such deprivation.[96] It is these discursive tensions that open up the possibilities for counter-hegemonic praxis to coalesce around the articulation of socioeconomic rights in opposition to neo-liberal globalisation.[97]

2.4 THE DEVELOPMENT OF SOCIOECONOMIC RIGHTS UNDER INTERNATIONAL LAW

Over recent decades the turn to socioeconomic rights discourse in global civil society has been enabled, or at any rate facilitated, in part by the elaboration and clarification of the normative content of socioeconomic rights by international human rights bodies, most notably the United Nations CESCR.

E/C.12/2002/11, para.11 ('water should be treated as a social and cultural good and not primarily an economic good') and para.12(c)(ii) ('water ... must be affordable for all').

[95] Thomas Humphrey Marshall, 'Citizenship and Social Class' in Thomas Humphrey Marshall and Thomas Bottomore (eds.) *Citizenship and Social Class* (Pluto Press, 1992) 7.

[96] Lisa Philips, 'Taxing the Market Citizen: Fiscal Policy and Inequality in an Age of Privatization' (2000) 63 *Law and Contemporary Problems* 111, 115–116.

[97] See generally, Lucie E White and Jeremy Perelman, *Stones of Hope: How African Activists Reclaim Human Rights to Challenge Global Poverty* (2010); Boaventura de Sousa Santos, *Towards a New Legal Common Sense: Law, Globalization and Emancipation* (Cambridge University Press, 2002) 271.

2.4.1 'IN FROM THE COLD': THE CLARIFICATION OF SOCIOECONOMIC RIGHTS STANDARDS

Socioeconomic rights have been recognised in international law since the adoption of the foundational document of international human rights law, the Universal Declaration of Human Rights (UDHR), in 1948. This document recognises, *inter alia*, the rights to social security, just and favourable conditions of employment, rest and leisure, food, clothing, housing, medical care and education.[98] The provisions in the UDHR were transformed into legally binding obligations by the United Nations' adoption of the International Covenant on Economic, Social and Cultural Rights (ICESCR) in 1966.[99] Socioeconomic rights are also contained in a number of regional human rights instruments[100] as well as international treaties aimed at protecting specific groups.[101] Despite the formal recognition of socioeconomic rights under international law, and repeated declarations of the equal footing of all human rights,[102] they have historically held a subordinate status in relation to their civil and political counterparts in the international framework of human rights protection.[103] Western Governments in particular have demonstrated

---

[98] UDHR (n 81) arts 22–26. For an analysis of the inclusion of socioeconomic rights in the UDHR see Jonas Morsink, *The Universal Declaration of Human Rights: Origins, Drafting and Intent* (University of Pennsylvania Press) 157–238.

[99] International Covenant on Economic, Social and Cultural Rights (adopted 16 December 1966, entered into force 23 March 1976) 993 UNTS 3 (ICESCR).

[100] E.g., European Social Charter (signed 18 October 1961, entered into force 26 February 1965) UNTS 89, ETS 155; Additional Protocol to the American Convention on Human Rights in the Area of Economic, Social and Cultural Rights (adopted 17 November 1988, entered into force 28 August 1991) OASTS 73 (Protocol of San Salvador); African Charter on Human and Peoples' Rights (adopted 27 June 1981, entered into force 21 October 1986) O.A.U. Doc. CAB/LEG/67/3 Rev 5 arts 15–17 (recognising the rights to work under satisfactory conditions, equal pay for equal work, the right to health and the right to education).

[101] E.g., Convention on the Elimination of All Forms of Discrimination against Women (adopted 18 December 1979, entered into force 3 September 1981) 1249 UNTS 13 arts 10–14; Convention on the Rights of the Child (adopted 20 November 1989, entered into force 2 September 1990) 1577 UNTS 3 arts 23–31; Convention on the Rights of Peoples with Disabilities (adopted 13 December 2006, entered into force 3 May 2008) G.A. Res. 61/106, 61 UN GAOR, Supp. (No.49), UN Doc. A/RES/61/106/AnnexII 65 arts 24–28.

[102] See, e.g., Vienna Declaration and Program of Action, UN GAOR, World Conference on Human Rights, 48th Sess., 22nd plen. Mtg., part 1, 18, UN Doc. A/CONF. 157/24 (1993), reprinted in 32 I.L.M. 1667 (1993) (stating that 'All human rights are universal, indivisible and interdependent and interrelated').

[103] David Marcus, 'The Normative Development of Socioeconomic Rights through Supranational Adjudication' (2006) 42 *Stanford Journal of International Law* 53–102.

much ambivalence, if not outright hostility, to the idea that socioeconomic rights give rise to legally binding obligations on States Parties.[104] This ambivalence dates back to the drafting of the ICESCR, when, as Daniel Whelan and Jack Donnelly point out:

> the understanding of economic and social rights as directive rather than justiciable [i.e. enforceable by courts] was shared by all states. No state ... seriously proposed – in the sense of being willing to adopt as a manner of enforceable national law – treating economic, social and cultural rights as matters of immediate rather than progressive realisation.[105]

This perception of state obligations in relation to socioeconomic rights is reflected in the language of article 2(1) of the ICESCR, which states:

> Each State Party to the present Covenant undertakes to take steps, individually and through international assistance and co-operation, especially economic and technical, to the maximum of its available resources, with a view to achieving progressively the full realization of the rights recognized in the present Covenant by all appropriate means, including particularly the adoption of legislative measures.[106]

Whilst the ICESCR's twin treaty, the International Covenant on Civil and Political Rights (ICCPR), requires states to adopt law and other measures to give effect to the rights in the Covenant and to ensure the provision and enforcement of remedies for breaches of the rights,[107] the obligations contained in article 2(1) of the ICESCR are clearly more qualified. The formation of state obligations under article 2(1) is widely regarded in the scholarship to be unsatisfactory due to its 'convoluted phraseology and numerous qualifying sub-clauses', which seem to 'defy any sense of obligation ... giving states almost total freedom of choice and action as to how rights should be implemented'.[108] The term 'maximum of its available resources' has been described as 'a difficult phrase – two warring

---

[104] Steiner, Alston and Goodman (n 62) 280–282.
[105] Daniel J Whelan and Jack Donnelly, 'The West, Economic and Social Rights, and the Global Human Rights Regime: Setting the Record Straight' (2007) 29(4) *Human Rights Quarterly* 908, 935.
[106] ICESCR (n 99) art 2(1).
[107] International Covenant on Civil and Political Rights (ICCPR) (adopted 16 December 1966, entered into force 23 March 1976) 999 UNTS 171, reprinted in 6 ILM – 368 (1967) art 2(1).
[108] Matthew Craven, 'The Justiciability of Economic, Social and Cultural Rights' in R Burchill, D Harris and A Owers (eds.), *Economic, Social and Cultural Rights: Their Implementation in United Kingdom Law* (1999) 5.

adjectives fighting over an undefined noun'[109] and the vague obligation on states to 'progressively realise' socioeconomic rights has been described as 'of such a nature as to be legally negligible'.[110] The meaning of resources, the timescale permitted for realisation and the nature of legislative action required have historically been 'stumbling blocks to interpretation'.[111]

Buttressing the problems associated with the vague and indeterminate formation of States Parties' obligations under article 2(1) has been the historic lack of enforcement and interpretive mechanisms for the rights contained within the ICESCR. Unlike the ICCPR, the ICESCR did not establish a human rights committee to review periodic state reports and consider individual communications of violations.[112] The vague nature of the rights contained within the ICESCR coupled with the historic lack of interpretive and enforcement mechanisms has produced a 'negative feedback cycle' whereby socioeconomic rights are seen as vague because they are non-justiciable, and in turn this non-justiciability has prevented judges and others from applying them and developing a jurisprudence to clarify their content.[113] Academic opinion has also often matched judicial scepticism in dismissing socioeconomic rights as mere non-binding 'aspirations' at best and denying that they possess the intrinsic character of rights at worst.[114]

---

[109] Robert E Robinson, 'Measuring Compliance with the Obligation to Devote the "Maximum Available Resources" to Realising Economic, Social and Culture Rights' (1994) 16 *Human Rights Quarterly* 693, 694.

[110] Egbert W Vierdag, 'The Legal Nature of the Rights Granted by the International Covenant on Economic, Social and Cultural Rights' (1978) 9 *Netherlands Year Book of International Law* 69, 105.

[111] Chris Downes, 'Must the Losers of Free Trade Go Hungry? Reconciling WTO Obligations and the Right to Food' (2007) 47 *Virginia Journal of International Law* 619, 626.

[112] ICCPR (n 107) art 40; Optional Protocol to the International Covenant on Civil and Political Rights (adopted 16 December 1966, entered into force 23 March 1976) 999 UNTS 171, art 1.

[113] Marcus (n 103) 59–60.

[114] It is beyond the scope of this chapter to explore the nature of these critiques. See, for example: Maurice Cranston, 'Human Rights Real and Supposed' in David Daiches Raphael (ed.), *Political Theory and the Rights of Man* (Indiana University Press, 1967); Maurice Cranston, *What Are Human Rights?* (Bodley Head, 1973) 37–40; Charles Fried, Right and Wrong (Harvard University Press, 1978) 113; Kenneth Minogue, 'The History of the Idea of Human Rights' in Walter Laquer and Barry Rubin (eds.), The *Human Rights* Reader (New Amsterdam Library, 1979); Michael Ignatieff, *Human Rights as Politics and Idolatry* (Princeton University Press, 2001) 88; Aryeh Neier, 'Social and Economic Rights: A Critique' (2006) 13(2) *Human Rights Brief* 1.

THE DEVELOPMENT OF SOCIOECONOMIC RIGHTS   71

However, 'an increasingly expansive array of international instances (has) generated social rights "jurisprudence"' since the adoption of the ICESCR, despite the initial lack of interpretative and enforcement mechanisms.[115] Since the beginning of the 1990s international human rights institutions and UN agencies have taken steps to assist the development of more rigorous definitions, monitoring and implementation of socioeconomic rights.[116] However, for the sake of brevity this section will focus only on the statements of the CESCR, the international body with the most relevance and authority in the monitoring and interpretation of the international law of socioeconomic rights.

The CESCR was formed in 1985 as a panel of experts responsible for monitoring States Parties' compliance with the ICESCR and for helping to clarify the treaty's content.[117] Since the CESCR was established it has attempted to delineate the normative content of state obligations through its 'Concluding Observations' in relation to state reports but more significantly through the adoption of 'General Comments'. While General Comments are non-binding they are persuasive interpretations of the rights and obligations set out in the Covenant. Through the General Comments the CESCR has been able to develop jurisprudence with regard to the concrete obligations of States Parties. It is worth briefly outlining several key facets of this jurisprudence.

Although the formulation of state obligations in relation to the rights in the ICESCR is couched in terms of 'progressive realisation' the CESCR have noted that there are a number of exceptions to this requirement where a state must act immediately. First, the obligation 'to take steps' requires that deliberate, concrete and targeted steps towards the implementation of the right are taken within a reasonably short time after the Covenant's entry into force.[118] Second, the requirement of non-discrimination in the enjoyment of ICESCR rights contained in article 2(2) is 'an immediate and

---

[115] Philip Alston, 'Foreword' in Malcolm Langford (ed.), *Social Rights Jurisprudence: Emerging Trends in International and Comparative Law* (Cambridge University Press, 2008) 3.

[116] Paul J Nelson and Ellen Dorsey, *New Rights Advocacy Changing Strategies of Development and Human Rights NGOs* (Georgetown University Press, 2008) 45–46.

[117] Office of the High Commissioner for Human Rights (OHCHR), 'Review of the composition, organization and administrative arrangements of the Sessional Working Group of Governmental Experts on the Implementation of the International Covenant on Economic, Social and Cultural Rights' (28 May 1985) ECOSOC Res 1985/17.

[118] CESCR, 'General Comment No. 3: The Nature of State Parties Obligations (Art. 2, par.1)' (14 December 1990) UN Doc. E/1991/23.

cross-cutting obligation' on states to refrain from discriminatory practices and to take 'concrete, deliberate and targeted measures to ensure that discrimination in the exercise of Covenant rights is eliminated'.[119] This obligation requires, *inter alia*, that states ensure the satisfaction of socioeconomic rights is available and affordable for all and that poorer households are not disproportionately burdened with expenses.[120]

Third, it is incumbent on States Parties to guarantee the satisfaction of 'at the very least, minimum essential levels of each of the rights' contained in the Covenant.[121] For example, 'a State party in which any significant number of individuals is deprived of essential foodstuffs, of essential primary health care, of basic shelter and housing, or of the most basic forms of education is, *prima facie*, failing to discharge its obligations under the Covenant'.[122]

In relation to 'progressive realization', the CESCR state that this 'imposes an obligation to move as expeditiously and effectively' towards ensuring the fulfilment of socioeconomic rights as possible.[123] It also implies a strong presumption against the adoption of retrogressive measures in relation to socioeconomic rights.[124] The burden is placed on the State Party to prove that all other alternatives were considered, the measure taken was justified in relation to the totality of the rights provided for in the ICESCR and in the context of the full use of the State Party's maximum available resources.[125] The CESCR have also used the General Comments to adopt the tripartite typology of State Obligations as developed by Henry Shue and Asbjørn Eide.[126] This imposes three

---

[119] CESCR, 'General Comment No. 20: Non-discrimination in economic, social and cultural rights (Art. 2, para.2, of the International Covenant on Economic, Social and Cultural Rights)' (2 July 2009) UN Doc. E/C.12/GC/20, paras.7 and 36.

[120] See, e.g., CESCR, 'General Comment No. 14 (2000) The right to the highest attainable standard of health (article 12 of the International Covenant on Economic, Social and Cultural Rights)' (11 August 2000) UN Doc. E/C.12/2000/4, para.12(b)(iii); CESCR, 'General Comment No. 15 (n 94) para.27.

[121] CESCR, 'General Comment No. 3' (n 118) para.10.   [122] ibid.   [123] ibid, para.9.

[124] ibid.

[125] See, for example, CESCR, 'General Comment No. 13: The right to education (article 13 of the Covenant)' (8 December 1999) UN. Doc. E/C.12/1999/10, para.45; General Comment 14 (n 120) para.32; General Comment 15 (n 94) para.19.

[126] Henry Shue, 'Rights in Light of Duties' in Peter G Brown and Douglas MacLean (eds.), *Human Rights and U.S. Foreign Policy* (Lexington, 1979) 51–64 (arguing that human rights impose three core duties on States: the duty to avoid depriving, the duty to protect from deprivation and the duty to aid the deprived); Asbjørn Eide, 'The International Human Rights System' in Asbjørn Eide, Catarina Krause and Allan Rosas (eds.), *Economic, Social and Cultural Rights: A Textbook* (Martinus Nijhoff Publishers, 1995)

THE DEVELOPMENT OF SOCIOECONOMIC RIGHTS   73

types or levels of obligations on states parties: to *respect, protect* and *fulfil*.[127] The duty to respect simply requires that states refrain from interfering with the enjoyment of a right.[128] The duty to protect requires the adoption of measures to ensure that third parties do not interfere with the socioeconomic rights of individuals and collectives under the State Party's jurisdiction.[129] Lastly, the duty to fulfil requires states to 'adopt appropriate legislative, administrative, budgetary, judicial, promotional and other measures towards the full realization of' rights within the ICESCR.[130] In contrast to the neo-liberal rejection of socioeconomic rights as 'positive rights', it follows from the CESCR's framework that that socioeconomic rights, just like civil and political rights, impose a mixture of positive and negative duties.[131]

A final noteworthy development has been the adoption of the Optional Protocol to the International Covenant on Economic, Social and Cultural Rights (OP-ICESCR or Optional Protocol) on 10 December 2008.[132] The Optional Protocol allows the CESCR to receive and consider individual, group and inter-State communications claiming violations of socioeconomic rights contained within the Covenant.[133] The unanimous adoption by the United Nations General Assembly perhaps indicates the final step in the mainstreaming of socioeconomic rights and their placement on formally equal terms with civil and political rights. Furthermore, these developments have been supported by a wide range of national courts in Latin America, South Asia, Africa, Eastern Europe and some Western Countries that have adopted positions in relation to socioeconomic

---

(arguing that that human rights obligations can be classified into three categories: the State's obligations to respect, protect and fulfil).

[127] See, e.g., CESCR, 'The Right to Adequate Food (Art. 11)' (12 May 1999) UN Doc. E/C.12/1999/5, para.15; CESCR 'General Comment 13' (n 125) para.46; CESCR, General Comment 14 (n 120) para.33; CESCR 'General Comment 15' (n 94) para.20.

[128] Asbjørn Eide 'Realization of Social and Economic Rights and the Minimum Threshold Approach' (1989) 10 *Human Rights Law Journal* 35, 35.

[129] CESCR, General Comment No. 14 (n 120) para.33.   [130] ibid.

[131] See, for example, Magdalena Sepulveda, *The Nature of Obligations under the International Covenant on Economic, Social and Cultural Rights* (Intersentia, 2003) 7.

[132] Optional Protocol to the International Covenant on Economic, Social and Cultural Rights (adopted 10 December 2008, entered into force 5 May 2013) G.A. Res. 63/117, UN GAOR, 63rd Sess., UN Doc. A/RES/63/117 (OP-ICESCR). For an account of the development of the OP-ICESCR see Catrina de Albuquerque, 'Chronicle of an Announced Birth: The Coming into Life of the Optional Protocol to the International Covenant on Economic, Social and Cultural Rights – The Missing Piece of the International Bill of Human Rights' (2010) 32(1) *Human Rights Quarterly* 144.

[133] OP-ICESCR ibid, ibid arts 2, 8, 9 and 10.

rights[134] as well as a wealth of legal and philosophical scholarship aimed at establishing the nature and scope of parties' duties under the ICESCR and providing theoretical justification for the legal protection of socioeconomic rights.[135]

In summary, the normative content and scope of socioeconomic rights has been increasingly clarified over the last two decades. This is not to suggest that oppositional tendencies towards socioeconomic rights do not continue to exist, that ambiguities are not still present, nor that significant imbalances between the international and constitutional protection and enforcement of civil and political rights and socioeconomic rights do not remain, but there is nevertheless an increasing realisation that socioeconomic rights are 'real rights' and that they give rise to certain legal and moral obligations and entitlements.[136]

2.4.2 THE UN HUMAN RIGHTS FRAMEWORK AND GLOBALISATION

Following a series of pioneering studies conducted under the auspices of the Sub-Commission on Prevention of Discrimination and Protection of Minorities, beginning in the early 1990s, human rights bodies within the UN began expressing concern about the negative impact of trends of globalisation for socioeconomic rights.[137] A number of key studies drew attention to the problematic role played by the Bretton Woods

---

[134] Roberto Gargarella, Pilar Domingo and Theunis Roux, 'Courts, Rights and Social Transformation: Concluding Reflections' in Roberto Gargarella, Pilar Domingo and Theunis Roux (eds.), *Courts and Social Transformation in New Democracies: An Institutional Voice for the Poor?* (Ashgate, 2006) 255 and Alston 'Foreword' in Langford ed. (n 115) x.

[135] See, e.g. Cécile Fabre, *Social Rights under the Constitution: Government and the Decent Life* (Oxford University Press, 1999); Frank I Michelman, 'The Constitution, Social Rights, and Liberal Political Justification' (2003) 1(1) *International Journal of Constitutional Law* 13; Donnelly (n 6) 29–31; Raymond Plant, 'Social and Economic Rights Revisited' (2003) 14 *Kings College Law Journal*, 1; David Bilchitz, *Poverty and Fundamental Rights: The Justification and Enforcement of Socio-Economic Rights* (Oxford University Press, 2007); Paul O' Connell, *Vindicating Socio-Economic Rights* (n 64); Katherine G Young, *Constituting Economic and Social Rights* (Oxford University Press, 2012).

[136] Paul O'Connell, 'The Death of Socio-Economic Rights' (2011) 74(4) *Modern Law Review* 532, 532.

[137] Asbjørn Eide, 'The Importance of Economic and Social Rights in the Age of Globalization' in Asbjørn Eide, 'Wenche Barth Eide and Uwe Kracht (eds.) *Food and Human Rights in Development Volume 1: Legal and Institutional Dimensions and Selected Topics* (Intersentia, 2005) 20.

institutions, the negative impact of increasing income differences for the enjoyment of human rights and the adverse impact of the liberalisation of trade services, agricultural reform and enhanced intellectual property rights (IPRs) on the realisation of human rights.[138] Andrew Lang argues that these bodies 'adopted and re-articulated many of the critiques of trade emanating from the global justice movement – but they did so selectively, incrementally, and usually in a moderated and modified form'.[139] This section will examine some of the relevant jurisprudence of the CESCR in relation to globalisation. As Lang suggests, the CESCR's criticisms of the policies and trends associated with neo-liberal globalisation are more moderated than those expressed by the global justice movements. In contrast to the demands for an alternative development paradigm to neo-liberal globalisation, the CESCR have adopted what might be termed a compensatory approach which is concerned with correcting or mitigating the perceived malfunctions of the existing international system.

One of the earliest engagements the CESCR had with the policies associated with neo-liberal globalisation was in its 1990 General Comment on International Technical Assistance. The CESCR noted that the adverse impact of debt burden and 'relevant adjustment measures' (i.e. IMF and World Bank mandated SAPs) were a 'particular concern' in the realization of socioeconomic rights.[140] Whilst the Committee accepted that adjustment programmes are often unavoidable and frequently involve a major element of austerity, it urged States Parties and relevant UN agencies to incorporate the protection of the rights of the poor and vulnerable into the basic objectives of economic adjustment. The CESCR termed this approach 'adjustment with a human face'.[141] Despite adopting what appeared to be a cautious yet critical stance on trends associated with globalisation, the Committee has been keen to emphasise that it is not a political organisation. At the end of 1990 it stressed that the ICESCR does not endorse or oppose any form of government or economic system and that:

> in terms of political and economic systems the Covenant is neutral and its principles cannot accurately be described as being predicated exclusively

---

[138] ibid, 20–21.   [139] Lang (n 5) 104.
[140] CESCR, 'General Comment No. 2: International Technical Assistance Measures' (2 February 1990) UN Doc. E/1990/23 para.9.
[141] ibid.

upon the need for, or the desirability of a socialist or a capitalist system, or a mixed, centrally planned, or *laisser-faire* economy, or upon any other particular approach.[142]

This ostensibly neutral position was maintained in the CESCR's 1998 statement on the impact of globalisation on the enjoyment of ESC rights. This statement – arising out a of 'day of discussion' with participants from the Office of the High Commissioner for Human Rights (OHCHR), UN bodies, specialised agencies, NGOs and individual experts – identified a number of trends within globalisation that could impact upon the enjoyment of socioeconomic rights, including the increased reliance on the free market, the growth of international finance, the diminution in the role of the state, the privatisation of former government functions and the growth of private actors such as TNCs in civil society.[143] Whilst the CESCR suggested that none of these trends are necessarily incompatible with the principles of the Covenant, it argued that their combined impact risked undermining socioeconomic rights unless they were complemented by 'new and innovative policies' and 'necessary safeguards'.[144] The CESCR expressed concern that the trends and policies associated with globalisation were overtly fixated with competitiveness, efficiency and economic rationalism whilst paying little attention to the realisation of ESC rights. The Committee also called upon the IMF, World Bank and WTO to pay enhanced attention in their activities to respect ESC rights, including by incorporating human rights impact assessments in relation to their loan, credit, trade and investment policies.[145] The following year, in November 1999, the CESCR issued a similar statement directed towards the WTO in the aftermath of the Seattle Ministerial, urging that the WTO ensure 'human rights principles and obligations are fully integrated in future negotiations in the WTO'.[146]

In its periodic monitoring and supervision procedure the CESCR has repeatedly emphasised that influential States Parties – as members of the IMF, World Bank and WTO – should ensure that the policies and decisions of those organisations are in conformity with the obligations

---

[142] CESCR, 'General Comment No. 3' (n 118) para.9.
[143] CESCR, 'Globalization and Economic, Social and Cultural Rights' (n 86) para.2.
[144] ibid, para.3.   [145] ibid, para.7.
[146] CESCR, 'Statement of the UN Committee on Economic, Social and Cultural Rights to the Third Ministerial Conference of the World Trade Organization (Seattle 30 November to 3 December 1999)' (26 November 1999) UN Doc. E/C.12.1999/9.

under the ICESCR.¹⁴⁷ It has also strongly recommended that the obligations of the Covenant are taken into account by a state in all aspects of its negotiations with international financial institutions (IFIs), like the IMF, the World Bank and the WTO, to ensure that ESC rights, particularly of the most vulnerable groups of society, are not undermined.¹⁴⁸ It has repeatedly expressed concerns that aspects of structural adjustment programmes and economic liberalisation policies have impeded and undermined the implementation of the ICESCR, particularly with regard to the most vulnerable groups in society.¹⁴⁹ The CESCR has also highlighted the potential or actual negative impact of international trade agreements on the realisation of ESC rights on numerous occasions.¹⁵⁰

The CESCR has also repeatedly stressed in its General Comments that States Parties should ensure that the rights in the ICESCR are given due attention in relevant international instruments¹⁵¹ and that States Parties have an obligation to ensure that their actions as members of IFIs take due account of the covenant rights. States Parties should also pay greater attention to the protection of ESC rights in influencing the lending policies, credit agreements and international measures of these institutions.¹⁵² In General Comment 15 on the right to water, the CESCR state that 'agreements concerning trade liberalization should not curtail or inhibit a country's capacity to ensure

---

[147] See Concluding Observations of the CESCR: UK (5 June 2002) UN Doc. E/C.12/1/Add. 79, para.26; Ireland (5 June 2002) UN Doc. E/C 12/1/Add.77, para.37; Italy (23 May 2000) UN Doc. E/C.12/1/Add.43, para.20; Belgium (1 December 2000) UN Doc. E/C.12/1/Add.54, para.31Germany (24 September 2001) UN Doc. E/C.12/1/Add.68, para.31.

[148] See CESCR, 'Concluding Observations of the CESCR: Morocco' (1 December 2000) UN Doc. E/C.12/1/Add.55, para.38.

[149] See CESCR, 'Concluding Observations of the CESCR: Egypt' (23 May 2000) UN Doc. E/C.12/1/Add.44, paras.10 and 28; Algeria (30 November 2001) UN Doc. E/C.12/1/Add.71, para.43; Venezuela. (21 May 2001) UN Doc. E/C.12/1/Add.56, para.8.

[150] CESCR, 'Concluding Observations of the Committee on Economic Social and Cultural Rights: Jamaica' (30 November 2001) UN Doc. E/C.12/1/Add.75, para.15; Ecuador (7 June 2004) UN Doc. E/C.12/1/Add.100, para.55; India (8 August 2008) UN Doc. E/C.12/CRI/CO/4, paras.29, 46.

[151] CESCR, 'General Comment No. 12: The Right to Adequate Food (Art. 11 of the Covenant)' (12 May 1999) UN Doc. E/C.12/1999/5, para.36.

[152] CESCR, 'General Comment No. 13' (n 125) para.56; CESCR, 'General Comment No. 14' (n 125) para.39; CESCR 'General Comment No. 15' (n 94) para.36; CESCR 'General Comment No. 18: The Right to Work (Art. 6 of the Covenant)' (6 February 2006) UN Doc. E/C.12/GC/18, para.30; CESCR 'General Comment No. 19: The Right to Social Security (Art. 9 of the Covenant)' (4 February 2008) UN Doc. E/C.12/GC/19, para.58.

the full realization of the right to water'.[153] In addition to these obligations for State Parties the CESCR have begun to suggest that the IMF, World Bank and WTO have certain forms of obligations themselves. It has stated that IFIs promoting measures of structural adjustment should ensure that such measures do not compromise the enjoyment of the rights in the ICESCR.[154] In particular the CESCR has stipulated that IFIs such as the IMF and World Bank should pay greater attention to the protection of ICESCR rights in their lending policies and credit agreements and in international measures to deal with the debt crisis. It has also argued that 'international organizations concerned with trade such as the WTO, should cooperate effectively with States Parties, building on their respective expertise, in relation to the implementation of human rights'.[155]

To what extent does the CESCR contest neo-liberalism? On the one hand, it is clear that the CESCR's statements do challenge the simplistic focus on competitiveness, efficiency and economic rationalism that underpins much of neo-liberal discourse and invites a wider human rights-based approach to development grounded in human dignity and wellbeing. Indeed, behind the façade of political neutrality, it is arguable that the CESCR's approach does prescribe a particular type of politics that emphasises the state's proactive and protective role in ensuring the material wellbeing of its citizens in ways that challenge the neo-liberal calls for withdrawal of the state from economic activity. On the other hand, it was suggested in the previous chapter that moves towards a 'Post-Washington Consensus' (PWC) in global governance have involved preserving the fundamental tenets of neo-liberal governance whilst incorporating aspects of subaltern criticisms of the *status quo*, including the acknowledgement that limited social safety nets and other mechanisms may be required to offset some of the problems associated with global free trade and the market mechanism. It may be that certain overlaps between the CESCR's compensatory approach and the PWC approach could open up possibilities for the discursive incorporation of socioeconomic rights into the neo-liberal hegemonic

---

[153] CESCR, 'General Comment 15' (n 94) para.35.
[154] CESCR, 'General Comment 12' (n 151) para.41; CESCR 'General Comment 13' (n 125) para.60; CESCR 'General Comment 14' (n 125) para.64; General Comment 15' (n 94) para.60; CESCR 'General Comment 18' (n 152) para.30.
[155] CESCR, 'General Comment 15' (n 94) para.60.

formation. Nevertheless, it is arguable that the jurisprudence of the CESCR does provide a basis for interrogating neo-liberalism that subaltern social forces can build upon. It is likely therefore that the CESCR's commentary and jurisprudence will be a terrain for dispute between hegemonic and counter-hegemonic social forces.

2.5 CRITICALLY EVALUATING THE COUNTER-HEGEMONIC POTENTIAL OF SOCIOECONOMIC RIGHTS

Before concluding this chapter, this section will briefly consider some of the potential strengths and limitations of subaltern movements turning to the language of socioeconomic rights as part of their counter-hegemonic praxis. This section does not attempt to provide a comprehensive account of the counter-hegemonic potential – or lack thereof – of socioeconomic rights discourse. Rather, it simply seeks to flag up some preliminary issues in anticipation of deeper consideration in the forthcoming case study chapters.

In Section 2.3 a number of discursive tensions were identified between the normative prescriptions of socioeconomic rights and the theoretical foundations of neo-liberalism. Whilst this might indicate that socioeconomic rights could form the basis for counter-hegemonic praxis, it does not necessarily imply that this discourse would be the most effective one to deploy. There are a multitude of other discourses and ideologies that also (potentially) stand opposed to neo-liberalism: socialism, environmentalism, third-worldism, post-structuralism, nationalism and so forth. It may be the case that these discourses more effectively interrogate neo-liberal processes and their underlying conceptions as well as provide more useful visions to move towards alternative models of development and societal organisation.

The first point to note here is that whilst some have argued that human rights strategies may compete with, and preclude, other emancipatory programmes,[156] there is no reason to view the discourses that counter-hegemonic praxis can draw upon in terms of competing, either-or categories.[157] Instead, socioeconomic rights discourse might

---

[156] Kennedy, *The Dark Side of Virtue* (n 39) 4; Wendy Brown '"The Most We Can Hope For ..."' Human Rights and the Poverty of Fatalism' (2004) 103(2/3) *South Atlantic Quarterly* 451, 461–462.

[157] Oriol Mirosa and Leila M Harris, 'Human Right to Water: Contemporary Challenges and Contours of a Global Debate' (2011) 44(3) *Antipode* 932, 933.

be thought of as a component in a 'portfolio' of discourses aimed at contesting neo-liberal globalisation.[158]

Nevertheless, even if it is accepted that socioeconomic rights can play a complementary role in tandem with other oppositional discourses, it must be asked: exactly what do they bring to the table? For whilst, as this chapter has sought to demonstrate, human rights are open to competing interpretations, this does not imply that they are merely empty signifiers entirely devoid of any discursive content. Human rights are 'inscribed' with at least basic meanings as a result of their past and present deployments.[159] With that in mind, this section will identify three potential advantages and three possible limitations of socioeconomic rights discourse as counter-hegemonic strategy. The potential advantages are that socioeconomic rights discourse transcends particular and sectorial interests, implies strong normative obligations, and provides the basis for 'imminent critique' of the extant order. The disadvantages are the historic emphasis on economic liberalism in human rights discourse, the limitations of 'narrow legalist' interpretations of rights and the historic failure of human rights law to adequately address the responsibilities of private and transnational actors.

2.5.1 POTENTIAL STRENGTHS OF SOCIOECONOMIC RIGHTS DISCOURSE

2.5.1.1 SOCIOECONOMIC RIGHTS TRANSCEND PARTICULAR AND SECTORIAL INTERESTS

Recourse to human rights discourse allows social movements to frame claims in ways that transcend particular and sectorial interests. Human rights are commonly understood – as the name implies – as 'the rights one has because one is human'.[160] In other words, re-articulating a given interest as a human right engages the process of *universalisation*. Universalisation involves 'taking an interpretation of the interests of some group, less than the whole polity, and [arguing] that it corresponds to the interests or the ideals of the whole'.[161] This is precisely what human rights discourse does in that it reinstates 'the interests of the group as the characteristics of all people'.[162]

In the previous chapter, it was argued that counter-hegemony emerges out of a series of 'moments' in the political consciousness of the subaltern

---

[158] ibid. [159] Bartholomew and Hunt (n 50) 52. [160] Donnelly (n 6) 7.
[161] Duncan Kennedy, 'The Critique of Rights in Critical Legal Studies' (n 39) 188.
[162] ibid.

classes, ranging from the 'economic-corporate' moment, in which individuals recognise and begin to articulate their interests as being linked to that of others from within their narrow social grouping, through to the 'counter-hegemonic' moment, in which individuals realise that their interests must become the interests of subaltern allies as well.[163] Resistance to neo-liberal globalisation becomes counter-hegemonic at the point at which it transcends various forms of defensive particularism and becomes part of an alternative ethico-political world viewpoint 'expressed in universal terms'.[164] Given that the idea of 'universalism' is a key principle upon which human rights are founded, human rights praxis can be understood as a potential integral component of counter-hegemonic struggles. The re-articulation of particular needs, interests or wants as 'rights' that inhere in individuals or collectives on the basis of their belonging to 'humanity' rather than a more particular interest or rights-generating category marks a passage from the 'corporate' to the 'universal' plane upon which the construction of counter-hegemony takes place.[165] For example, if a particular group is deprived of access to affordable medicines due to the enforcement of stringent IPRs protection, this becomes a matter of universal concern once it is demonstrated that it constitutes a violation of the right to health. Human rights can therefore potentially facilitate the integration of geographically dispersed movements with divergent ideological, political and cultural references into unified global campaigns expressed in universal terms.[166]

2.5.1.2 SOCIOECONOMIC RIGHTS IMPLY STRONG NORMATIVE OBLIGATIONS

The second important aspect of articulating demands in terms of socioeconomic rights is that they imply strong normative obligations. Rights discourse differs from other expressions of interests – such as needs, wants or desires – in the sense that it implies a social relationship in which a right holder has an enforceable legal or moral claim against a

---

[163] See Chapter 1 Section 1.2.3.
[164] Robert Cox, 'The Way Ahead: Toward a New Ontology of World Order' in Catherine Eschle and Brice Maiguashea (eds.), *Critical Theory and World Politics* (Boulder, 2001) 58.
[165] Hunt (n 45) 320–321.
[166] Priscilla Claeys, 'From Food Sovereignty to Peasants' Rights: An Overview of Via Campesina's Struggle for New Human Rights' (*La Via Campesina*, 15 May 2013) 2. http://viacampesina.org/downloads/pdf/openbooks/EN-02.pdf accessed 3 January 2014.

duty bearer.[167] Individuals in need of essential goods and services are transformed from 'supplicants' to 'claimholders'. This constitutes an important paradigm shift in social struggles against poverty and exclusion. As Shareen Hertel and Lanse Minkler argue:

> By invoking the normative force of human rights in defence of their own needs ... grassroots protesters can change the nature of their interaction with powerful government and private sector representatives. Instead of offering petitions for help, they can demand that rights be fulfilled – explicitly referencing documents such as the UDHR [and] the ICESCR ... which codify key [socio-economic] rights. In doing so, they transform their status from that of supplicants to claimants.[168]

Furthermore, 'by moving to a human rights framework, the elimination of poverty becomes more than just a desirable, charitable and even moral goal. It becomes [a] ... duty of all states'.[169] This shifts opposition to poverty from the purely *subjunctive* terrain about how the world ought to be, to an *injunctive* terrain in which legal and other institutional avenues are opened up to demand change. This progression corresponds with Gramsci's distinction between discourses that 'are arbitrary [and] ... only create individual "movements", polemics and so on' and those that '"organise" human masses ... and create the terrain on which [they] acquire consciousness of their own position, struggle etc.'.[170] Of course, this is not to imply that demands can only be expressed in terms of rights, but the claim-holder/duty-holder relationship implied in rights discourse makes it a particularly strong language to rely on for that purpose.

Despite initial uncertainties about the nature of States Parties' obligations under the ICESCR, the jurisprudence of the CESCR and other international sources has provided a normative framework against which state and other behaviour can be measured, critiqued and challenged. The presumption against retrogressive measures can be used to challenge the logic of austerity; the prohibition against discrimination

---

[167] See generally, Wesley Newcomb Hohfield, 'Some Fundamental Legal Concepts as Applied to Judicial Reasoning' (1913) 23 *Yale Law Journal* 16; Wesley Newcomb Hohfield, 'Some Fundamental Legal Concepts as Applied to Judicial Reasoning' (1917) 26 *Yale Law Journal* 710.

[168] Shareen Hertel and Lanse Minkler, 'Economic Rights: The Terrain' quoted in Paul O'Connell, 'Let Them Eat Cake: Socio-Economic in an Age of Austerity' in Colin Harvey, Aoife Nolan and Rory O'Connell (eds.), *Human Rights and Public Finance* (Hart 2013) 73.

[169] ibid.   [170] Gramsci (n 7) 377.

can challenge privatisation measures that disproportionately impact on poor and marginalised groups; and the goal of progressive realisation of universal access to certain material entitlements condemns widespread poverty and material deprivation and opens up legal and other institutional channels to challenge them.[171] In the context of neo-liberal globalisation, which seeks to remove or neutralise the effectiveness of social protections inscribed within the welfare/developmentalist state, the task of rejuvenating the *redistributivist* state and making it more ethical, progressively just and increasingly accountable is central to a counter-hegemonic rights strategy today.[172]

2.5.1.3 SOCIOECONOMIC RIGHTS AS IMMANENT CRITIQUE

Today, as Costas Douzinas notes, human rights rhetoric has become a new global *lingua franca* 'adopted by left and right, the north and the south, the state and the pulpit, the minister and the rebel'.[173] In other words, human rights have become part of the 'common sense' of political and ethical discourse today. As argued in the previous chapter, common sense is not a monolithic phenomenon but rather is open to multiple interpretations and 'potentially supportive of very different kinds of social visions and political projects'.[174] Common sense therefore constitutes the starting point for counter-hegemonic activity. The widespread acceptance of human rights discourse, and the equally widespread disagreement over its meaning, makes it an important battleground for a counter-hegemonic *war of position*.

Immanent critique is premised upon the idea that 'social criticism should be based on concrete, existing and accepted politico-legal standards

---

[171] See Ariranga G Pillay, Chairperson, Committee on Economic, Social and Cultural Rights, 'Letter to States Parties' (16 May 2012) CESCR/48th/SP/MAB/SW (reminding states that even in times of crisis they must use their maximum available resources to fulfil socioeconomic rights obligations); CESCR, 'Concluding observations on the fourth report of Iceland' (30 November 2012) UN Doc. E/C.12/ISL/CO/4, para.6; CESCR; CESCR, 'Concluding observations of the Committee on Economic, Social and Cultural Rights: Spain' (6 June 2012) UN Doc. E/C.12/ESP/CO/5, para.8.

[172] Upendra Baxi, *The Future of Human Rights* (Oxford University Press, 2002) 41. See also Mark Boden, 'Primitive Accumulation, Neo-Liberalism and Counter-Hegemony in the Global South: Re-Imagining the State' (n/d) www.nottingham.ac.uk/shared/shared_cssgj/Documents/smp_papers/boden.pdf accessed 3 July 2013.

[173] Costas Douzinas, *Human Rights and Empire: The Political Philosophy of Cosmopolitanism* (Routledge, 2007) 33.

[174] Mark Rupert, 'Reading Gramsci in an Era of Globalising Capitalism' in Andreas Bieler and Adam David Morton (eds.), *Images of Gramsci: Connections and Contentions in Political Theory and International Relations* (Routledge, 2006) 488.

(rather than simply contested and abstract moral and ethical criteria)'.[175] As discussed in the previous chapter, this form of critique is a key component of counter-hegemonic activity. Hegemony is not simply a dominant ideology that shuts out all alternative visions and political projects. One of the important aspects of a hegemonic project is the way that it absorbs counter-hegemonic ideas through at least partially addressing some of the concerns and demands of the subaltern classes. The task of a counter-hegemonic strategy must therefore be to 'tease out' and supplement aspects of the hegemonic apparatus that embody the 'good sense' of the subaltern classes and refashion them into an ideologically mature alternative to the extant hegemony.[176] The contradictions that exist between the hegemonic message on the one hand and the 'lived reality' of the subaltern classes on the other, makes hegemonic discourses particularly vulnerable to immanent critique. As Ralph Miliband argued:

> ... the main reason why the struggle for hegemony-as-consent can never be taken to be finally won ... is that there exists a vast discrepancy between the message which hegemonic endeavours seek to disseminate, and the actual reality which daily confronts the vast majority of the population for whom the message is mainly intended. The message speaks of democracy, equality, opportunity, prosperity, security, community, common interests, justice, fairness, etc. The reality, on the other hand, as lived by the majority, is very different, and includes the experience of exploitation, domination, great inequalities in all spheres of life, material constraints of all kinds, and very often great spiritual want .... A crucial purpose of hegemonic endeavours is to prevent such sentiments from turning into a generalised availability to radical thoughts.[177]

Today, every Member State of the UN has either signed or ratified at least one international treaty that contains some socioeconomic rights obligations. Yet despite 'the sheer volume of voluntarily assumed standards and the regularity with which human rights are endorsed in political rhetoric',[178] there is nevertheless 'a growing perception that economic and social rights are increasingly being eroded by the momentous disruptions brought about by economic globalisation'.[179] Despite the formal

---

[175] Eoin Rooney and Colin Harvey, 'Better on the Margins? A Critique of Mainstreaming Economic and Social Rights' in Nolan et al. (eds.) (n 168) 130 (fn 39).
[176] Gramsci (n 7) 326–327.
[177] Ralph Miliband, 'Counter-Hegemonic Struggles' (1990) 26 *Socialist Register* 346, 347.
[178] Rooney and Harvey (n 175) 130 (fn 39).
[179] Fantu Cheru, 'Debt, Adjustment and the Politics of Effective Response to HIV/AIDS in Africa' (2002) 23(2) *Third World Quarterly* 299, 299.

commitment of governments and international institutions to protecting and promoting socioeconomic rights, in reality many have been responsible for enforcing agreements and implementing policies that have eroded social protections and expropriated local communities and primary producers of the means of subsistence.[180] This glaring contradiction – between the aspirations towards universal socioeconomic security on the one hand and the dynamics of neo-liberal capitalism's drive towards 'accumulation by dispossession' on the other – constitutes a pressure point in neo-liberalism's hegemonic framework. The task of counter-hegemonic movements is to draw attention to these discrepancies, highlight the contradictions and channel grievances into struggles for modes of social organisation where purported universal aspirations can be more fully realised.

2.5.2 POSSIBLE LIMITATIONS OF SOCIOECONOMIC RIGHTS DISCOURSE

2.5.2.1 THE PRESENCE OF ECONOMIC LIBERALISM IN HUMAN RIGHTS DISCOURSE

The historic emphasis of certain strands of human rights on economic liberty – understood as individual appropriation and exclusive ownership of resources – at the expense of the equitable distribution of resources has been well documented.[181] Nevertheless, many commentators argue that this excessive emphasis on economic liberty in early expressions of natural rights has been displaced by a more balanced vision of human rights contained in the UDHR and subsequent human rights instruments.[182] In particular, the inclusion of socioeconomic rights in these instruments is argued to militate against the property-orientated expressions of human rights found in both classical and neo-liberal thought.

Whilst neo-liberal doctrine and the ideal of socioeconomic rights appear to be in discursive tension, following a neo-Gramscian analysis

---

[180] Isabella Bakker and Stephen Gill, 'Global Political Economy and Social Reproduction' in Isabella Bakker and Stephen Gill (eds.), *Power, Production and Social Reproduction* (Palgrave, 2003) 4.

[181] See, e.g. Karl Marx, 'On the Jewish Question' reproduced in Joseph O'Malley (ed.), *Marx: Early Political Writings* (Cambridge University Press, 1993) 28; John Charvet and Elisa Kaczynska-Nay, *The Liberal Project and Human Rights: The Theory and Practice of a New World Order* (Cambridge University Press, 2008) 11–12.

[182] Stephen P Marks, 'From the "Single Confused Page" to the "Decalogue for Six Billion Persons": The Roots of the Universal Declaration of Human Rights in the French Revolution' (1998) 20(3) *Human Rights Quarterly* 459, 475.

it can be suggested that neo-liberal hegemony is achieved not through the imposition of a coherent and unified doctrine on social reality but rather through an on-going process of contestation that involves incorporating subaltern concerns into 'ever more refined but basically unchanged versions' of neo-liberal governance.[183] In addition to this, there is a rich collection of human rights scholarship that draws our attention to the fact that human rights discourse is a double-edged sword in relation to its ability to contest power.[184] Whilst human rights tend to begin their existence as subversive challenges to extant relations and structures of power, they are also likely to develop a more ambiguous relationship with dominant power structures as they become sedimented as sources of positive law (including public international law).[185] In short, the necessary reflexivity of a given hegemonic project coupled with the ambiguous relationship of rights discourse to power means that the tensions that appear to exist between socioeconomic rights and neo-liberalism can be overcome.

Indeed, the discursive impact of neo-liberalism on socioeconomic rights has been documented by a number of scholars. In the judicial setting, Paul O'Connell has surveyed an underlying trend in the jurisprudence of apex courts in Canada, India and South Africa that reflects an 'atomistic, "market friendly"' reading of socioeconomic rights congruent with neo-liberalism.[186] Similarly, Upendra Baxi has noted a tendency within the UN in which 'human rights standards ... lend themselves to a whole variety of foreign power and global corporate uses and abuses under the cover of "international consensus"'.[187] Indeed Baxi has gone so far as to argue that the materiality of neo-liberal globalisation has

---

[183] Richard Peet, 'Ideology, Discourse and the Geography of Hegemony: From Socialist to Neoliberal Development in Postapartheid South Africa' (2002) 34(1) *Antipode* 54, 65.

[184] See, e.g., Morton J Horwitz, 'Rights' (1988) 23 *Harvard Civil Rights-Civil Liberties Law Review* 393–406; Douzinas, *The End of Human Rights* (n 42) 1; Baxi *The Future of Human Rights* (2002) (n 172) 40–41; de Sousa Santos (n 97) 257–280; Balakrishnan Rajagopal 'Counter-Hegemonic International Law: Rethinking Human Rights and Development as a Third World strategy' (2006) *Third World Quarterly* 767, 768; Stammers, *Human Rights and Social Movements* (n 56) 3.

[185] Stammers (n 56) 3 ('human rights initially emerge as "struggle concepts" to support social movement challenges to power ... once institutionalised ... human rights stand in a much more ambiguous relation to power'); Douzinas (n 42) 7 (arguing that a significant shift occurs when rights 'are turned from a discourse of rebellion and dissent into that of state legitimacy').

[186] Paul O' Connell, 'The Death of Socioeconomic Rights' (2011) 74(4) *Modern Law Review* 532.

[187] Baxi, *The Future of Human Rights* (2002) (n 172) 9.

resulted in a shift towards a 'trade-related market friendly paradigm' within human rights discourse.[188] Within this alternate paradigm the 'promotion and protection of some of the most cherished contemporary human rights becomes possible only when the order of rights for global capital stands fully recognized'.[189] Socioeconomic rights are thus reconceptualised as derivative of the rights of private businesses and TNCs. The corporate friendly reading of socioeconomic rights finds its expression in arguments such as the 'right to food (now reconceptualised by the Rome Declaration as the right to food security systems) is best served by the protection of the rights of agribusiness corporations';[190] the 'right to health is best served, in a variety of contexts, by the protection of the research and development rights of pharmaceutical and diagnostic industries'[191] and the right to water is best served by granting 'corporate rights to withdraw water globally for private profit'.[192]

Within the trade-related market-friendly paradigm, socioeconomic rights are also dependent upon the creation of a neo-liberal macroeconomic framework. As Robert Anderson and Hannu Wager, two counsellors to the WTO Secretariat, put it:

> Trade liberalization, by enhancing possibilities for voluntary exchange according to the principles of competitive advantage, creates wealth for all participants and thereby generate the resources needed for the fuller realization of ... economic social and cultural rights.[193]

Indeed, the core international institutions associated with neo-liberal globalisation – the WTO, the IMF and the World Bank – whilst resisting formally integrating concern for human rights into their constitutions, maintain that their overall significance for socioeconomic rights is a positive one.[194] The formally neutral stance adopted by the CESCR

---

[188] Upendra Baxi, *The Future of Human Rights* (3rd edn., Oxford University Press, 2008) 234.
[189] ibid, 257.   [190] ibid.   [191] ibid (n 188) 256.
[192] Rada De Souza, 'Liberal Theory, Human Rights and Water-Justice: Back to Square One?' (2008) *Law, Social Justice & Global Development Journal* 1, 5.
[193] Robert D Anderson and Hannu Wager, 'Human Rights, Development and the WTO: The Cases of Intellectual Property and Competition Policy' (2006) 9(3) *Journal of International Economic Law* 707, 708.
[194] World Bank, *Development and Human Rights: The Role of the World Bank* (World Bank, 1999) 3 ('[t]hrough its support of primary education, health care and nutrition, sanitation, housing, and the environment, the Bank has helped hundreds of millions of people attain crucial economic and social rights'); François Gianviti, 'Economic, Social and

makes it possible for neo-liberals to argue that 'markets maximise economic growth, providing the optimal basis for the enjoyment of (socioeconomic rights) therefore states committed to (the) ICESCR will invariably privilege market solutions'.[195] Socioeconomic rights discourse is therefore not immune from appropriation by neo-liberal forces, a threat that the subaltern movements struggling for such rights must be aware of.

2.5.2.2 RISKS OF NARROW LEGALISM

The appropriation of socioeconomic rights discourse and the co-option of movements for socioeconomic rights can occur in far subtler forms than those that were identified in the previous section. There are dangers that human rights discourse can become focused on a 'narrow legalism' that can potentially exclude both wider structural critique of neo-liberalism and mass participation in transformative action. For Gramsci, counter-hegemony entails activity aimed at 'the reorganisation of the structure ... of the economy and production'.[196] There is a concern that human rights are too narrow and legalistic as a discourse to be used to challenge the systematic and material bases of social deprivation that are governed by the structural logic and organisation of the global political economy.[197] Human rights challenges, particularly in the form of litigation,

---

Cultural Rights and the International Monetary Fund' (2002) para.57 ('by promoting a stable system of exchange rates and a system of current payments free of restrictions ... the Fund contributes to providing the economic conditions that are a precondition for the achievement of the rights set out in the [ICESCR]') www.imf.org/external/np/leg/sem/2002/cdmfl/eng/gianv3.pdf accessed 12 December 2013; Pascal Lamy, 'Towards Shared Responsibility and Greater Coherence: Human Rights, Trade and Macroeconomic Policy' (2010) www.wto.org/english/news_e/sppl_e/sppl146_e.htm accessed 13 December 2013 ('The opening of markets creates efficiency, stimulates growth and helps spur development, thereby contributing to the implementation of the fundamental human rights that are social and economic rights'). For an account of how the IFIs have adopted a hegemonic interpretation of human rights congruent with neo-liberalism see Nlerum O Okogbule, 'Modest Harvests: Appraising the Impact of Human Rights Norms on International Economic Institutions in Relation to Africa' (2011) 15(5) *The International Journal of Human Rights* 728.

[195] Rooney and Harvey (n 175) 129.   [196] Gramsci (n 7) 263.
[197] Klein (n 43) 118–121 (arguing that human rights NGO activity in Chile and Argentina in the 1970s failed to identify and address the underlying political and structural factors that generated repressive state activities); Susan Marks, 'Human Rights and Root Causes' (2011) 41(1) *Modern Law Review* 57, 70–74 (suggesting that much human rights analysis still treats abuses as contingent phenomena rather than the necessary outcomes of particular systematic or material arrangements).

often revolve around relatively narrow issues, and underlying structural factors (political, social, cultural, economic etc.) are generally left unaddressed.[198] The identification of a violator, violation and remedy is foregrounded in human rights analysis, whilst the underlying background structures that give rise to such violations are neglected or minimised.[199] Indeed, Samuel Moyn has argued that human rights discourse emerged as 'the last utopia' in history in the 1970s as disillusionment grew with the revolutionary projects of national liberation and communism.[200] In this sense, the rise of human rights coincided with the decline of the language of radical political and economic transformation. There may be dangers that reliance on socioeconomic rights discourse could result in excluding more radical demands of social movements that lack a firm basis in international law, potentially draining social movement struggles of their subversive energy in the process. Indeed, as this chapter has highlighted, the overall orientation of the CESCR and the UN human rights system is significantly less radical than the stance adopted by sections of the subaltern movements currently resisting neo-liberal globalisation.

These tendencies towards narrow legalism can be encouraged by the distribution of power within global civil society. As noted in the previous chapter, the domain of global civil society should not be understood as a unitary agent of emancipation but rather as terrain upon which hegemonic discourses are produced and reproduced as well as contested. There are always risks that even well-meaning human rights movements may inadvertently help to reproduce the dominant values of neo-liberal globalisation. Transnational coalitions are usually dependent upon 'the ability of indigenous groups, grassroots movements, and activists in impoverished countries to establish discursive and organizational ties with relatively resourceful experts and often with highly professionalized NGOs'.[201] As a result, human rights organisations in the south are often formed by 'English-language speaking, cosmopolitan local activists who know how to relate to western donors ... and write fund raising

---

[198] Kennedy (n 41) 10–13.
[199] Alicia Ely Yamin, 'The Future in the Mirror: Incorporating Strategies for the Defense and Promotion of Economic, Social and Cultural Rights into the Mainstream Human Rights Agenda' (2005) 27 *Human Rights Quarterly* 1200, 1221.
[200] Samuel Moyn, *The Last Utopia: Human Rights in History* (Harvard University Press, 2010) 120–175.
[201] Ronen Shamir, 'Corporate Social Responsibility: A Case of Hegemony and Counter-Hegemony' in De Sousa Santos and Rodriguez-Garavito (eds.) (n 45) 113.

appeals'.²⁰² In order to be able to secure funding and gain a hearing at international forums it is often necessary for organisations to demonstrate their professional credentials by articulating concerns and demands in the technical vocabulary of human rights law.²⁰³ The exclusionary potential of these pressures could undermine the counter-hegemonic capacity of socioeconomic rights discourse. As Gramsci argued, counter-hegemony involves the construction of 'an intellectual-moral bloc which can make politically possible the intellectual progress of the mass and not only of small intellectual groups'.²⁰⁴

2.5.2.3 HUMAN RIGHTS STANDARDS FAIL TO ADEQUATELY ADDRESS THE RESPONSIBILITIES OF PRIVATE AND TRANSNATIONAL ACTORS

Some critics have called into question whether the international human rights framework's focus on the state as a primary or sole duty-bearer can adequately address the types of violations of socioeconomic rights associated with neo-liberal globalisation.²⁰⁵ The traditional human rights paradigm imposes obligations on States Parties to respect, protect and fulfil the human rights of those subjects within their jurisdiction. However, the capacity to regulate certain aspects of economic and social affairs within their own borders has been significantly weakened by developments in the financial and commodity markets, the consolidation of global productive capacity by TNCs and the economic and ideological leverage of IFIs like the IMF and World Bank.²⁰⁶ In addition, the policies of northern states – ranging from their economic protectionism to their roles vis-à-vis international lending institutions in the imposition of structural adjustment – stand accused of undermining the socioeconomic rights of the poor in the Global South.²⁰⁷ This has led some human rights advocates to call for the imposition of direct human rights obligations on

---

[202] Balakrishnan Rajagopal, *International Law from Below: Development, Social Movements and Third World Resistance* (Cambridge University Press, 2004) 261.

[203] Stammers (n 56) 126–127.   [204] ibid, 332–333.

[205] See, e.g., Gideon Baker, 'Problems in the Theorisation of Global Civil Society' (2002) 50 *Political Studies* 928–943; Tony Evans and Alison J Ayers, 'In the Service of Power: The Global Political Economy of Citizenship and Human Rights' (2006) 10(3) *Citizenship Studies* 289, 295–299.

[206] Leslie Sklair, *Globalization: Capitalism & Its Alternatives* (3rd edn., Oxford University Press, 2002) 309.

[207] Thomas Pogge, *World Poverty and Human Rights* (Polity, 2002) 15–20; Ricardo Faini and Enzo Grill, 'Who Runs the IFIs?' (*Centro Studi Luca D'Agliano*, October 2004). www.dagliano.unimi.it/media/WP2004_191.pdf accessed 7 January 2014; Adouharb and Cingranelli (n 89) 135–149.

IFIs and TNCs[208] and argue that northern states should be held accountable for 'extra-territorial' violations of socioeconomic rights.[209] Such calls have been consistently opposed by northern states, who argue that extra-territorial commitments under the ICESCR are, at best, moral obligations of a non-legal nature,[210] whilst IFIs and TNCs have also resisted the imposition of binding human rights standards.[211]

Nevertheless, the CESCR has used its General Comments to delineate certain extra-territorial obligations in relation to human rights obligations. In some instances, it has used robust language, asserting that states 'must ... do everything possible to protect at least the core content' of socioeconomic rights and that states 'have to respect the enjoyment' of those rights of peoples in other countries.[212] On other occasions the CESCR has used more cautious language, merely expressing States Parties' obligations in terms of what they *should* do.[213] For example, as members of IFIs – such as the IMF and World Bank – it is only asserted that they '*should* pay greater attention' to the protection of the relevant

---

[208] Manisuli Ssenyonjo, 'The Applicability of International Human Rights Law to Non-State Actors: What Relevance to Economic, Social and Cultural Rights?' (2008) 12(5) *The International Journal of Human Rights* 725-760; Ziegler et al. (ibid) 84-100.

[209] See, e.g., Jean Ziegler, Christopher Golay, Claire Mahon and Sally-Anne Way, *The Fight for the Right to Food* (Palgrave, 2010) 78-84.

[210] For a critical analysis see Matthew Craven, 'The Violence of Dispossession: Extra-Territoriality and Economic, Social and Cultural Rights' in Mashood A Baderin and Robert McCorquodale (eds.), *Economic, Social and Cultural Rights in Action* (Oxford University Press, 2007) 71-88.

[211] Gianviti (n 194) 10-30 (arguing that the ICESCR does not apply to the IMF); Ibrahim FI Shihata (ed.), *The World Bank Inspector Panel in Practice* (2nd edn., Oxford University Press, 2000) 241 ('There is no legal obligation on behalf of the Bank or its staff to guarantee that the project it finances will succeed or will not cause harm to any party.'); Ernst-Ulrich Petersmann, 'Time for a United Nations "Global Compact" for Integrating Human Rights into the Law of Worldwide Organizations: Lessons from European Integration' (2002) 13 *European Journal of International Law* 621 (noting that the WTO does not integrate human rights standards into its law and practice); Steiner, Alston and Goodman (n 62) 1404-1405 (commenting on the widespread corporate and governmental opposition to the 2003 UN Sub-Commission prepared Norms of the Responsibilities of Transnational Corporations and other Business Enterprises with Regard to Human Rights).

[212] CESCR, 'General Comment No. 8: The relationship between economic sanctions and respect for economic, social and cultural rights' (12 December 2997) UN Doc. E/C.12/ 1997/8, para.7; CESCR, General Comment No. 14 (n 125) para.39; CESCR, 'General Comment No. 15' (n 94) paras.31 and 36.

[213] See, e.g., CESCR, 'General Comment No. 2' (n 140) para.9; CESCR, 'General Comment No. 12' (n 151) para.36; CESCR, General Comment No. 13 (n 125) para.56; CESCR, 'General Comment No. 15' (n 94) paras.33-36; CESCR, 'General Comment No. 18' (n 152) para.30.

rights in their influencing of lending policies, credit agreements and international measures.[214] With regard to non-State actors, such as TNCs and IFIs, the CESCR can only acknowledge that such entities are not directly bound by the ICESCR[215] and are consequently restricted to encouraging them to 'pay greater attention' to socioeconomic rights concerns in the countries that they affect.[216]

The CESCR's jurisprudence on the extraterritorial scope of socioeconomic rights obligations is less radical than the obligations suggested by other commentators.[217] Nevertheless, it is arguable that a normative (if not legally binding) framework for assessing extraterritorial socioeconomic rights obligations is emerging that can be marshalled by activists in global civil society to challenge global neo-liberal institutional structures.[218] That being said, critical ambiguities in the understanding of extraterritorial obligations, coupled with the opposition of northern states to the imposition of such standards, could undermine appeals by social movements to socioeconomic rights standards, when many of the (moral) violations of such standards may lack a firm or clear basis in international law.

With regard to the obligations of non-state actors (e.g. TNCs) and inter-state actors (e.g. IFIs, the WTO), whilst the current framework does not impose direct obligations upon these entities, it does establish obligations upon states to regulate the former under its obligation to protect human rights[219] and imposes obligations upon the conduct of individual states parties when taking part in the latter.[220] Nevertheless, there are

---

[214] CESCR, 'General Comment 14' (n 125) para.39; CESCR, 'General Comment No. 15' (n 94) paras.31 and 36; CESCR, 'General Comment No. 19' (n 152) paras.53 and 58.
[215] See, e.g., CESCR, General comment No. 12 (n 151) para.20.
[216] See, e.g., CESCR: General Comment No. 2 (n 140); General Comment No. 13 (n 125) para.60; General Comment No. 14 (n 122) para.39; General Comment 15 (n 94) para.60; General Comment No. 18 (n 152) para.53.
[217] See, for example, Maastricht Principles on Extraterritorial Obligations of States in the area of Economic, Social and Cultural Rights (2011), www2.lse.ac.uk/humanRights/articlesAndTranscripts/2011/MaastrichtEcoSoc.pdf accessed 7 January 2014.
[218] On this see generally, Malcolm Langford et al. (eds.), *Global Justice, State Duties: The Extraterritorial Scope of Economic, Social, and Cultural Rights in International Law* (Cambridge University Press, 2013).
[219] UNHCR, 'Report of the Special Representative of the Secretary- General on the issue of human rights and transnational corporations and other business enterprises, John Ruggie: Guiding Principles on Business and Human Rights: Implementing the United Nations "Protect, Respect and Remedy" Framework' (21 March 2011) UN Doc. A/HRC/17/31, Annex.
[220] See n 151.

myriad practical difficulties in applying these norms. In relation to TNCs, the very real power discrepancies between these powerful actors and weak southern states often make it very difficult for the latter to sufficiently regulate the former. In relation to IFIs, it can be difficult to identify particular state conduct within the often opaque institutional arrangements of institutions such as the IMF and World Bank, and even where it is possible these institutions do not possess any mechanisms whereby these states can be held to account for failing to take socioeconomic rights considerations into account.[221]

2.6 CONCLUSION

This chapter has argued that socioeconomic rights constitute a potentially important discourse to contest the economic basis and ethicopolitical foundations of neo-liberalism. In contrast to the discourses of neo-liberal governance that emphasise competitiveness, efficiency, economic rationalism and limited government, socioeconomic rights discourse promotes a broader conception of governance and development based on the principle of human dignity. The clarification of the normative content of socioeconomic rights by UN human rights bodies over the past two years has in some senses buttressed the opportunities for social movements to mobilise around socioeconomic rights in global civil society. Nevertheless, given the incorporative dimension of hegemony and the malleability of rights discourse, it cannot be assumed that all articulations of socioeconomic rights necessarily have a discrete and oppositional stance vis-à-vis neo-liberalism. Whilst socioeconomic rights discourse has a number of advantages in that it can universalise particular claims, articulate strong normative obligations and be used as a form of immanent critique of the extant order, it also has potential limitations – most notably in the context of the historic liaison between human rights and economic liberalism, the potential for narrow legalist understandings of human rights and also a lack of clarity on the extraterritorial obligations in relation to socioeconomic rights.

---

[221] For a discussion of the limitations of TNC and IFI legal accountability for violations of socioeconomic rights see Smita Narula, 'International Financial Institutions, Transnational Corporations and Duties of States' in Langford et al. (eds.) (n 218) 116–152.

3

FOOD SECURITY VS. FOOD SOVEREIGNTY
THE RIGHT TO FOOD AND GLOBAL HUNGER

> That the objective possibilities exist for people not to die of hunger and that people do die of hunger has its importance, or so one would have thought. But the existence of objective conditions ... is not enough: it is necessary to "know" them, and know how to use them. And to want to use them.[1]

> To recognize the right to food is not without dangers ... [it is to recognize] the right of peasants to organize themselves against the taking over of land by a few, the right of workers to organize themselves in defence of their livelihood, the right of the unemployed to organize themselves to loot food stores in order to survive. It is not unusual for these movements to meet the opposition of those who had, a little hastily, in the euphoria of some conference, recognized solemnly the right to food and the right to a livelihood ... but suddenly decide to put the rights of ownership first.[2]

3.1 INTRODUCTION: THE RIGHT TO FOOD AND WORLD HUNGER

Despite repeated international declarations committed to the reduction and eradication of global hunger since 1974, the shocking reality is that hunger has doubled since that date.[3] In 1974 there were 500 million

---

[1] Antonio Gramsci, *Selections from the Prison Notebooks of Antonio Gramsci* (Quintin Hoare and Geoffrey Nowell Smith eds. and trans.) (Lawrence & Wishart, 1971) 275–276 (hereafter 'Gramsci') 360.

[2] Pierre Spitz, Former Director of the Office of Evaluation and Studies, International Fund for Agricultural Development. Quoted in Philip Alston, 'International Law and the Human Right to Food' in Philip Alston and Katarina Tomasevki (eds.), *The Right to Food* (Martinus Nijhoff, 1984) 18.

[3] In 1974 the World Food Conference pledged to eradicate child hunger in ten years; in 1996 the World Food Summit pledged to reduce the number of people living in hunger by half in 2015; in 2000 world leaders pledged at the Millennium Summit to reduce extreme poverty by half in 2015. In 2002 at the World Food Summit (five years later), Governments were forced to admit that progress towards the Millennium Goals had been poor. See Eric Holt-Giménez and Raj Patel, *Food Rebellions! Crisis and the Hunger for Justice* (Pambazuka Press, 2009) 9.

hungry people in developing countries[4] but in 2009 the United Nations Food and Agriculture Organization (FAO) announced that, for the first time, there were more than one billion undernourished people worldwide.[5] The year 2009 represented the peak of the global hunger crisis, but today there are still an estimated 795 million undernourished people globally.[6] Whilst the vast majority of the global hungry live in the Global South – notably sub-Saharan Africa, Asia and the Caribbean – hunger and food poverty are becoming increasingly prevalent in northern states as well.[7] For example, there are now close to a quarter of a million people dependent upon food banks in the United Kingdom.[8] Paradoxically, half of the world's hungry people are themselves food producers, usually smallholder farmers who live off small plots of land without access to adequate resources.[9] The rural and urban poor – and in particular children, elderly people, women and refugees – are identified as being the most vulnerable to hunger.[10]

The causes of world hunger and other related injustices associated with the global production and distribution of food are complex and multifarious. However, it is not the case that hunger and malnutrition are simply the products of fate, geographic location or climatic phenomena. In the past the FAO has suggested that enough food is produced to feed 12 billion people: double the world's current population.[11] It was recently estimated that as much as half of all the food produced in the world – equivalent to 2 billion tonnes – ends up as waste every year.[12] Furthermore, hunger and undernourishment co-exist with 'obesity

---

[4] ibid.
[5] FAO, 'More People than Ever Are Victims of Hunger' (2009) www.fao.org/fileadmin/user_upload/newsroom/docs/Press%20release%20june-en.pdf accessed 5 November 2012.
[6] FAO, IFAD and WFP, *The State of Food Insecurity in the World: Meeting the 2015 International Hunger Targets: Taking Stock of Uneven Progress* (2015) 4.
[7] Olivier de Schutter, 'The Right to Food in Times of Crisis' in Just Fair Report, *Freedom from Hunger: Realising the Right to Food in the UK* (Just Fair 2013) 8–9 www.edf.org.uk/blog/wp-content/uploads/2013/03/Freedom-from-Hunger.Just-Fair-Report.FINAL_..pd> accessed October 7 2013.
[8] Just Fair (ibid) 2.
[9] UNGA, 'Res 28/10: The Right to Food' (2 April 2015) UN Doc. A/HRC/RES/28/10) Para.19.
[10] FAO, 'People at Grave Risk of Hunger' (1996) available at www.fao.org/FOCUS/E/WFDay/WFGra-e.htm accessed November 5 2012.
[11] UNCHR, 'Report of the Special Rapporteur on the right to food, Jean Ziegler' (10 January 2008) UN Doc. A/HRC/7/5 2.
[12] Rebecca Smithers, 'Almost Half of the World's Food Thrown Away, Report Finds' *Guardian* (Manchester, 10 January 2013).

pandemics' in North America and Europe.[13] The roots of the problem of hunger and malnutrition are therefore not concerned with the lack of food but rather the lack of access to available food.[14]

For the purposes of this chapter it is not necessary to attempt a detailed explanation of all the factors that contribute towards hunger, nor is it necessary to embark upon the reductive exercise of attempting to explain all of the world's food ills with reference to a single causal determinant. Instead, this chapter will limit itself to considering the impact of the economic, agricultural and trade policies associated with what will be termed 'the neo-liberal food regime'. In relation to this, this chapter will explore attempts by subaltern social movements to contest the neo-liberal food regime by mobilising around articulations of the right to food. The chapter will begin by identifying the political-institutional arrangements of the neo-liberal food regime and the ways in which they have been criticised for contributing towards world hunger and other food-related injustices. The next section will explore the legitimising framework of 'food security' that underpins the neo-liberal food regime. Following this, the chapter will introduce an outline of an alternative ethico-political framework of 'food sovereignty', a paradigm for food and agriculture developed by the international peasant movement, La Via Campesina, and subsequently taken up by small-scale farmers, pastoralists, indigenous peoples, landless peasants and urban food campaigners. The chapter will then examine how articulations of the right to food within global civil society – particularly in the contexts of the UN Human Rights System and the FAO – have interacted with the hegemonic frame of 'food security' and the counter-hegemonic framework of 'food sovereignty'. The chapter will conclude by critically evaluating the extent to which the food sovereignty model has succeeded in establishing a counter-hegemonic basis for articulations of the right to food.

3.2 THE POLITICAL-INSTITUTIONAL SETTING: THE NEO-LIBERAL FOOD REGIME

In order to understand global neo-liberal food policy this chapter adopts a 'food regime' analytical framework as developed by Harriet Friedmann

---

[13] See generally, Raj Patel, *Stuffed and Starved: Markets, Power and the Hidden Battle for the World Food System* (Black, 2007).
[14] Committee on Economic, Social and Cultural Rights (hereafter the CESCR or 'the Committee'), 'General Comment 12: The Right to Adequate Food (Art. 11)' (12 May 1999) UN Doc. E/C.12/1999/5, para.5.

and Philip McMichael.[15] A food regime is conceptualised as 'a rule-governed structure of production and consumption on a world scale'.[16] It constitutes a constellation of class and interstate power relations, norms and institutional structures that link global relations of production and consumption to specific periods of capital accumulation.[17] The emergence of the neo-liberal food regime is intimately related to the restructured global economy over the last thirty years, and the general trends towards financial deregulation, trade liberalisation, public spending reductions and privatisation.[18] Whilst patterns of food production have long been dependent on capitalist social relations, the 1990s witnessed a shift away from a nationally-centred development paradigm towards a corporate globalisation project with transnational corporations (TNCs) exercising a central role in global food chains.[19]

The most relevant organs of global political society that comprise the political-institutional structure of the neo-liberal food regime can be identified as the private investment arms of the International Monitory Fund (IMF) and the World Bank, the World Trade Organization (WTO) and Northern States – particularly core powers within the European Union (EU) and the United States of America (USA).[20] The third world debt crisis of the 1980s provided the basis for IMF and World Bank mandated 'structural adjustment' requirements that indebted southern countries open their markets up to agricultural imports.[21] This was closely followed by the negotiation of the Agreement on Agriculture (AoA)

---

[15] See, e.g., Harriet Friedmann, 'International Regimes of Food and Agriculture since 1870', in Theodor Shanin (ed.) *Peasants and Peasant Societies* (Basil Blackwell, 1987) 258–276; Harriet Friedmann and Philip McMichael, 'Agriculture and the State System: The Rise and Decline of National Agricultures, 1870 to the Present' (1989) 29(2) *Sociologia Ruralis* 93–117.

[16] Harriet Friedmann, 'The Political Economy of Food: A Global Crisis' (1993) 197(1) *New Left Review* 29, 30–31.

[17] Harriet Friedmann, 'Discussion: Moving Food Regimes Forward: Reflections on Symposium Essays' (2009) 26(4) *Agriculture and Human Values* 335–344; Philip McMichael 'A Food Regime Genealogy' (2009) 36(1) *Journal of Peasant Studies* 139–169.

[18] Valeria Sodano, 'Food Policy Beyond Neo-Liberalism' in Dennis Erasga (ed.) *Sociological Landscape: Theories, Realities and Trends* (InTech, 2012) 375, 377.

[19] Philip McMichael, 'Food Security and Social Reproduction: Issues and Contradictions' in Isabella Bakker and Stephen Gill (eds.), *Power, Production and Social Reproduction* (Palgrave Macmillan, 2003) 171.

[20] Eric Holt-Giménez and Annie Shattuck, 'Food Crises, Food Regimes and Food Movements: Rumblings of Reform or Tides of Transformation?' (2011) 38 *Journal of Peasant Studies* 109–144.

[21] Susan Soederberg, 'The Transnational Debt Architecture and Emerging Markets: The Politics and Paradoxes of Punishment' (2003) 26(6) *Third World Quarterly* 927–949.

during the Uruguay Round of the General Agreement on Tariffs and Trade (GATT), which entered into force with the establishment of the WTO in 1995.[22] The AoA brings agriculture within the general WTO principles of trade liberalisation, thereby fully integrating Southern agricultural systems into the global market. Northern States have facilitated the development of the corporate food regime by granting multi-billion dollar subsidies to the agricultural sectors of the USA and Europe, establishing intellectual property regimes favourable to the commercial expansion of Genetically Modified (GM) food and creating new markets for financial speculation in food commodities.[23]

3.2.1 CRITICISMS OF THE NEO-LIBERAL FOOD REGIME

Having sketched the political-institutional basis of the neo-liberal food regime, this section will briefly identify three aspects of these institutional arrangements that have been criticised for contributing to the problem of global hunger: structural adjustment, trade liberalisation and the domination of the food chain by TNCs and global finance capital.

Structural adjustment measures imposed on the Global South since the 1980s by the IMF and the World Bank have led to privatisation and the dismantling of government support for the domestic agricultural sector.[24] The removal of subsidies on agricultural products, the abolition of domestic stabilisation institutions, the dismantling of food price controls and budgetary cutbacks in rural support services have undermined low-income farmers who cannot compete with capital-intensive large-scale farms.[25] These measures have also threatened the food security of the poor, particularly in times of high food prices and natural disasters where access to available and affordable food may be threatened.[26]

---

[22] Agreement on Agriculture (AoA), April 15, 1994, Marrakesh Agreement Establishing the World Trade Organization, Annex 2, Legal Instruments – Results of the Uruguay Round, 33 I.L.M 1125 (1994).

[23] Nicholas John Rose, 'Optimism of the Will: Food Sovereignty as Transformative Counter-Hegemony of the 21st Century' (PhD Thesis, RMIT University 2013) 117.

[24] Farshad Araghi, 'The Invisible Hand and the Visible Foot: Peasants, Globalization and Dispossession' in A Haroon Akram-Lodhi and Cristóbal Kay (eds.), *Peasants and Globalisation: Political Economic, Rural Transformation and the Agrarian Question* (Routledge, 2009) 133.

[25] ibid.

[26] Rolf Kunnermann and Sandra Epal-Ratjen, *The Right to Food: A Resource Manual for NGOs* (AAAS Science and Human Rights Program, 1999) 49; FIAN International, *Parallel Report: The Right to Adequate Food in Cameroon* (FIAN, 1999) 16.

Furthermore, structural adjustment measures have required that developing countries expand their export-orientated agriculture (so-called cash crops) in order to earn foreign exchange to pay off their debts. The cash crops strategy has been criticised for actually devaluing the price of developing countries' food exports whilst creating dependency on food imports and undermining national self-sufficiency.[27]

The liberalisation of trade in agriculture promoted by the AoA and regional free trade agreements like the North American Free Trade Agreement (NAFTA)[28] has been identified as entrenching and deepening inequalities between northern and southern states. For example, the AoA requires the progressive elimination of barriers to agricultural trade by reducing barriers to agricultural imports,[29] reducing export subsidies[30] and not increasing support for domestic agriculture.[31] It has been noted that these requirements are drafted in such a way that actually allow the EU and USA to protect their agricultural markets to a greater degree than the countries of the Global South.[32] These imbalances have led to the importation of cheap subsidised goods from northern industrialised countries by southern states – a practice often termed 'dumping'.[33] These measures, combined with World Bank and IMF mandated structural adjustment, have had a devastating impact on smallholder farmers in the Global South who are unable to compete with cheap imported food.[34] For example, the FAO has estimated that at least 20–30 million peasants were displaced in the 1990s following the institution of the WTO, and in Mexico upward of two million peasants lost land after the national market was flooded with subsidised US maize in accordance with the NAFTA agreement.[35] The undermining of domestic production in the

---

[27] See generally, Peter Robbins, *Stolen Fruit: The Tropical Commodities Disaster* (Zed Books, 2003).

[28] North American Free Trade Agreement, U.S.-Can.-Mex., 17 December 1992, 32 I.L.M. 289 (1993).

[29] ibid, articles 4 and 5 and annex 5.   [30] ibid, articles 8, 9, 10 and 11.

[31] ibid, article 6 and annexes 2, 3 and 4.

[32] See, for example, Carmen G Gonzalez, 'Institutionalizing Inequality: The WTO Agreement on Agriculture, Food Security and Developing Countries' (2002) 27 *Colombia Journal of Environmental Law* 433; Actionaid, *The WTO Agreement on Agriculture* www.actionaid.org.uk/sites/default/files/doc_lib/51_1_agreement_agriculture.pdf accessed 12 October 2013.

[33] Philip McMichael, 'The Land Grab and Corporate Food Regime Restructuring' (2012), 39 (3–4) *The Journal of Peasant Studies* 681, 682.

[34] Walden Bello, 'How to Manufacture a Global Food Crisis' (2008) 51(4) *Development* 450–455.

[35] Cited in McMichael, 'The Land Grab' (n 33) 682.

Global South results in these countries becoming increasingly dependent on food imports for their food security.[36]

Both contributing to the processes of structural adjustment and trade liberalisation as well as benefiting from them, TNCs and other powerful private interests now exert enormous control over international agriculture, often to the detriment of the global poor.[37] In 1977 Susan George noted that unless TNC control of the food chain was challenged 'continued malnutrition and starvation will be the only prospect for those hundreds of millions who will never be "consumers": the invisible poor.'[38] Corporate power has grown exponentially since George's remarks and has negatively impacted on access to food in a number of ways. First, the concentration of power in agricultural trade by an increasingly small number of TNCs means that these companies now dominate markets for agricultural inputs, such as seeds and fertiliser, which in turn drives up production costs for small farmers.[39] Second, corporate land grabs in the Global South for biofuels, 'cash crops' and other exports are depriving poor countries of acres of arable land that could have been used to provide jobs and grow affordable food for local communities.[40] Third, the increased global deregulation and financialisation of the global markets has enabled powerful economic actors like hedge funds, pension funds and investment banks to exert inordinate control over the food commodities markets in ways that threaten price volatility for the poor.[41] Finally, TNCs engaged in resource extraction

---

[36] Jacqueline Mowbray, 'The Right to Food and the International Economic System: An Assessment of the Rights-Based Approach to the Problem of World Hunger' (2007) 20(3) *Leiden Journal of International Law* 545, 552.

[37] UNCHR, 'Report of the Special Rapporteur (n 11) 17–19.

[38] Quoted in Philip Alston, 'International Law and the Human Right to Food' in Philip Alston and Katarina Tomasevki (eds.) (n 2) 42.

[39] ActionAid International, *Power Hungry: Six Reasons to Regulate Global Food Corporations* (2005), 31 www.actionaid.org.uk/_content/documents/power_hungry.pdf accessed November 7 2012.

[40] A Haroon Akram-Lodhi, 'Land Grabs, the Agrarian Question and the Corporate Food Regime' (2015) 2(2) *Canadian Food Studies* 233.

[41] The food price crisis of 2008 – in which at least 40 million extra people around the world were driven into hunger as a direct result of increased food prices – is widely recognised to have been strongly influenced by speculative activity in food commodity derivative markets. Noemi Pace, Andrew Seal and Anthony Costello, 'Food Commodity Derivatives: A New Cause of Malnutrition?' (2008) 371 *The Lancet* 1648–1650; United Nations, World Economic Situation and Prospects 2009 (United Nations, 2009) 48; Olivier De Schutter 'Food Commodity Speculation and Food Price Crises: Regulation to Reduce the Risks of Price Volatility' (United Nations Office for the High Commissioner for Human Rights, 2010).

have often interfered with food production in their areas of operation. Oil companies, for example, have been found to have destroyed and contaminated crops, land and water sources that are necessary to grow and store food.[42]

In summary, the global neo-liberal agricultural regime stands accused by its critics of failing to address, and even of exacerbating, the problems of hunger and malnutrition, particularly in the Global South, by undermining national support structures for agriculture, imposing an asymmetric trading regime rigged in the interests of the industrialised north and reducing food to a commodity in a market dominated by powerful TNCs and other private interests at the expense of smallholder farmers and the rural and urban poor.

3.3 THE POLITICS OF FOOD: FOOD SECURITY VS. FOOD SOVEREIGNTY

It is widely recognised that there are two broad competing ethico-political frameworks for global food governance: *food security* and *food sovereignty*.[43] A certain conception of food security has become hegemonic in global governance and currently serves as the legitimising framework for the neo-liberal food regime. Food sovereignty, by contrast, is a counter-hegemonic discourse advanced by an array of subaltern forces as an alternative to the neo-liberal food regime. This section will explore these counter-posed frameworks before going on to assess the extent to which international formulations around the right to food interact with them, and hence either bolster or contest the dominant neo-liberal discursive framework. Whilst these competing discourses are fluid and changing,[44] the following section will use Weberian 'ideal type' analysis outlined in the introduction for comparative analytical purposes.

---

[42] See, for example, *Social and Economic Rights Action Centre (SERAC) and the Centre for Economic and Social Rights (CESR) v. Nigeria,* Communication No. 155/96, (2003) 10(1) IHRR 282 (2001) AHRLR 60 (15th Annual Activity Report) paras.58 and 64; *Kichwa Indigenous People of Sarayuku v. Ecuador* IACtHR Series C. No. 245 (June 27, 2012) paras. 174 and 249.

[43] See generally, William S Schanbacher, *The Politics of Food: The Global Conflict between Food Security and Food Sovereignty* (Praeger, 2010); Hannah Wittman, 'Food Sovereignty: A New Rights Framework for Food and Nature?' (2011) 2 *Environment and Society: Advances in Research* 87–105.

[44] See Lucy Jarosz, 'Comparing Food Security and Food Sovereignty Discourses' (2014) 4(2) *Dialogues in Human Geography* 168.

3.3.1 THE NEO-LIBERAL ETHICO-POLITICAL FRAMEWORK: FOOD SECURITY

As a concept in international politics, 'food security' has evolved over the last thirty years. In the past it was understood at a macro-economic level relating to food supply and price stability, but over the past two decades the concept has shifted from the overall availability of food to the availability of food to individuals and groups.[45] Food security is defined as 'a situation that exists when all people, at all times, have physical, social and economic access to sufficient, safe and nutritious food that meets their dietary needs and food preferences for an active and healthy life'.[46] During the consolidation of the neo-liberal food regime in the 1980s and 1990s, the private sphere increasingly became identified as the setting for the achievement of food security. Whilst still formally a state responsibility, food security was transformed into a policy goal to be realised principally through market mechanisms and private sector actors, with the state playing an enabling role in establishing the legal frameworks in which markets can efficiently function.[47]

Since 1996, a number of international food conferences and high level food summits have been held to address problems of world hunger and food security. These various gatherings have produced a certain conception of food security that promotes the further consolidation and expansion of the neo-liberal food regime.[48] It is possible to identify three main components of this ethico-political framework. First, there is the theory of comparative advantage, which, as Carmen G. Gonzalez argues, 'plays a central role in legitimating both the ideology of free trade and the economic policy recommendations of the WTO, the World Bank and the IMF.'[49] This theory, first developed by David Ricardo, is premised upon the idea that each country should specialise in the goods that it produces relatively more efficiently and should import the goods that it produces relatively less efficiently.[50] For the countries of the Global South

---

[45] Michael Windfuhr and Jennie Jonsen, *Food Sovereignty: Towards Democracy in Localized Food Systems* (ITDG, 2005) 21–22.
[46] Definition from the FAO Committee on World Food Security (CFS) www.fao.org/cfs/en/ accessed 12 October 2013.
[47] McMichael, 'Food Security and Social Reproduction' (n 19) 171–177.
[48] See generally, Section 3.4 of this chapter.
[49] Carmen G Gonzalez, 'An Environmental Justice Critique of Comparative Advantage: Indigenous Peoples, Trade Policy and the Mexican Neoliberal Reforms' (2011) 32(3) *University of Pennsylvania Journal of International Law* 723, 736–737.
[50] Alan O Sykes, 'Comparative Advantage and the Normative Economics of International Trade Policy' (1998) 1 *Journal of International Economic Law* 49, 50–53.

it is argued that they should specialise in the non-grain crops they have a comparative advantage in producing, whilst purchasing their grains and dairy products from northern countries that have surpluses of these products.[51] Its promoters argue that subsidies, tariffs and government-controlled food stocks are inefficient because they distort the market mechanism and encourage countries to produce goods in which they do not have a comparative advantage and which might be produced more cheaply elsewhere.[52] Hence, principle 7(e) of the Declaration of the High-Level Conference on World Food Security in 2008 states that:

> We encourage the international community to continue its efforts in liberalizing international trade in agriculture by reducing trade barriers and market distorting policies. Addressing these measures will give farmers, particularly in developing countries, new opportunities to sell their products on world markets and support their efforts to increase productivity and production.[53]

A second component of the neo-liberal ethico-political framework stresses the importance of corporate-led research and development to enhance the efficiency and output of farming methods. Hunger and the plight of smallholder farmers in the Global South are identified as stemming from problems with unproductive soil, unreliable water supplies, low-quality seeds and scarce markets for their crops. The solution to these problems lies with bio-technological developments in fertilisers, pesticides and GM seed varieties that can increase crop yields, developing better quality seeds and improving soil fertility.[54] For example, the agribusiness TNC Monsanto warns that low tech agriculture 'will not produce sufficient crop yield increases and improvements to feed the worlds burgeoning population', and states that 'biotechnological innovations will triple crop yields without requiring any additional farmland, saving valuable rainforests and animal habitats'.[55] It concludes

---

[51] Gonzalez (n 49) 737.
[52] This nominal commitment to 'free trade' should not, however, obscure the concrete reality of asymmetrical protectionism and subsidies afforded to Northern States through the AoA as well as regional trade agreements like NAFTA and the EU Common Agricultural Policy.
[53] Declaration of the High-Level Conference on World Food Security: The Challenges of Climate Change and Bioenergy (5 June 2008) www.fao.org/fileadmin/user_upload/food climate/HLCdocs/declaration-E.pdf accessed 13 October 2013.
[54] Kees Jansen and Aarti Gupta, 'Anticipating the Future: "Biotechnology for the Poor" as Unrealized Promise?' (2009) 41 *Futures* 436–445.
[55] Quoted in McMichael, 'The Land Grab' (n 33) 182.

'biotechnology can feed the world ... let the harvest begin'.[56] These arguments are also forwarded by philanthropic organisations, like the Bill and Melinda Gates Foundation and the Rockefeller Foundation, that have invested hundreds of millions of dollars to promote a new 'Green Revolution' in Africa.[57]

Finally, there are reformist trends within the neo-liberal food regime that seek to address subaltern critiques of the neo-liberal food regime. This was most strikingly apparent in the World Bank's 2008 World Development Report on agriculture.[58] In the report the Bank frankly admits that its own past policies had:

> left smallholders exposed to extensive market failures, high transaction costs and risks, and service gaps. Incomplete markets and institutional gaps impose huge costs in forgone growth and welfare losses for smallholders, threatening their competitiveness and, in many cases, their survival.[59]

To address these past 'shortcomings' the World Bank offers a number of reformist measures to ensure that trade liberalisation 'reaches the poor', including commitments to prioritising the needs of small farmers, redressing inequalities experienced by women in agriculture and introducing social safety nets for the poorest sections of society. However, Philip McMichael argues that these reforms all operate within the boundaries of neo-liberalism's market-orientated, corporate-friendly framework.[60] For example, the World Bank argues in favour of limited social safety nets to 'the chronic and transitory poor' because they 'can increase both efficiency and welfare. Efficiency gains come from reducing the cost of risk management and the risk of asset decapitalization in response to

---

[56] ibid.
[57] Eric Holt-Giménez and Miguel A Altieri, 'Agroecology, Food Sovereignty, and the New Green Revolution' (2013) 37(1) *Agroecology and Sustainable Food Systems* 90, 92. This technological solution to World Hunger is also reflected in the Declaration of the High-Level Conference on World Food Security, which states that 'We urge the international community, including the private sector, to decisively step up investment in science and technology for food and agriculture. Increased efforts in international cooperation should be directed to researching, developing, applying, transferring and disseminating improved technologies and policy approaches'. Declaration of the High-Level Conference on World Food Security (n 53) principle 7(d).
[58] The World Bank, *World Development Report 2008: Agriculture for Development* (World Bank, 2008).
[59] ibid, 138.
[60] Philip McMichael, 'Banking on Agriculture: A Review of the World Development Report 2008' (2009) 9(2) *Journal of Agrarian Change* 235, 239–241.

shocks'.[61] Thus social safety nets are considered a limited and temporary measure designed primarily to promote economic rationality. Indeed, economic rationality is understood by the Bank as delineating the nature and scope of social assistance: 'These programs have to be organized so that they do not undermine the local labor market and food economy and do not create work disincentives for beneficiaries, but do reach those most in need "just in time."'[62] As Eric Holt-Giménez and Miguel A. Altieri note, whilst the reformist wing of the neo-liberal food regime aims to 'mitigate the excesses of the market, its "job" is ... the reproduction of the corporate food regime'.[63] The limited reforms proposed and sometimes implemented by the World Bank constitute a Gramscian 'passive revolution': elite-led reform aimed at political and economic stabilisation and the forestalling of potentially transformative social movements from below.

3.3.2 THE COUNTER-HEGEMONIC ETHICO-POLITICAL FRAMEWORK: FOOD SOVEREIGNTY

The concept of food sovereignty was first discussed by the international peasant organisation La Via Campesina in 1996 as a response to what they perceived as the failure of the neo-liberal food security paradigm to ensure sustainable access to nutritious and culturally appropriate food.[64] La Via Campesina was created by peasant organisations from around the world and defines itself as 'the international movement which brings together millions of peasants, small and medium-size farmers, landless people, women farmers, indigenous people, migrants and agricultural workers from around the world'.[65] In Chapter 2 it was argued that a global counter-hegemonic movement is one that organises globally around an alternative ideology expressed in universal terms. La Via Campesina satisfies these criteria: it represents around 200 million farmers organised in 150 national and local organisations in 70 countries

---

[61] The World Bank, *World Development Report 2008* (n 58) 18.   [62] ibid.
[63] Holt-Giménez and Altieri (n 57) 92.
[64] Hannah Wittman, Annette Desmarais and Nettie Wiebe, 'The Origins and Potential of Food Sovereignty', in Hannah Wittman, Annette Aurélie Desmarais, Nettie Wiebe (eds.), *Food sovereignty: Reconnecting Food, Nature and Community* (Fernwood, 2013) 2–3.
[65] La Via Campesina, 'The International Peasant's Voice' (*La Via Campesina* 9 February 2011) http://viacampesina.org/en/index.php/organisation-mainmenu-44 accessed 13 October 2013.

from every continent.⁶⁶ Since 1996 it has organised and taken part in an array of international gatherings of transnational civil society and in the process has formed an international network of social movements, research institutions and NGOs committed to developing and implementing food sovereignty as an alternative to the dominant neo-liberal food paradigm.⁶⁷ It is committed to 'globalizing the struggle' by building 'unity and solidarity between small and medium-scale agricultural producers from the North and South' to 'realize food sovereignty and stop the destructive neoliberal process'.⁶⁸

Food sovereignty is defined (in part) as 'the right of peoples to healthy and culturally appropriate food produced through ecologically sound and sustainable methods, and their right to define their own food and agriculture systems'.⁶⁹ Advocates of food sovereignty explain the paradox that the majority of the world's hungry are themselves food producers by reference to the political and economic disempowerment and marginalisation of small-scale food-producing communities: 'In a neo-liberal capitalist structure, a people that does not produce its own food (or a great part of it), is a people that can easily be subjugated to pressure, extortion or domination imposed by the trans-national empire and will end up losing its sovereignty'.⁷⁰ Whilst the 'Food Sovereignty model' is still developing and subject to differing interpretations, two common threads run through all the different approaches to food sovereignty: (1) analysis of food injustice must begin from the perspective of those facing hunger and malnutrition and; (2) it is vital that nations and communities that produce food have the ability to democratically determine their own food systems.⁷¹ In that

---

[66] ibid. [67] Schanbacher (n 43) 53–54.i. [68] La Via Campesina (n 65).

[69] This is the first sentence from the long definition contained within the 2007 Nyéléni Declaration of the Forum for Food Sovereignty www.nyeleni.org/spip.php?article290 accessed 1 December 2012. There have been many definitions of food sovereignty since the concept was first articulated in 1996 but the Nyéléni declaration is widely regarded as the most representative, as more than 500 representatives of organisations of peasants/family farmers, artisanal fisherfolk, indigenous peoples, landless peoples, rural workers, migrants, pastoralists, forest communities, women, youth, consumers, and environmental and urban movements from more than eighty countries agreed on this definition. For a friendly critique of some of the tensions and contradictions within the Nyéléni Declaration see Raj Patel, 'What Does Food Sovereignty Look Like?' (2009) 36(2) *Journal of Peasant Studies* 663, 666.

[70] Juana Curio, 'Seed at the Centre of Food Sovereignty' (2007). http://viacampesina.net/downloads/PDF/seed_heritage_of_the_people_for_the_good_of_humanity.pdf accessed 1 December 2012.

[71] Windfuhr and Jonsen (n 45) xii.

sense, food sovereignty can be understood as a meta-right or the *right to have rights* over food.[72]

Food sovereignty can be understood as providing the counter-hegemonic basis for alternative food governance for a number of reasons. First, in opposition to the neo-liberal conception of food as a commodity to be traded like any other, the food sovereignty model regards food primarily as a fundamental human right because of its social and cultural dimensions.[73] Second, in contrast to the neo-liberal view that food security is achieved through the comparative advantage of importing food from where it is cheapest, the food sovereignty model insists that food security is greatest when food production is in the hands of the hungry, or where food is locally produced.[74] Third, whereas the neo-liberal model proposes industrial, monocultural, chemical-intensive and GMO-based farming technology, proponents of food sovereignty advocate sustainable farming methods based upon 'agroecological' principles and local knowledge.[75] Fourth, while the neo-liberal model regards government subsidies to farmers in the developing world as constituting 'market distortions', the food sovereignty model advocates the provision of public support for family farmers in terms of price and income support, soil conversion, conversion to sustainable farming and research.[76]

The food sovereignty model is not simply a different set of policy proposals to the dominant neo-liberal model; it is also founded upon alternative values and principles. As William Schanbacher points out, food sovereignty discourse expressly contests the core underlying assumptions of the neo-liberal model in that it 'emphasizes mutual wellbeing over self-interest, cooperation over competition, the survival

---

[72] Amartya Sen, 'The Right Not to Be Hungry' in Alston and Tomasevki eds. (n 2) 70 (defining a meta-right as 'the right to have policies p(x) that genuinely pursue the objective of making the right to x realizable'); Patel (n 69) 663 ('In many ways, Via Campesina's call for food sovereignty is precisely about invoking a right to have rights over food').

[73] 'Food cannot be considered a commodity because of its social and cultural dimension' NGO/CSO Forum for food sovereignty, 'Food Sovereignty: A Right for All Political Statement of the NGO/CSO Forum for Food Sovereignty' (June 8–13, 2002 Rome NGO/CSO Forum for Food Sovereignty 2002), principle 6; La Via Campesina, 'Food Sovereignty: A Future without Hunger' (Rome, 1996) www.voiceoftheturtle.org/library/1996%20Declaration%20of%20Food%20Sovereignty.pdf accessed 13 October 2013) principle 1.

[74] La Via Campesina, 'Food Sovereignty: A Future without Hunger' ibid, principle 4.

[75] Schanbacher (n 43) 55–60.

[76] La Via Campesina, 'Food Sovereignty: A Future without Hunger' (n 73) principle 2.

of communities, traditions, and cultural values over efficiency and profiteering and sustainable development over unfettered consumption'.⁷⁷ Similarly, Walden Bello argues:

> [t]he paradigm of food sovereignty challenges at every point the pillars of capitalist industrial agriculture, emphasizing, among other principles, food self-sufficiency, the right of a people to determine their patterns of agricultural production, farming that is not based on chemical-intensive agriculture or biotechnology, equality in land distribution, and agricultural production and distribution resting mainly on small farmers and cooperative enterprises.⁷⁸

This chapter will not examine attempts to implement food sovereignty at the national or community level. Instead, it will be exploring the extent to which food sovereignty activists have both drawn upon the international legal framework around the right to food to support demands for food sovereignty and sought to influence the shape of that international legal framework by relating it to the ideas of food sovereignty. In Gramscian terminology, this chapter is concerned with the extent to which La Via Campesina and others have successfully engaged in a counter-hegemonic war of position within the domain of global civil society by relating their struggle for food sovereignty to the right to food under international law. At the international level, food sovereignty activists have raised a number of demands: a code of conduct on the human right to food; an international convention on food sovereignty to replace the current AoA and other WTO agreements; a reformed and strengthened United Nations; an independent dispute settlement mechanism – integrated within an International Court of Justice; and an international legally binding treaty that defines the rights of smallholder farmers.⁷⁹ It is clear from this that La Via Campesina regards the contestation of global norms and institutions as a key component of its mission.

Having established the existence of these two perspectives on global food governance, the following section will examine the development of the right to food within the UN human rights system and the FAO, and the extent to which these two competing conceptions have influenced or are present in international articulations of the right to food.

---

[77] Schanbacher (n 43) 99. [78] Walden Bello, *The Food Wars* (Verso, 2009) 148–149.
[79] Windfuhr and Jonsen (n 45) 15–16.

Table 3.1 *Food security and food sovereignty paradigms compared*

	Food security paradigm	Food sovereignty paradigm
Conceptualisation of food	Commodity	Fundamental right
Food principally produced for	Global markets	Local communities
Mode of production	Industrial, chemical-intensive and GMO technologies	Agro-ecology
Economic model	Free market	Interventionist
Approach to plant genetic resources	Private property rights	Anti-patent/communal
Lead instrument	Agreement on agriculture	Nyéléni Declaration
Lead organisation	World Trade Organization	La Via Campesina

3.4 DISCURSIVE CONTESTATION OVER THE RIGHT TO FOOD UNDER INTERNATIONAL LAW

How does the right to food under international law relate to the contestation discussed above? In one sense, it can be argued that the mere assertion of the right to food is *inherently* opposed to neo-liberal food security discourse and closer to the food sovereignty approach. The neo-liberal approach adopts a utilitarian framework that seeks to maximise aggregate welfare through the market mechanism. Within this framework, food is viewed primarily, if not exclusively, as a commodity to be bought and traded like any other. The tension between the concept of food as a commodity and food as a right has long been recognised. For example, in 1975 Addeke H. Boerma, the Director-General of the FAO 1968–1975, argued that:

> food, like any other commodity, is bought and sold internationally as well as on domestic markets. But food, I maintain, is not like any other commodity. If human beings have a right to life at all, they have a right to food. So the mechanism of the international free market, while valuable in adjusting supply and demand under normal circumstances, simply cannot apply across the board when you have literally hundreds of millions of people who, collectively, cannot afford to pay the going prices for the most basic need of life itself.[80]

---

[80] Addeke H Boerma, *A Right to Food: A Selection from Speeches* (FOA, 1976) 153.

By contrast, the food sovereignty model relies heavily on a rights-based framework, including the idea that everybody has a fundamental right to food and the right to define their own food systems. Within this viewpoint, the rights and dignity of individuals and communities cannot be overridden by economic efficiency, competitiveness or other utilitarian considerations.[81] Moreover, the idea of the right to food implies that governments have obligations in relation to those under its jurisdiction to ensure that they have access to adequate food. This runs contrary to the neo-liberal approach to food security, which regards food as a private commodity to be distributed by the market. Nevertheless, as it was argued in the last chapter, despite the apparent discursive incompatibility of socioeconomic rights and neo-liberalism, hegemonic incorporative strategies can be deployed to bring socioeconomic rights into line with the ethico-political framework of neo-liberal governance. The rest of this chapter will critically evaluate the extent to which attempts to advance the right to food at the international level have been able to bolster the counter-hegemonic discourse of food sovereignty and contest the dominant neo-liberal food security paradigm.

3.4.1 THE LEGAL BASIS FOR THE RIGHT TO FOOD

The right to adequate food is set out in Article 25 of the Universal Declaration of Human Rights (UDHR) in the context of the right to an adequate standard of living.[82] This formulation is repeated in the International Covenant on Economic, Social and Cultural Rights (ICESCR),[83] which also recognises the 'fundamental right of everyone to be free from hunger'.[84] The exact content and interrelationship of the right to adequate food and the fundamental right to be free from hunger is not entirely clear, although the inclusion of the word 'fundamental' in

---

[81] Flavio Luiz Schieck Valente and Ana Maria Suarez Franko, 'Human Rights and the Struggle against Hunger: Laws, Institutions, and Instruments in the Fight to Realize the Rights to Adequate Food' (2010) 13 *Yale Human Rights and Development Law Journal* 435, 438–439.

[82] Universal Declaration of Human Rights (adopted 10 December 1948 UNGA Res 217 A (III) (UDHR) article 25.

[83] International Covenant on Economic, Social and Cultural Rights (adopted 16 December 1966, entered into force 23 March 1976) 993 UNTS 3 (ICESCR) article 11(1).

[84] ibid, article 11(2).

relation to the latter suggests that state action directed towards the right to adequate food should focus primarily on the enjoyment of minimum levels of subsistence.[85] The ICESCR also links the right to food with the obligation of states to take measures individually and through international co-operation to improve methods of food production, conservation and distribution, make use of technical and scientific knowledge and develop or reform agrarian systems to achieve the most efficient use of resources.[86] Furthermore, it requires state measures to take account of the problems of both food-importing and food-exporting countries to 'ensure an equitable distribution of world food supplies in relation to need'.[87]

Before considering the ways in which the international right to food has been drawn upon within the context of global civil society, this section will briefly consider the extent to which the obligations in relation to the right to food in the ICESCR support or contest the two different ethico-political frameworks discussed above. In discussing the relationship between the right to food under the ICESCR and global food policy, it is important to recall from the previous chapter that the Committee on Economic, Social and Cultural Rights (CESCR) has consistently stressed that the ICESCR is a neutral instrument with regard to political and economic systems. Nevertheless, there are a number of provisions within the ICESCR that have implications for the way in which global food governance should be organised. One important aspect of the Covenant is the right to self-determination contained in article 1. This reads:

---

[85] Matthew Craven, *The International Covenant on Economic, Social and Cultural Rights: A Perspective on Its Development* (Clarendon Press, 1995) 307.

[86] ICESCR (n 83) article 11(2)(a).

[87] ibid 11(2)(b). Rights to food are also contained in a number of international human rights treaties aimed at protecting specific groups and in regional human rights instruments. For example: the Convention on the Elimination of All Forms of Discrimination against Women, 1979, UN Doc. A/34/46, article 12 (2); The Convention on the Rights of the Child, 1989, UN Doc. A/44/49, articles 24(2)(c) and (e) and 27(3); The Convention on the Rights of Persons with Disabilities, 2007, UN Doc. A/RES/61/106, articles 25(f) and 28(1); Additional Protocol on Economic, Social and Cultural Rights (San Salvador Protocol), OAS Treaty Series No. 69 (1988) articles 12 and 17(a); African Charter on the Rights and Welfare of the Child, OAU Doc. CAB/LEG?24.9/49 (1990) articles 14 (2) (c), (d) and (h); Protocol to the African Charter on Human and People's Rights on the Rights of Women in Africa, Maputo, 11 July 2003, available at www.achpr.org/instruments/women-protocol/ articles 15 and 14(2) (b).

Article 1

1. All peoples have the right of self-determination. By virtue of that right they freely determine their political status and freely pursue their economic, social and cultural development.

2. All peoples may, for their own ends, freely dispose of their natural wealth and resources without prejudice to any obligations arising out of international economic co-operation, based upon the principle of mutual benefit, and international law. In no case may a people be deprived of its own means of subsistence.

Philip Alston notes that despite the complexity and vagueness of the concept of self-determination enshrined in article 1, it is arguable that 'a government would be in violation of article 1 if it permitted the exploitation of the country's food-producing capacity (natural resources) in the exclusive interests of a small part of the population or of foreign (public or private) corporate interests while a number of State's inhabitants are starving or malnourished'.[88] Whilst the concept of sovereignty used in food sovereignty discourse may be wider and more multi-scalar in scope than the concept of self-determination enshrined in the ICESCR – in the sense that it involves 'the right of peoples, *communities* and countries to define (their agricultural policies)'[89] – the concept contained within the ICESCR nevertheless seems congruent with the food sovereignty model's insistence that national agricultural policies must prioritise production for domestic consumption and food self-sufficiency.

However, a recurring theme in the food sovereignty critique of the neo-liberal food regime is that countries, particularly those in the Global South, are constrained in their ability to 'freely dispose of their natural wealth and resources' because of the external pressures exerted upon them as a result of the food dumping practices of the Northern States and the structural adjustment and trade liberalisation measures they are compelled to introduce. Indeed, the Human Rights Committee has noted that the freedom to deal with natural resources contained in common

---

[88] Philip Alston, 'International Law and the Human Right to Food' in Alston and Tomasevki (eds.) (n 2) 23.

[89] NGO/CSO Forum for food sovereignty, 'Food Sovereignty: A Right for All Political Statement of the NGO/CSO Forum for Food Sovereignty' (June 8–13, 2002 Rome NGO/CSO Forum for Food Sovereignty 2002) 2 (emphasis added) www.foodsovereignty.org/Portals/0/documenti%20sito/Resources/Archive/Forum/2002/political%20statement-eng.pdf accessed 14 October 2013.

article 1(2) of the Covenants involves a correlative duty on other states not to interfere with such freedom.[90]

A question that must therefore be raised is whether the right to food under the ICESCR gives rise to 'extraterritorial' obligations on states, or even multilateral organisations. In other words, does it impose obligations upon states in relation to human rights beyond their own national borders? This is a hotly contested topic within the legal literature.[91] However, it is worth bearing two points in mind. First, it should be noted that the ICESCR, unlike the majority of international human rights treaties, makes no explicit mention of the scope of its territorial application.[92] Whereas article 2(1) of the International Covenant on Civil and Political Rights imposes obligations on States Parties to respect and ensure the rights of all individuals 'within its territory and subject to its jurisdiction', there is no mention of territory or jurisdiction in the wording of article 2(1) of the ICESCR.[93] Second, there is an explicit reference within article 2(1) of the ICESCR to international assistance and co-operation as a means to achieve the full realisation of the rights provided by the Covenant. This reference to international assistance and co-operation is reiterated in several other articles, including article 11(2) in relation to the fundamental right to be free from hunger.[94]

The exact nature of the legal obligation to co-operate internationally is far from clear. Matthew Craven argues that 'it does not appear to go very

---

[90] UN Human Rights Committee (HRC), 'CCPR General Comment No. 12: Article 1 (Right to Self-Determination), The Right to Self-Determination of Peoples' (13 March 1984), para.5.
[91] For some accounts of the extraterritorial obligations in relation to the right to food see: Asbjørn Eide, 'The Right to Adequate Food and to Be Free from Hunger, Updated Study on the Right to Food', UN Doc. E/CN.4/Sub.2/1999/12 para.130; Sigrun Skogly, 'The Obligation of International Assistance and Co-Operation in the International Covenant on Economic, Social and Cultural Rights' in Morten Bergsmo (ed.), *Human Rights and Criminal Justice for Downtrodden: Essays in Honour of Asbjørn Eide* (Kluwer Law International, 2003); FIAN, Brot fur die Welt and the Evangelischer Enterwicklungsdienst, *Extraterritorial State Obligations* (FIAN 2004); Maastricht Principles on Extraterritorial Obligations of States in the Area of Economic, Social and Cultural Rights (2011). www2.lse.ac.uk/humanRights/articlesAndTranscripts/2011/MaastrichtEcoSoc.pdf accessed 2 November 2013.
[92] On the implications of this see Margot E Salomon, *Global Responsibility for Human Rights: World Poverty and the Development of International Law* (Oxford University Press, 2007) 76–77.
[93] International Covenant on Civil and Political Rights (adopted 16 December 1966, entered into force 23 March 1976) 999 UNTS 171, article 2(1).
[94] ICESCR (n 83) articles 11(2), 15(4), 22 and 23.

far beyond a commitment to participate in certain types of international activity of a humanitarian character',[95] while Alston suggests that the most significant aspects of the obligation to co-operate internationally are the duties to avoid depriving people in other countries of their right to food and to protect against such deprivations.[96] Alston's interpretation would suggest that the right to food gives rise to duties on states to refrain from taking decisions within the WTO, IMF or the World Bank that can lead to violations of the right to food in other countries as well as to ensure that TNCs under their jurisdiction do not violate the right to food extraterritorially.

An understanding of the right to food as imposing some international obligations is supported by the wording of article 11(2)(b) which requires states 'to ensure equitable distribution of world food supplies in relation to need'.[97] Clearly the reference to 'world food supplies' suggests an extraterritorial element to this obligation and the notion of 'equitable distribution ... in relation to need' elevates food above the status of a mere commodity. Moreover, the drafting history of article 11(2)(b) indicates that the framers saw the right to food as imposing international obligations that transcend economic interest. The reference to 'problems' rather than 'interests' was included to indicate that 'the distribution of food supplies should be based not solely on the interests of the countries involved or on purely economic grounds, but also on social and humanitarian considerations'.[98] Furthermore, during the drafting of article 11(2)(b) 'it was pointed out that freedom from hunger should not be interpreted as freedom to dispose of agricultural surpluses to the detriment of the economies of the less developed countries'.[99] This lends weight to the argument that many food dumping practices of northern states are incompatible with the fundamental right to be free from hunger contained in the ICESCR.

In summary, whilst the right to adequate food contained within the ICESCR is open to widely varying interpretations, there are a number of provisions contained within it – freedom from hunger giving rise to duties upon governments, countries having a right to freely pursue their

---

[95] Matthew Craven, 'The Violence of Dispossession: Extra-Territoriality and Economic, Social and Cultural Rights' in Mashood Baderin and Robert McCorquodale (eds.) *Economic, Social and Cultural Rights in Action* (Oxford University Press, 2007).
[96] Alston (n 2) 41.   [97] ICESCR (n 83) article 11(2)(b).
[98] Eluchans (Chile), UN Doc. A/C.3/SR.1268, para.16 (1963) cited in Craven (n 95) 301.
[99] ibid.

economic, social and cultural development, and certain extraterritorial obligations with regard to the right to food – that can be construed as broadly supportive of some of the key demands of the food sovereignty movement as well as problematising certain assumptions that underpin the neo-liberal food security model.

3.4.2 THE WORLD FOOD SUMMIT, ROME 1996

With the exception of some general preparatory efforts to get the right to food on the international development agenda, very little attention had been paid to the meaning of the right within global governance until the mid-1990s.[100] The year 1996 was a turning point, with the FAO organising the World Food Summit (WFS) in Rome. This event constituted a 'milestone in the efforts to bring attention to the right to food and nutrition as a human right in wider development circles'.[101] The WFS produced the Rome Declaration on World Food Security, which reaffirmed 'the right of everyone to have access to safe and nutritious food, consistent with the right to adequate food and the fundamental right of everyone to be free from hunger'.[102] It also adopted a Plan of Action which included a commitment towards clarifying the content of these rights.[103] To achieve this, governments were encouraged to ratify and implement the ICESCR, and the UN High Commissioner for Human Rights was invited to better define the right to food contained in article 11 of the ICESCR, in consultation with relevant treaty bodies and specialised agencies.[104]

The WFS has been lauded as constituting an important milestone in efforts to recognise the right to food as an international norm.[105] This achievement was undoubtedly the product of intensive lobbying and negotiations preceding and during the Summit by a growing right to

---

[100] Margret Vidar, 'The Right to Food in International Law' (2003) 5 www.actuar-acd.org/uploads/5/6/8/7/5687387/fao_the_right_to_food_in_international_law.pdf accessed 15 October 2013; Kerstin Mechem, 'Food Security and the Right to Food Discourse in the United Nations' (2004) 10 *European Journal of International Law* 631, 637.

[101] Arne Oshaug and Wenche Barth Eide, 'The Long Process of Giving Contention to an Economic, Social and Cultural Right: Twenty Five Years with the Case of the Right to Adequate Food' in Morten Bergsmo (ed.) (n 90) 330 and 346.

[102] Rome Declaration of World Food Security (World Food Summit, Rome 13–17 November 1996), preamble.

[103] ibid, objective 7.4.   [104] ibid, objective 7.4(a),(b) and (e).

[105] Jean Ziegler, Christopher Golay, Claire Mahon and Sally-Anne Way, *The Fight for the Right to Food* (Palgrave, 2010) 5.

food movement.[106] Nevertheless, some have expressed concerns about the broader discursive shift that took place in food security discourse at the WFS.[107] For example, the Declaration identifies the major causes of food insecurity as poverty, war and conflict, environmental degradation and gender inequality.[108] Whilst these are undoubtedly significant challenges to achieving food security, the Declaration does not identify the inequalities within the global political economy as a significant producer of food insecurity. This is in sharp contrast to earlier international declarations, which recognised that violations of the right to food were the product of global inequalities.[109] In failing to address the contradiction between growing numbers of malnourished people and a world that produces an abundance of food, the WFS declaration helps to naturalise the organisation of the global political economy and the political-institutional arrangements that sustain it. Indeed, the Declaration goes further and suggests that these arrangements are part of the solution to world hunger. Commitment Four states the need to ensure that 'food, agricultural trade and overall trade policies are conducive to fostering food security for all through a fair and market-orientated world trade system'.[110] It is therefore 'essential that all members of the World Trade Organisation (WTO) respect and fulfil the totality of the undertakings of the Uruguay Round'.[111]

The 1996 Plan of Action has been criticised by the food sovereignty movement because it is perceived to support 'policies that lead to hunger, policies that support economic liberalization for the South and cultural homogeneity'.[112]

Another recommendation in the Rome Declaration that appears to undermine the potency of the right to food as a counter-hegemonic

---

[106] Oshaug and Eide (n 101) 347.
[107] See generally, NGO/CSO Forum for food sovereignty (n 89).
[108] Rome Declaration of World Food Security (n 102) preamble.
[109] For example, the 1974 Universal Declaration on the Eradication of Hunger and Malnutrition recognised that 'hunger and malnutrition arises from ... historical circumstances, especially social inequalities, including in many cases alien and colonial domination, foreign occupation, racial discrimination, apartheid and neo-colonialism in all its forms' See UNGA Res 3348 (17 December 1974) principle c. The 1992 World Declaration and Plan of Action for Nutrition stated that 'We recognize that globally there is enough food for all and that inequitable access is the main problem', FOA and World Health Organization, 'World Declaration and Plan of Action for Nutrition' (Rome, December 1992) article 1 http://whqlibdoc.who.int/hq/1992/a34303.pdf accessed 13 October 2013.
[110] Rome Declaration of World Food Security (n 102) preamble.
[111] ibid Commitment 4, para.37.
[112] NGO/CSO Forum for food sovereignty (n 89) 1. For critical remarks see also Upendra Baxi, *The Future of Human Rights* (Oxford University Press, 2002) 141.

challenge to neo-liberalism is its invitation to the UN High Commissioner for Human Rights to take account of 'the possibility of formulating voluntary guidelines for food security for all'.[113] The recommendation to adopt voluntary guidelines rather that legally binding instruments reinforces the neo-liberal assertion that socioeconomic rights are not legal obligations but rather mere 'aspirations'. In stark contrast to this, the parallel NGO Forum on Food Sovereignty presented the demands for more 'effective instruments' to implement the right to food, such as a Code of Conduct to govern the activities of those involved in achieving the Right to Food, including private actors such as TNCs and a Global Convention on Food Security that gives precedence to the right to food over other international agreements such as the AoA.[114]

In spite of the rather cautionary approach to the right to food within the declaration, its express inclusion was clearly a matter of concern for the US delegation, who issued an interpretive statement after the adoption of the Plan of Action that sought to stymie any potential impact it might have on international legal norms:

> In joining consensus on this and other similar paragraphs of the Rome Declaration on World Food Security and the World Food Summit Plan of Action, the United States does not recognize any change in the current state of conventional or customary international law regarding rights related to food. The United States believes that the attainment of any "right to food" or "fundamental right to be free from hunger" is a goal or aspiration to be realized progressively that does not give rise to any international obligations nor diminish the responsibilities of national governments toward their citizens.[115]

What we witness in the Rome Declaration on Food Security and in the Statement by the NGO forum at the World Food Summit is two counterposed conceptions of the right to food. On the one hand, the conception that is present in the official Declaration regards the right to food as a policy goal to be achieved progressively through compliance with the WTO trading regime and complemented by domestic poverty reduction policies. The contradiction here is that it is WTO trading rules, as well as IMF and World Bank diktats, that are identified by the food sovereignty movement as undermining domestic poverty reduction strategies. Alternatively, the conception of the right to food articulated in the NGO

---

[113] ibid, objective 7.4(e). [114] NGO/CSO Forum for food sovereignty (n 89) 3.
[115] FAO 'Report of the World Food Summit', 13–17 November, Rome, annex II www.fao.org/docrep/003/w3548e/w3548e00.htm#Scalfaro accessed 14 October 2013.

forum was one that argued that the right to food should be a legally binding international obligation not only upon states but also international institutions and TNCs, and furthermore an obligation that takes precedence over WTO agreements.

The Rome Declaration marks the beginning of the process of the mainstreaming of the right to food into global governance. This was undoubtedly a response to pressure being exerted by the human rights movement concerned about some of the negative impacts of the globalisation of agricultural trade. Nevertheless, what we also witness in the Rome Declaration is the process of *trasformismo*, the co-option of the right to food agenda by tying it to the need to comply with WTO agreements. Notwithstanding this, the call in the Plan of Action for governments, relevant treaty bodies and appropriate specialised agencies of the UN to consider how they might contribute to the elaboration of the right to food opened up future opportunities to develop alternative understandings of the right.

3.4.3 THE DRAFT INTERNATIONAL CODE OF CONDUCT ON THE HUMAN RIGHT TO ADEQUATE FOOD

In response to the call of the NGO forum at the WFS for a Code of Conduct on the human right to adequate food, three human rights organisations – FIAN International (FoodFirst Information and Action Network), WANAHR (World Alliance for Nutrition and Human Rights) and Institute Maritain International – took the lead in preparing such a document. A draft was produced in 1997 that was supported by more than 800 NGOs.[116] This draft was addressed to states and other relevant actors, including civil society actors and international organisations that are responsible for securing the right to food.[117] The draft code begins by fleshing out the meaning of adequate food as incorporating the dimensions of food safety, nutrition and cultural preferences.[118] In relation to state obligations at the national level, the Code draws upon the tripartite framework for human rights obligations developed by Asbjørn Eide to suggest that states have three levels of obligation in relation to the right

---

[116] Vidar (n 100) 6.
[117] International Code of Conduct on the Human Right to Adequate Food (1997) Draft endorsed by FIAN International, Human Rights Organization to Feed Oneself, World Alliance for Nutrition and Human Rights and Institute Jacques Maritain International, preamble www.iatp.org/files/International_Code_of_Conduct_on_the_Human_Rig.htm accessed 21 October 2013.
[118] ibid, article 4.

to food: to respect, protect and fulfil.[119] This means that states must refrain from interfering with existing access to food, prevent third parties under their jurisdiction – including TNCs – from interfering with individuals' access to adequate food and develop long-term strategies to ensure the ability of everybody to access adequate food.[120] Furthermore, states are under a minimum core obligation to ensure that vulnerable populations are protected, through social programmes, from hunger.[121]

The Code also recognised state obligations at the international level. States must not violate or assist in violating the right to adequate food of persons not in their jurisdiction and their international policies and programmes must respect the full realisation of the right to adequate food.[122] This has implications for their trade and finance policies as well as their domestic agricultural policies that have international implications (e.g. food dumping).[123] Finally, the draft code also identified direct obligations on international organisations – such as the IMF and World Bank – and civil society actors – including TNCs.[124] In relation to the former, they are under obligations not to take measures that presuppose a violation of the ICESCR and to respect and protect people's access to adequate food.[125] In relation to the latter, they are prohibited from contributing to violations of the right to adequate food.[126]

The Code provided a robust interpretation of the obligations associated with the right to adequate food. It advocates binding duties to be imposed upon an array of actors, including those traditionally conceived as being outside of the framework of directly binding human rights law. It contained the key demands of the food sovereignty movement to subordinate international trade practices and structural adjustment conditionalities to the right to adequate food and freedom from hunger for all. It also acted as a springboard for increased collaboration between specialist human rights NGOs and rural peasant movements. In 1999 FIAN joined forces with La Via Campesina to launch 'the Global Campaign for Agrarian Reform':

> Under the campaign banner "food, land and freedom", peasants joined human rights activists in 12 countries in Asia, the Americas and Europe in mobilizations, land occupations and other public events to demand the

---

[119] ibid, article 6. See Asbjørn Eide, 'The Right to Adequate Food as a Human Right: Special Rapporteur's Report on the Right to Adequate Food' (1987) UN. Doc. E/CN.4/Sub.2/1987/23, para.66.
[120] ibid, articles 6(1) to 6(3).   [121] ibid, article 6(4).   [122] ibid, article 7(4).
[123] ibid.   [124] ibid, articles 8 and 10.   [125] ibid, article 8(2).   [126] ibid, article 10.

right to land and security of land tenure as a prerequisite to the human right to food stipulated in Article 11 of the International Covenant on Economic, Social and Cultural Rights.[127]

This Global Campaign involved work on a number of different fronts, including the lobbying of international institutions for the integration of the right to food into global governance practices. The implementation of an international Code of Conduct in relation to the right to food would be one key component of this strategy.

3.4.4 GENERAL COMMENT 12

The CESCR was one of the first UN bodies to respond to the call of the WFS for greater clarity on the meaning of the right to food. In 1999 it issued General Comment 12 on the right to adequate food under article 11 of the ICESCR.[128] The Comment, whilst not imposing model substantive solutions to the problem of global hunger, nevertheless provides a rights-based approach to adequate food and freedom from hunger that is congruent with many aspects of the food sovereignty-based framework. Importantly, the Comment recognises that the 'roots of the problem of hunger and malnutrition are not lack of food but lack of access to available food'.[129] The Comment thus politicises the question of unequal access to available resources rather than adopting a narrow technological approach aimed at improving existing agricultural productive methods.

In many respects, the Comment was influenced in its form and content by the Code. Like the Code it recognises the dimensions of nutrition, food safety and cultural preference as being essential components of the right to adequate food.[130] It also follows the Code in recognising the three levels of state obligations in relation to the right to food.[131] However, the Comment goes one step further in recognising two sub-categories to the obligation to fulfil: the obligation to facilitate and the obligation to provide. The former requires the proactive engagement of the state in activities to strengthen people's access to land and other resources required for food security and the latter requires the state to directly provide food for those individuals

---

[127] Annette-Aurelie Desmaraisi, 'The Vía Campesina: Consolidating an International Peasant and Farm Movement' (2002) 29(2) *Journal of Peasant Studies* 91, 108–109.
[128] General Comment 12 (n 16).   [129] ibid, para.5.   [130] ibid, paras.9–11.
[131] ibid, para.15.

who are unable to provide for themselves.[132] Like the Code, the Comment also recognises that whatever the level of resources a state has, there is a 'core obligation to take necessary action to mitigate and alleviate hunger'.[133] The Comment therefore provides a useful corrective to the insistence of some, including the US government, that the right to food merely expresses programmatic aims rather than binding obligations. Conversely, the Comment legitimises the Food Sovereignty movement's demand for binding obligations in relation to the right to food by providing further detail about violations of the right to adequate food and implementation at the national level.

There are, however, a number of ways in which the Comment differs from the Code. First, whilst both the Code and the Comment recognise the responsibilities of civil society and the private sector in the full realisation of the right to adequate food, the latter recognises that only states are parties to the ICESCR and thus are ultimately responsible for complying with it.[134] In this context the Comment refers to the need for private businesses to pursue their activities within the framework of a code of conduct agreed upon jointly with the Government and civil society, whilst recognising that such obligations are not currently legally binding.[135] The same applies to international financial institutions such as the IMF and World Bank, which the Comment states 'should pay greater attention to the protection of the right to food in their lending policies and credit agreements and in international measures to deal with the debt crisis'.[136] This is in contrast to the Code, which unequivocally states that international organisations 'must not take measures which would presuppose a violation of the ICESCR.[137] The Comment also differs from the Code in relation to the international obligations in relation to the right to food. The Comment does recognise that there are international dimensions to state party obligations with regard to appropriate economic, environmental and social policies.[138] At the international level states have extraterritorial obligations that mirror their domestic obligations to respect, protect and fulfil the right to food:

> States parties should take steps to respect the enjoyment of the right to food in other countries, to protect that right, to facilitate access to food

---

[132] ibid.   [133] ibid, paras.6 and 14.   [134] ibid, para.20.   [135] ibid.
[136] ibid, para.41.   [137] International Code of Conduct (n 117) article 8(2).
[138] General Comment 12 (n 16) para.4.

and to provide the necessary aid when required. States parties should, in international agreements whenever relevant, ensure that the right to adequate food is given due attention and consider the development of further international legal instruments to that end.[139]

In relation to these international obligations it is important to note that the CESCR uses the recommendatory language of 'should' to describe states' and IFI's duties. This is in contrast to a state's domestic obligations which are expressed in the more mandatory terms of what they are *required* to do and *must* do. This indicates that the CESCR did not regard the extraterritorial commitments in relation to the right to food as having as firm a basis in international law as domestic obligations. However, whilst the legal basis of extraterritorial and non-state actor obligations is not as forcefully expressed in the Comment, it is clear that the Comment lends moral weight and legitimacy to these aspirations of the food sovereignty movement. Given that General Comments are persuasive in nature rather than legally binding, the task of enshrining the recommendations of the Comment into a binding legal agreement would remain a central task for the right to food movement.

3.4.5 THE WORLD FOOD SUMMIT: FIVE YEARS LATER

As a result of poor progress on the reduction of global hunger since the 1996 WFS, a follow up summit entitled World Food Summit: five years later (WFS:fyl) was convened by the FAO in 2002.[140] The Summit was concerned with identifying the resources, means and actions necessary for states to be in compliance with the commitments made in the 1996 Plan of Action and to ensure that everybody has access to sufficient and safe food. One hundred and eighty countries participated in the Summit, alongside a host of inter-governmental agencies and NGOs.[141] Many of the present states, including Norway, Venezuela and Chile, were in favour of a Code of Conduct in relation to the right to food. Nevertheless, the USA, with the support of the United Kingdom and other countries, was successful in removing all references to a Code of Conduct

---

[139] ibid, para.36.
[140] H.E. Carlo Azeglio Ciampi, President of the Italian Republic (10 June 2002), FAO Annex 1: Statements – Inaugural Ceremony www.fao.org/docrep/MEETING/005/Y7106E/Y7106E02.htm#TopOfPage accessed 22 October 2013.
[141] FAO, Annex V: List of Participating Countries and Organizations www.fao.org/docrep/MEETING/005/Y7106E/Y7106E06.htm#TopOfPage accessed 22 October 2013.

from the text of the Summit's final declaration.¹⁴² Instead, the declaration calls for the FAO to establish an Intergovernmental Working Group (IGWG) to elaborate 'a set of voluntary guidelines to support the progressive realization of the right to adequate food in the context of national food security'.¹⁴³

The outcome of the WFS:fyl presented a number of challenges to the counter-hegemonic potential of the right to food within the international domain. First, the proposed Guidelines were 'voluntary', that is, they would not create any new legally binding obligations. As A. Clare Cutler argues, the global promulgation of non-binding voluntary codes can be understood as a key component of the construction of neo-liberal hegemony:

> To the extent that juridification is taking a non-binding, "soft" form of law, one must consider whether law is operating dialectically to juridify certain relations in hard legal disciplines (enforcement under WTO, NAFTA, the EU), and de-juridify others (corporate social responsibility; corporate environment and labour practices) in 'soft law' and voluntary legal regimes.¹⁴⁴

Indeed, when one considers that the property and investment rights of transnational capital are protected in 'exquisite detail' under extensive NAFTA, GATT and WTO regulations and articles,¹⁴⁵ while social rights norms remain comparatively meagre and lack effective monitoring and enforcement mechanisms, it becomes easy to see how the latter become instruments of *trasformismo*: 'They are promoted as the efficient and rational means for giving globalisation a "human face", but this mythology conceals their nature as safety valves for capital'.¹⁴⁶ In other words, socioeconomic rights standards, and the promise they contain, can be marshalled to legitimate world order, but these very same standards lack

---

¹⁴² See the report of the International Service for Human Rights 'World Food Summit: five years' later (Rome, 10–13 June 2002)' http://olddoc.ishr.ch/hrm/archive/FAO/FAO-WFSfyl.pdf accessed 22 October 2013.

¹⁴³ FAO, Declaration of the World Food Summit: five years later, International alliance against hunger, para.10, Report of the World Food Summit: five years later, part one, Appendix, Rome, 2002 www.fao.org/docrep/MEETING/005/Y7106E/Y7106E09.htm#TopOfPage accessed 22 October 2013.

¹⁴⁴ A Claire Cutler, 'Gramsci, Law and the Culture of Global Capitalism' (2005) 8(4) *Critical Review of International Social and Political Philosophy* 527, 537.

¹⁴⁵ Tony Evans and Alison J Ayers, 'In the Service of Power: The Global Political Economy of Citizenship and Human Rights' (2006) 10(3) *Citizenship Studies* 289, 293.

¹⁴⁶ Cutler (n 144) 539.

the 'bite' of hard law regimes to be able to mount effective challenges to the injustices associated with neo-liberalism.

The second aspect of concern in relation to the Voluntary Guidelines proposal was that they were framed in the language of 'progressive realisation', which may have suggested that there are no immediate obligations in relation to the right to food. Whilst it is clearly the case that the standard of 'progressive realisation' is an important component of the obligation to fulfil the right to adequate food for all, the CESCR has noted, since General Comment 3, that there are a number of exceptions to the requirement of progressive realisation – for example, in relation to non-discrimination and minimum core obligations.[147] Furthermore, General Comment 12 states that the ICESCR recognises that some obligations of States Parties in relation to the right to food 'are of a more immediate nature, while other measures are more of a long-term character, to achieve progressively the full realization of the right to adequate food'.[148] By framing the right to adequate food solely in terms of 'progressive realisation' the WFS:fyl declaration had the potential to run roughshod over the calibrated state duties that have been identified by the CESCR and scholars such as Asbjørn Eide. This, in turn, could have the potential to downgrade the status of the right to food from an *existing* right to an *emerging* right and bolster the neo-liberal position advanced by the USA and other northern states that socioeconomic rights are 'aspirational' in character and do not give rise to legally binding obligations.[149]

Third, the proposed guidelines were framed in the 'context of national food security'. By limiting the scope of the guidelines to the national context, the international dimensions of the right to food are obscured. From the perspective of the food sovereignty movement this is undesirable given the negative impact that the activities of TNCs, Northern States and IFIs can have on the right to food in the Global South. By limiting the possibility of shining a critical light on neo-liberal global political society, the Voluntary Guidelines could

---

[147] CESCR, 'General Comment No. 3: The Nature of State Parties Obligations (Art. 2, par.1)' (14 December 1990) UN. Doc. E/1991/23 para.9; CESCR, 'General Comment No. 20: Non-Discrimination in Economic, Social and Cultural Rights (Art. 2, para.2, of the International Covenant on Economic, Social and Cultural Rights)' (2 July 2009) UN. Doc. E/C.12/GC/20, para.7.

[148] General Comment 12 (n 16) para.16.

[149] Tony Evans, 'The Human Right to Health?' (2002) 201 23(2) *Third World Quarterly* 197, 199.

potentially help to obscure key structural impediments to the elimination of global hunger. This concern would appear to be supported by the final text of the WFS:fyl declaration. During the WFS:fyl negotiations, a number of developing countries expressed concern about the negative impact of the food dumping practices of the OCED countries, but this did not find a voice in the final declaration, which only contained a vague commitment to the need for co-operation in order to enable developing countries 'to cope with the challenges and reap the benefits of globalization'.[150]

Finally, the guidelines were to be drafted 'with the participation of stakeholders'. This will include international human rights NGOs that advocate the right to food. From a Gramscian perspective what is interesting here is the potential for *trasformismo*, that is to say the process whereby leaders and potential leaders of subordinate groups are co-opted into the dominant hegemonic project. Analysts of neo-liberal globalisation have long identified the tactic of encouraging the participation of civil society within global governance as being a central strategy aimed at reproducing neo-liberal hegemony.[151] Often the basis of this participation amounts to little more than the dissemination of information concerning a set of pre-existing policies or possibly consultation concerning a narrow range of policy options.[152] It was clear from the framework adopted in the WFS:fyl that the future guidelines process would be heavily circumscribed in terms of what can be achieved. In contradistinction to the demands of the food sovereignty movement, the right to food is already framed as a domestic concern, reduced to an emerging right rather than an existing right and is a voluntary commitment to be achieved progressively. Even after the USA was successful in watering down the final declaration it issued a reservation reiterating its understanding of the right to food as 'a goal or aspiration to be realized progressively that does not give rise to any international obligation' and described the Voluntary Guidelines as a distraction 'from the real work of

---

[150] FAO, *Declaration of the World Food Summit: five years later* (2002) para.26 www.fao.org/docrep/meeting/005/y7106e/y7106e09.htm accessed 30 January 2014. See Peter Russet, 'The US Gets Its Way' (*APRN*, 30 June 2002) www.aprnet.org/concerns/50-issues-a-concerns/146-us-opposes-right-to-food-at-world-summit accessed 3 November 2013.

[151] Stephen Gill, 'New Constitutionalism, Democratisation and Global Political Economy' (1998) 10(1) *Pacifica Review* 23, 38.

[152] Arne Ruckert, 'Producing Neoliberal Hegemony? A Neo-Gramscian Analysis of the Poverty Reduction Strategy Paper (PRSP) in Nicaragua' (2007) 79 *Studies in Political Economy* 91, 106.

reducing poverty and hunger'.[153] This polarisation led the then UN Special Rapporteur on the Right to Food, Jean Ziegler, to observe that there are:

> profound internal contradictions in the United Nations system. On the one hand, the UN agencies emphasize social justice and human rights ... on the other, the Bretton Woods Institutions, alongside the Government of the United States of America and the World Trade Organisation oppose the right to food, preferring the Washington Consensus, which emphasises liberalization, deregulation, privatization and the compression of State domestic budgets.[154]

Given the final declaration's adoption of a weak voluntary guidelines strategy, as well as its failure to address the inequities of global trade and structural adjustment, it is not surprising that it was widely rejected by subaltern social movements concerned with food justice. The International NGO/CSO Forum for Food Sovereignty, comprising organisations of farmers, fisherfolk and indigenous peoples, environmentalist and women's organisations as well as trade unions and NGOs, expressed their 'collective disappointment in, and rejection of' the official Declaration, arguing that it prescribed 'more of the same failed medicine'.[155] It further noted that there would be no progress towards the goal of eliminating hunger until the neo-liberal policies of liberalisation, privatisation and commodification were reversed.[156] Nevertheless, some NGOs viewed the Guidelines strategy as opening up a potential vista to continue to advance the right to food in the international arena.[157]

3.4.6 THE VOLUNTARY GUIDELINES ON THE RIGHT TO FOOD

The IGWG began work on the Voluntary Guidelines in March 2003 at the FAO headquarters in Rome and subsequently held two further meetings in October and February.[158] The negotiations broke down in July 2004 due to a lack of consensus around issues of international trade

---

[153] United States of America 'Reservation' FAO, Annex II: Explanatory Notes and Reservations www.fao.org/docrep/MEETING/005/Y7106E/Y7106E03.htm#TopOfPage accessed 22 October 2013.
[154] Quoted in Russet (n 150).   [155] NGO/CSO Forum for food sovereignty (n 89).
[156] ibid, 2.   [157] See Windfuhr and Jonsen (n 45) 20-21.
[158] Arne Oshaug, 'Developing Voluntary Guidelines for Implementing the Right to Adequate Food: Anatomy of an Intergovernmental Process' in Wenche Barch Eide and Uwe Kracht (eds.), *Food and Human Rights in Development* (Intersentia, 2005) 259.

and assistance. These disputes were finally resolved in a 'friends of the chair' meeting, where the Voluntary Guidelines were approved and submitted to the FAO Council for final approval in November 2004.[159] During the negotiation process, many proposals were submitted and several drafts were produced until the final document was adopted. The negotiations were characterised by heated contestation on a broad array of issues including the legal status of the right to food, the obligations that it entailed, the array of actors that had duties in relation to it, its international dimensions and what policies should be adopted to progressively realise it.[160] Clear divisions could be identified between the various social forces participating in the discussions.[161] On the one hand, social movements such as the All African Farmers Network, the South Asian Peasants Coalition and the World Forum of Fisher Peoples as well as human rights NGOs such as FIAN and the Brazilian Action for Nutrition and Human Rights (ABRANDH) seized the opportunity to promote the human right to food as a fundamental challenge to neo-liberal globalisation.[162] On the other hand, Northern States, including the USA, Canada, Japan, the UK, the Netherlands and Sweden as well as the European Community, favoured using the Guidelines to advance a policy-based approach to food security that would be largely congruent with the policies of the neo-liberal food regime.[163]

The presence of these two perspectives can be witnessed in the submissions of various participants and stakeholders at the first IWGW working group meeting in March 2003. A number of social movements and human rights NGOs collaborated in issuing a 'Joint North-South Contribution' (JNSC).[164] The JNSC relates the right to food to the

---

[159] ibid, 275.
[160] See, for example, FAO, Intergovernmental Working Group for the Elaboration of a Set of Guidelines to Support the Progressive Realization of the Right to Adequate Food in the Context of National Food Security: Updated Synthesis Report of Submissions Received From Governments and Stakeholders (First Session, Rome 24–26 March 2003) ftp://ftp.fao.org/docrep/fao/meeting/006/y8803e.pdf accessed 29 October 2013.
[161] For an excellent critical account of the competing positions taken in the negotiations see Julian Germann, 'The Human Right to Food: "Voluntary Guidelines" Negotiations' in Yildiz Atasoy (ed.), *Hegemonic Transitions, the State and Crisis in Neoliberal Capitalism* (Routledge, 2009) 126.
[162] ibid, 131.   [163] Oshaug, 'Developing Voluntary Guidelines' (n 158) 274.
[164] A Joint North-South Contribution 'Voluntary Guidelines for the Implementation of the Right to Adequate Food: A Joint North-South Civil Society Contribution' (March 2003) www.fian.org/fileadmin/media/publications/Voluntary-Guidelines-for-the-Implementation-of-the-Right-to-Adequate-Food-2003.pdf accessed 29 October 2013.

concept of food sovereignty, arguing that 'the right to adequate food describes the minimum standards of policy space needed by each state to implement its obligations under the right to adequate food'.[165] It follows General Comment 12 in identifying the roots of hunger and malnutrition as lying not in the lack of food but in the lack of access to food as well as in 'inadequate policies at the national and international level'.[166] The JNSC specifically identifies the liberalisation of agricultural markets, the erosion of public services, the privatisation of natural resources, the effects of industrialised agriculture by TNCs, food dumping practices, WTO rules and structural adjustment conditions imposed by IFIs as areas of concern that contributed to food insecurity.[167] The submission includes a draft version of the voluntary guidelines that follows the Draft Code of Conduct in elaborating detailed national and international human rights-based obligations for states to respect, protect and fulfil the right to food.[168]

By contrast, the US government submission contains no reference to the right to food and instead focuses exclusively on the concept of food security.[169] Its central recommendations were that:

> The Intergovernmental Working Group (IGW) should focus on practical steps that can be taken by Member States to achieve national food security – increasing agricultural productivity, creating an enabling environment, promoting transparent and accountable government, boosting agricultural science and technology, developing domestic market and international trade opportunities, securing property rights and access to finance, enhancing human capital, and protecting the vulnerable.[170]

The US submission equates the right to food with the concept of food security and therefore reduces it to a policy goal that does not give rise to any binding obligations. Moreover, the means for achieving food security are entirely consistent with neo-liberal ideals of food governance – focusing on trade liberalisation, technological advancement and limited social protections.[171] The JNSC and US submissions represent two

---

[165] ibid, 8. [166] ibid, 9 and 14. [167] ibid, 10. [168] ibid, 13–31.
[169] US Submission to the IGWG (Rome, 28 February) www.fao.org/righttofood/en/ accessed 29 October 2013.
[170] ibid, 2.
[171] ibid, 7. With regard to the latter, the US submission stipulates that emergency food aid and social safety nets should be integrated into a development strategy through 'targeting the most needy and limiting possible negative effects of building competitive and open markets for food and other agricultural projects'. Consistent with the approach of the World Bank discussed in Section 3.2.1, social safety nets are legitimate to the extent that they support and do not conflict with market rationality.

counter-posed interpretations of the right to food. The JNSC framework presents food as a fundamental human right that takes precedence over trade obligations. It advocates concrete obligations for states to ensure that every citizen has access to adequate food as a matter of right. By contrast, the US submission treats the right to adequate food as a policy goal or aspiration that can be achieved through the voluntary adoption of food security measures congruent with the imperatives of neo-liberal globalisation. The proposals respectively represent counter-hegemonic and hegemonic interpretations of the right to food.

The final adopted version of the Voluntary Guidelines (VGs), it must be concluded, represents something in between the JNSC and US proposals.[172] In contrast to the US position, the VGs do contain aspects of the human rights framework with regard to the right to food, but these were substantially watered down so as so be compatible with, rather than subversive of, the neo-liberal food security model. The VGs are divided into three sections. Part 1 includes a preface and an introduction which provides a background to the VGs, recounts the relevant international instruments under which the right to adequate food is enshrined and establishes the content of the right to food in relation to food security. Part 2 contains nineteen guidelines which create an enabling environment in which the right to food can be progressively realised. Finally, part 3 provides a further sixteen paragraphs on international measures, actions and commitments in relation to the realisation of the right to food.

Concerning part 1, it is noticeable that, in contrast to the suggestions made in the JNSC, the preface does not address the causes of global vulnerability and hunger. Instead, it simply reiterates the WFS's and the Millennium Development Goal's (MDGs) commitments to reducing world hunger by half by 2015.[173] Not only does this approach contribute to the continued depoliticisation of hunger, it is also problematic from a human rights perspective.[174] This incremental approach treats hunger

---

[172] FAO, 'Voluntary Guidelines to Support the Progressive Realisation of the Right to Adequate Food in the Context of National Food Security' (Adopted by the 127the Session of the RAO Council November 2004) www.fao.org/docrep/meeting/009/y9825e/y9825e00.htm accessed 29 October 2013.
[173] ibid, Preface, para.1.
[174] For a human rights-based critique of the Millennium Goals see: Thomas Pogge, 'The First UN Millennium Development Goal: A Cause for Celebration?' in Andrease Follesdal and Thomas Pogge (eds.), *Real World Justice* (Springer, 2005) 318; Margot E Salomon, 'Poverty, Privilege and International Law: The Millennium Development Goals and the Guise of Humanitarianism' (2008) 51 *German Yearbook of International Law* 39.

as a problem to be overcome progressively through technical and technological means rather a violation of human rights that gives rise to immediate obligations on actors to stop contributing engaging in actions that prevent people from accessing adequate food. Whereas the counter-hegemonic potential of the right to food stems from its ability to critically interrogate the global institutions and policies that produce and reproduce hunger, the VGs begin by framing hunger in a manner that obscures and depoliticises these structures.

Next, the guidelines identify a number of 'basic instruments' which enshrine the right of everyone to an adequate standard of living, including adequate food. This list includes the UDHR, the ICESCR, articles 55 and 56 of the UN Charter and other international instruments, resolutions and General Comments.[175] However, it is noticeable that there is no explicit acknowledgement of General Comment 12 as an authoritative text on the content of the right to food, even though this was recommended by most submissions during the drafting process.[176] In the absence of an acknowledged pre-existing authority on the normative content of the right to adequate food and its concomitant obligations, the VGs themselves address this point in the third section of part 1 entitled 'the right to adequate food and the achievement of food security'.[177] It is noteworthy that this section does not contain a legal definition of the right to adequate food, but instead contains a definition of food security.[178] Whilst these two definitions are almost identical there are nonetheless fundamental differences between the languages of food security and the language of rights: 'food security implies a desirable state of affairs which governments claim to work for – however there exists no legally binding obligations or legal mechanisms linked to it that could be used by the malnourished to defend themselves'.[179] Julian Germann argues that the discursive imbrication of the right to food and food security within the VGs amounts to a hegemonic reframing strategy aimed at reducing the human right to food to a policy goal achieved through the adoption of neo-liberal policy prescriptions rather than a legally binding obligation which can be used as a standard to identify, critique and challenge such policies.[180]

---

[175] Voluntary Guidelines (n 172) Introduction, paras.10–19.
[176] Updated Synthesis Report of Submissions Received from Governments and Stakeholders (n 160) para.7.
[177] Voluntary Guidelines (n 172) Part 1, paras.15–19.   [178] ibid, para.15.
[179] Windfuhr and Jonsen (n 45) 22.   [180] Germann (n 160) 134.

Nevertheless, the VGs contain a delineation of states' obligations in relation to the progressive realisation of the right to food:

> States have obligations under relevant international instruments relevant to the progressive realization of the right to adequate food. Notably, States Parties to the International Covenant on Economic, Social and Cultural Rights (ICESCR) have the obligation to respect, promote and protect and to take appropriate steps to achieve progressively the full realization of the right to adequate food. States Parties should respect existing access to adequate food by not taking any measures that result in preventing such access, and should protect the right of everyone to adequate food by taking steps so that enterprises and individuals do not deprive individuals of their access to adequate food. States Parties should promote policies intended to contribute to the progressive realization of people's right to adequate food by proactively engaging in activities intended to strengthen people's access to and utilization of resources and means to ensure their livelihood, including food security. States Parties should, to the extent that resources permit, establish and maintain safety nets or other assistance to protect those who are unable to provide for themselves.[181]

It is immediately apparent upon reading this paragraph that in many respects it bears close resemblance to the States Party obligations established by the CESCR in General Comment 12. However, there are also a number of respects in which it differs. First, the widely used human rights tripartite formulation of *respect, protect and fulfil* has been replaced with the obligations to *respect, promote and protect*. The obligation to promote entails two components similar to the two sub-obligations of the duty to fulfil – the duties to facilitate and provide – but is qualified in terms of the *promotion of policies*.[182] Merely promoting policies is clearly a weaker commitment than implementing those policies and consequently imposes less stringent commitments on States Parties than the replaced duty to fulfil. Furthermore, the requirement to establish and maintain safety nets is limited by 'the extent that resources permit' rather than according to the 'maximum available resources' as contained in article 2(1) of the ICESCR.[183] Again, this

---

[181] Voluntary Guidelines (n 172) Part 1, para.17.
[182] Contrast CESCR, General Comment 12 (n 16) para.15.
[183] During the IWG negotiations FIAN attempted to have the 'maximum available resources' formulation included but were unsuccessful. See Intergovernmental Working Group for the Elaboration of a Set of Voluntary Guidelines to Support the Progressive Realisation of the Right to Adequate Food in the Context of National Food Security, First Session, Rome 24–26 March 2003 available at www.fao.org/unfao/bodies/rtf/igwg_e.htm accessed 8 November 2012.

lower threshold for state justification to not provide safety nets represents a weaker obligation for States Parties.

Second, the general formulation of all three levels of obligation is weaker. Whereas General Comment 12 uses mandatory verbs to indicate the legally binding nature of the obligations under the ICESCR (The obligation to respect ... *requires* States parties ... [t]he obligation to protect *requires* measures ... [t]he obligation to fulfil ... means the States *must* proactively engage ...),[184] the 'obligations' in the Voluntary Guidelines use 'should' (States parties should respect ... States parties should protect ... States parties should promote ...).[185] This reformulation, of course, reflects the voluntary nature of the guidelines, but it also serves as a reminder of the retrograde nature of the process, which has effectively reduced legally binding rights into policy options.

Third, these 'obligations' only apply to States Parties that have ratified the ICESCR. They therefore do not impose any additional duties on countries like the USA that have not ratified the treaty. Whilst states that are not party to the ICESCR are 'invited to consider' ratification,[186] they are not required to. The guidelines could have gone further than this and attempted to clarify the duties that States Parties have in relation to the right to food under Customary International Law. Such an exercise would have been useful in clarifying the food-related duties of states that have not ratified the ICESCR and could have helped at least clarify global minimum core obligations. The guidelines explicitly prevent such developments, stating that no provision in them is to be 'interpreted as amending, modifying or otherwise impairing rights and obligations under national and international law'.[187] In the absence of such clarification, the guidelines in many senses are regressive, given the extent to which they water down pre-existing more robust interpretations of the right to food under international law.

A final noteworthy point concerning the first section of the guidelines is that it emphasises:

> the achievement of food security as an outcome of the realization of existing rights and includes certain key principles: the need to enable individuals to realize the right to take part in the conduct of public affairs, the right to freedom of expression and the right to seek, receive and impart information.[188]

---

[184] CESCR, General Comment 12 (n 16) para.15 (my emphasis).
[185] Voluntary Guidelines (n 172) Part 1, para.17.   [186] ibid, Part 1, para.18.
[187] ibid, Part 1, para.9.   [188] ibid, Part 1, para.19.

Whilst it is undeniable that these rights are essential to the realisation of the right to adequate food, it is noteworthy that other relevant rights – such as the right to work, the right to social security, the right to development, or the right to self-determination – are not included, reflecting a neo-liberal bias towards civil and political rights at the expense of socio-economic and collective rights. Moreover, suggesting that the right to adequate food is simply an 'outcome of the realization of existing rights' implies that the right to adequate food is not a free-standing right with its own distinct content and obligations, but is rather a subsidiary product of the implementation of existing (civil and political) rights.

Having recast the human right to food largely in terms of a policy goal or aspiration in part 1 of the VGs, part 2 provides a series of guidelines relating to creating the enabling environment for the realisation of the right to food. These are not human rights obligations but rather are 'practical guidance' to assist states to implement the right.[189] The existence of neo-liberal discourse is clearly present in this section, which stipulates that states should 'ensure non-discriminatory access to markets',[190] 'prevent uncompetitive market practices',[191] 'be in conformity with WTO agreements',[192] 'benefit from opportunities created by competitive agricultural trade',[193] 'foster ... food security ... through a ... market orientated ... world trade system'[194] and operate 'within the framework of relevant international agreements, including those on intellectual property'.[195] As Jacqueline Mowbray argues:

> the Guidelines treat "the market", and the international trading system which is its embodiment at the international level, as mechanisms for addressing the right to food, and render any negative effect that these institutions may have on food security beyond the scope of discussion.[196]

In that way the guidelines support neo-liberal 'common sense' rather than contest it.

Having observed the existence of neo-liberal discourse within the VGs, it is important to also note that the guidelines contain elements of the subaltern critique of the neo-liberal paradigm. For example, it is noted that 'States should take into account that markets do not automatically result in everyone achieving a sufficient income ... and should therefore

---

[189] ibid, Part 1, para.9.   [190] ibid, Part 2, Guideline 4.2.   [191] ibid.
[192] ibid, Part 2, Guideline 4.4.   [193] ibid, Part 2, Guideline 4.6.
[194] ibid, Part 2, Guideline 4.7.   [195] ibid, Part 2, Guideline 8.4.
[196] Mowbray (n 36) 560.

seek to provide adequate social nets' and protect 'the environment and public goods'.[197] Furthermore, states should 'promote the participation of the poor in economic policy decisions',[198] 'advance land reform to enhance access to the poor and women',[199] 'invest in rural infrastructure, education and research'[200] and 'give priority to providing basic services to the poorest ... by ensuring access to primary education for all, basic health care ...'.[201]

Whilst these measures appear to address aspects of subaltern critique, they do not indicate that the VGs lend weight to counter-hegemonic visions of food governance. There are a number of reasons for this. First, these are policy recommendations rather than legal obligations. To the extent that these recommendations conflict with legally binding obligations under WTO agreements, it is clear that the former are 'trumped' by the latter. Second, the VGs do not address the ways in which many practices associated with the neo-liberal food regime – such as dumping practices, WTO rules and structural adjustment conditions imposed by IFIs – undermine the efforts of countries in the Global South to provide support systems to help the poorest parts of their population achieve food security. Indeed, in many ways the VGs are complicit in these practices by promoting the WTO and international free trade in agriculture as the means to achieve food security.

Finally, as already noted, the VGs are formulated in the context of *national* food security. As this chapter has documented, one of the radical claims asserted by the right to food movement in opposition to neo-liberal globalisation is that the right to food gives rise to obligations beyond the traditional nationally based, state-centric human rights paradigm: it also gives rise to international obligations for states as well as multilateral organisations and TNCs. It is for this reason that the JNSC recommends that states should be 'obliged to respect the right to food of persons living in other states ... the role of governments in Intergovernmental Organizations (IGOs) needs to be pursued in a way that these IGOs fully respect and contribute, as far as possible, to the protection and full implementation of the right to food'.[202] In contrast to these demands, the VGs adopt a weaker stance, stating that:

---

[197] Voluntary Guidelines (n 172) Part 2, Guidelines 4.9 and 4.10.
[198] ibid, Part 2, Guideline 2.6.   [199] ibid, Part 2, Guideline 8.10.
[200] ibid, Part 2, Guideline 2.6.   [201] ibid, Part 2, Guideline 3.6.
[202] A Joint North-South Contribution (n 164) 20.

States are strongly urged to take steps with a view to the avoidance of, and refrain from, any unilateral measures not in accordance with international law ... that impedes the full achievement of economic and social development by the populations of the affected countries and that hinders the progressive realization of the right to adequate food.[203]

Whilst the acknowledgement that there are international dimensions to the right to food undoubtedly represents progress from the vantage point of the right to food movement, and was the product of skilled negotiating from human rights NGOs and a number of southern states during the negotiations,[204] the formulation this takes in the VGs is unduly restrictive as a result of the limitation of 'unilateral measures'. Many extraterritorial violations of the right to food are not 'unilateral' but rather are the product of multilateral policies such as WTO rules or the EU Common Agricultural Policy.

Furthermore, the international measures, actions and commitments that are supposed to create the enabling environment for the progressive realisation of the right to food largely reflect the Post Washington Consensus neo-liberal agenda. The guidelines call for adherence to the Monterrey Consensus,[205] the WTO special and differential treatment measures for developing countries[206] and the Heavily Indebted Poor Countries initiative[207] in supporting states achieve food security in the national context. All of these approaches have been comprehensively critiqued as strategic instruments of neo-liberal *trasformismo* that seek to co-opt elites from peripheral states and absorb counter-hegemonic challenges from subaltern social forces whilst only making superficial or symbolic alternations of extant power relations.[208]

---

[203] Voluntary Guidelines (n 172) Part 3, para.3.
[204] See Arne Oshuag, 'The Netherlands and Making of the Voluntary Guidelines on the Right to Food' in Otto Hospes and Bernd Van Der Meulen (eds.), *Fed up with the Right to Food? The Netherlands Policies and Practices Regarding the Human Right to Adequate Food* (Wageningen, 2009) 91 (discussing the position of the various Countries and Stakeholders in relation to the international dimensions of the right to food).
[205] Voluntary Guidelines, (n 172) Part 3, para.4.   [206] ibid, Part 3, para.9.
[207] ibid, Part 3, para.11.
[208] See Susanne Soederberg, 'Recasting Neoliberal Dominance in the Global South? A Critique of the Monterrey Consensus' (2005) 30 *Alternatives* 325–364; Arne Ruckert, 'Towards an Inclusive-Neoliberal Regime of Development: From the Washington to the Post-Washington Consensus' (2006) 39(1) *Labour, Capital and Society* 35–67; Sarah Joseph, 'Trade to Live or Live to Trade: The World Trade Organization, Development, and Poverty' in Mashood Baderin and Robert McCorquodale (eds.) *Economic, Social and Cultural Rights in Action* (Oxford University Press, 2007) 392.

In summary, the VGs provide a striking illustration of the ways in which subaltern discourses that challenge the *status quo* can be recast and absorbed into the dominant hegemonic discursive framework. The right to food was adopted as a core component of the food sovereignty movement's challenge to the neo-liberal food regime in 1996. Within the VGs, the right to food becomes largely aligned with neo-liberal discourse by placing emphasis on the need for states to conform to WTO agreements and participate in competitive agricultural trade. The guidelines, however, are wider in terms of their ideological scope than the unilateral position of the US government, which unequivocally rejects any conception of the right to food. The VGs acknowledge the right to food and stipulate obligations in relation to it that recall and reflect more radical earlier interpretations, such as the Draft Code and General Comment 12, albeit in ways that render these obligations considerably weaker. Furthermore, the guidelines incorporate aspects of the counter-hegemonic critique of the neo-liberal food regime in that they recognise the need for states to engage in an array of corrective or compensatory measures to offset some of the failures associated with agricultural liberalisation. The guidelines are therefore framed in such a way as to be capable of generating universal appeal – the *raison d'être* of hegemony – rather than in the ideologically narrow terms advanced by the USA during the negotiations.

Nevertheless, the guidelines do not touch upon the fundamental tenets of the neo-liberal food regime that are contributing towards the very problems that the VGs claim to address, and indeed, the VGs tacitly approve of key aspects of them. In that sense the VGs can be seen as emblematic of the Post Washington Consensus, which seeks to wed the provision of social safety nets and other complimentary measures with traditional neo-liberal economic policy prescriptions. Indeed, during the negotiations on the guidelines, Margret Vidar from the FOA Legal Council sought to assuage concerns that the right to food would constitute an unacceptable interference in market activity by assuring that '[t]here are numerous instruments for ensuring the realisation of food rights that do not conflict with market liberalisation and deregulation and the principles of efficiency'.[209] Hence, although the VGs contain a number of ideological concessions that acknowledge aspects of counter-hegemonic critique of neo-liberal practices, these concessions do not

---

[209] Vidar, cited in Germann (n 160) 138.

touch upon the essential nucleus of neo-liberal economic relations, and thereby satisfy Gramsci's formula for ruling class hegemony.[210] It can be surmised therefore that the VGs are an instance of *trasformismo* that sustain, rather than contest, the dominant neo-liberal food security framework.[211]

3.4.7 THE RIGHT TO FOOD AFTER THE VOLUNTARY GUIDELINES: THE WORLD FOOD CRISIS OF 2008

The next key moment concerning the right to food in global governance came with the food crisis of 2008. In response to rising food prices and rising levels of hunger, food riots broke out in more than thirty countries by early 2008.[212] Two of the main causes of the steep rise in food prices were identified as the increased speculative investment into agricultural commodities and the accelerated global land grab for the production of agrofuels.[213] For many in the food sovereignty and right to food movements, the crisis was the product of the market-driven approach to agriculture which left many countries and peoples vulnerable to the vagaries of the international markets and unable to implement adequate domestic safeguards to protect against them.[214] In response to one of the greatest challenges to the legitimacy of the international globalised food market, many international and inter-governmental High Level

---

[210] See Chapter 1 Section 1.2.3.

[211] However, it is important to note that the VGs themselves can also be selectively drawn upon and evoked by counter-hegemonic forces as well. For a radical interpretation see Katja Albrecht, Julian Germann and Sandra Ratjen, *How to Use the Voluntary Guidelines on the Right to Food: A Manual for Social Movements, Community Based Organisations and Non-Governmental Organisations* (FIAN, 2007). See also Global Right to Food and Nutrition Watch, 'Ten Years of the Right to Food Guidelines: Gains, Concerns and Struggles (Right to Food and Nutrition Watch, 2014) www.fian.org/fileadmin/media/publications/10yearGuidelines_CivilSociety_SynthesisPaper_en.pdf accessed 21 February 2016.

[212] Raj Patel, 'Commentary: The Hungry of the Earth' (2008) *Radical Philosophy* 151 www.radicalphilosophy.com/commentary/the-hungry-of-the-earth accessed 31 October 2013.

[213] See Peter Wahl, 'The Role of Speculation in the 2008 Food Price Bubble' in FIAN (eds.), *The Global Food Challenge: Towards a Human Rights Approach to Trade and Investment Policies* (2009) www.fian.org/resources/documents/others/the-global-food-challenge/pdf accessed 2 November 2013.

[214] See, e.g., 'Civil Society Statement on the World Food Emergency: Sign on Letter' (2008) www.foodfirst.org/en/node/2155 accessed 31 October 2013.

Conferences and Summits relating to food security and agriculture were held in 2008 and 2009.[215] In April 2008 the UN Secretary General instituted the UN High Level Task Force on the Global Food Crisis, comprising fifteen UN specialised agencies, as well as the IMF, World Bank and WTO.[216]

A series of official declarations were produced which repeatedly emphasised the need for greater trade liberalisation in agriculture, whilst also formally recognising the right to food. For example, the 2008 Declaration of the High-Level Conference on World Food Security opened by recalling the 2004 Voluntary Guidelines on the Right to Adequate Food. However, the Declaration presents the solution to soaring food prices as lying with increased investment in science and technology for agriculture and increased efforts to liberalise international trade through reducing trade-distorting measures.[217]

Similarly, the Declaration of the 2009 World Summit on Food Security affirms the right of everyone to have access to safe, sufficient and nutritious food, consistent with the progressive realisation of the right to adequate food in the context of national food security.[218] However, the Declaration only addresses the structural global causes of the 2008 crisis in a very faint way: for example, by requesting 'relevant international organizations to examine possible links between speculation and agricultural price volatility'[219] and stressing the need to 'continue to address the challenges and opportunities posed by biofuels, in view of the world's food security, energy and sustainable development needs'.[220] The solution to food insecurity lies almost entirely with domestic government action, whilst internationally there is only the vague commitment 'to strengthening the multilateral system in the channelling of resources and in the promotion of policies dedicated to fighting hunger and malnutrition'.[221] Moreover, the realisation of food security is still framed in the context of increased trade liberalisation:

---

[215] Valente and Franko (n 81) 447.
[216] United Nations, 'The Secretary General's High Level Task Force on the Global Food Security Crisis: Background Information' www.un.org/en/issues/food/taskforce/background.shtml accessed 31 October 2013.
[217] Declaration of the High-Level Conference on World Food Security: The Challenges of Climate Change and Bioenergy (n 53) preamble and principles 7(d) and (e).
[218] FAO, 'Declaration of the World Summit on Food Security' (16–18 November 2009) WSFS 2009/2 para.2.
[219] ibid, para.24.   [220] ibid, para.30.   [221] ibid, para.32.

> We will pursue policies and strategies that improve the functioning of domestic, regional and international markets and ensure equitable access for all, especially smallholders and women farmers from developing countries. We support WTO-consistent, non-trade-distorting special measures aimed at creating incentives for smallholder farmers in developing countries, enabling them to increase their productivity and compete on a more equal footing on world markets. We agree to refrain from taking measures that are inconsistent with the WTO rules, with adverse impacts on global, regional and national food security. We reiterate support to a timely, ambitious, comprehensive and balanced conclusion of the Doha Development Round of trade negotiations that would be important to improving food security ...[222]

The World Food Crisis of 2008 therefore did not constitute the basis for a fundamental shift in neo-liberal 'common sense'. Trade liberalisation and participation in global markets is still presented as the framework in which food security will be achieved. The right to food, as developed in the VGs, is incorporated into this strategy, illustrating unambiguously that a version of the right to food now forms part of the neo-liberal hegemonic framework, albeit playing a fairly peripheral discursive role. The above declarations and statements can be contrasted with the CESCR 2008 statement on the world food crisis.[223] The statement recognises that 'the current food crisis represents a failure to meet the obligations to ensure an equitable distribution of world food supplies in relation to need'.[224] The food crisis also 'reflects failure of national and international policies to ensure physical and economic access to food for all'.[225] Closely mirroring the arguments of the food sovereignty movement, the CESCR recommend that the rapid rise in food prices should be limited by:

> encouraging production of local staple food products for local consumption instead of diverting prime arable land suitable for food crops for the production of agrofuels, as well as the use of food crops for production of fuel, and introducing measures to combat speculation in food commodities.[226]

The CESCR also recommend revising the WTO to ensure that agricultural trade promotes, rather than undermines, the right to food and also recommends increasing investment in small-scale agriculture.[227]

---

[222] ibid, para.22.
[223] CESCR, 'The World Food Crisis: Statement' (20 May 2008) UN Doc. E/C.12/2008/1.
[224] ibid, para.9. [225] ibid. [226] ibid, para.11. [227] ibid, para.13.

The marked difference between the CESCR statement and the 2009 WFS Declaration illustrates that, despite the increased mainstreaming of the norm of the right to food in global food governance, radical divergences about its content and associated obligations continue to persist.

3.4.8 THE RIGHT TO FOOD AND LA VIA CAMPESINA'S CAMPAIGNS AGAINST LAND GRABS

La Via Campesina, the movement that founded the concept of food sovereignty, also continues to critically engage with rights discourse in the international domain, playing a leading role in securing wide-ranging reforms to the newly revitalised FAO Committee on World Food Security (CFS) in 2009.[228] It subsequently has worked within the CFS's Civil Society Mechanism (CSM) through its International Planning Committee for Food Sovereignty (IPC)[229] and played a particularly noteworthy role in participating with government representatives and other civil society organisations during the negotiations that led to the CFS's 2012 Voluntary Guidelines on the Responsible Governance of Tenure of Land, Fisheries and Forests in the Context of National Food Security.[230] The guidelines place food security and the progressive realisation of the right to adequate food at the centre of its objectives.[231] They devote attention to the needs of indigenous peoples, small landowners and women, and enshrine the principle of 'free prior and informed consent' for indigenous communities vulnerable to losing their land tenure as a result of legislative or commercial arrangements.[232]

La Via Campesina has cautiously welcomed this document as representing a tool to be used to challenge corporate land grabs in the Global South. They note that the text importantly recognises customary ancestral rights, tenure systems for public or communal land use and legitimate occupation rights that are not yet legally protected.[233] However,

---

[228] Valente and Franko (n 81) 449.
[229] La Via Campesina, 'The Committee on World Food Security (CFS): A New Space for the Food Policies of the World: Opportunities and Limitations' (La Via Campesina Notebook Number 4 September 2012) 6 http://viacampesina.org/downloads/pdf/en/report-no.4-EN-2012-comp.pdf accessed 31 October 2013.
[230] FAO, 'Voluntary Guidelines on the Responsible Governance of Tenure of Land, Fisheries and Forests in The Context of National Food Security' (Rome 2012) www.fao.org/docrep/016/i2801e/i2801e.pdf accessed 31 October 2013.
[231] ibid, Part 1, Objective 1.1.   [232] ibid, Part 3, principle 9.9.
[233] La Via Campesina, 'The Committee on World Food Security' (n 229) 11.

demonstrating clear awareness of the dangers of *trasformismo* when participating in inter-governmental forums, La Via Campesina note that 'Of course, the final text does not correspond to the initial proposal by civil society'.²³⁴ Significant shortcomings include the voluntary nature of the text, the absence of any reference to communal access to water, the acceptance of the transfer of land tenure through market mechanisms without any administrative or legislative support mechanisms for the poor and the lack of a strong enforcement mechanism.²³⁵

In sharp contrast to legal formalist approaches to international law, La Via Campesina adopt a highly politicised and instrumentalist approach to the production of such documents: 'we will have to demonstrate tactical intelligence to be able to use this new international instrument against land grabs by highlighting the paragraphs of the text that explicitly refer to the UDHR, the UN Declaration on the Rights of Indigenous Peoples, and other rights of peasants and other rural persons'.²³⁶ Rather than fetishising the guidelines, La Via Campesina activists are concerned with the ways in which they can selectively evoke them to advance their strategic objectives. Furthermore, La Via Campesina is aware of the limitations of engaging in such processes:

> While there can be little doubt that the participation of La Via Campesina in the CFS is supporting a democratic moment of the public confrontation, it has also been acknowledged that this could: be extremely energy intensive; reduce the time and effort dedicated to grass-roots operations; . . . [and] legitimize an institutionalized space where people can only take limited action.²³⁷

Moreover, it is understood by many in the movement that developing norms in the international arena is a thoroughly political exercise that relies not only on the formal mechanisms of the negotiations processes themselves, but on wider forms of political praxis. As Sofia Monsave of FIAN observed concerning the guidelines process:

> When we participated in negotiations on the guidelines, it wasn't just a question of presenting good technical proposals or of carrying out lobbying work. Without the symbolic protests organized outside the FAO building, without the Dakar appeal signed by more than 1000 organizations around the world, without the 15,000 emails sent to Hilary Clinton by the US Food Sovereignty Alliance, without the forceful interventions by Angel and Kalissa (of La Via Campesina) at key moments that allowed

---

²³⁴ ibid. ²³⁵ Rose (n 23) 160. ²³⁶ ibid. ²³⁷ ibid, 1.

us to turn the negotiating table to our advantage, we would not have obtained what we did. So it was all a sophisticated mix of clear vision, the capacity to mobilize people and the charisma of leaders from social movements.[238]

3.4.9 LIMITATIONS OF THE RIGHT TO FOOD AND THE STRUGGLE FOR 'PEASANT RIGHTS'

Perhaps not surprisingly, given repeated successful co-option and subversion of the idea of the 'right to food' by neo-liberal actors, La Via Campesina has not adopted it as their rallying cry.[239] Priscilla Claeys suggests that a number of La Via Campesina activists are ambivalent to the 'right to food' because they perceive it as being a 'compensatory' entitlement aimed at curbing inequalities within capitalist economic markets rather than tackling those inequalities at the root by positing *a right to produce food*.[240]

To remedy this potential limitation, La Via Campesina have pursued the recognition of 'peasant rights' closely associated with food sovereignty at the international level. In March 2009 it adopted a Draft Declaration on the Rights of Peasants in Seoul, South Korea.[241] The Declaration, which locates the on-going violations of peasants' rights within the context of neo-liberal policies and transnational corporate power, represents one of the most radical human rights-based challenges to the neo-liberal food regime.[242] The Declaration contains a wide-ranging collection of rights. This includes a number of rights already recognised under international human rights law, such as gender equality,[243] the right to life,[244] the right to an adequate standard of living[245] and the right to freedom of association and expression.[246] The declaration also contains an array of 'new rights' such as the right to land,[247] the right to seeds,[248] the right to the protection of agricultural values,[249] the right to biological diversity,[250] the right to preserve the environment,[251] the right to the means of production[252] and

---

[238] ibid, 11.
[239] Priscilla Claeys, 'The Creation of New Rights by the Food Sovereignty Movement: The Challenge of Institutionalizing Subversion' (2014) 46(5) *Sociology* 844, 848.
[240] ibid, 848–849.
[241] La Via Campesina 'Declaration of the Rights of Peasants: Women and Men' (2009) http://viacampesina.net/downloads/PDF/EN-3.pdf accessed 31 October 2013.
[242] ibid, Sections I and II. [243] ibid, article II. [244] ibid, article III. [245] ibid.
[246] ibid, article XII. [247] ibid, article IV. [248] ibid, article V. [249] ibid, article IX.
[250] ibid, article X. [251] ibid, article XI. [252] ibid, article VI.

the 'right to reject' policies and practices that interfere with the aforementioned rights.[253] These rights go well beyond the existing right to adequate food framework and can be understood as a manifesto of rural resistance to the current practices associated with neo-liberal globalisation.

Later in 2009 La Via Campesina presented the Declaration as a response to the world food crisis before the United Nations Human Rights Council and the General Assembly.[254] In 2010 Olivier de Schutter declared that the Peasants Rights initiative was 'intertwining' with the right to food.[255] On 24 September 2012, the UN Human Rights Council adopted a resolution on the 'Promotion of the human rights of peasants and other people working in rural areas'.[256] The resolution led to the creation of an open-ended inter-governmental working group with the mandate of negotiating a draft United Nations Declaration on the Rights of Peasants and Other Persons Working in Rural Areas.[257] In light of the experience of previous attempts to impose a significantly less radical Code of Conduct on the right to adequate food, it is almost inconceivable that the content of La Via Campesina's Draft Declaration will remain in anything like its original form when the Declaration is finally adopted.[258] John Rose, who conducted interviews with La Via Campesina activists, argues that the campaign for peasant rights has 'less to do with legal niceties and judicial procedures, and much more to do with consciousness–raising'.[259] This involves a dual process of education,

---

[253] ibid, see *inter alia*, articles VI(9), V(2) and (3), IX(3), X(3),(4) and (6).

[254] UNHRC, Preliminary study of the Human Rights Council Advisory Committee on discrimination in the context of the right to food (22 February 2010) UN Doc. A/HRC/13/32, para.53.

[255] Olivier de Schutter, cited in Marc Edelman and Carwil James, 'Peasants' Rights and the UN System: Quixotic Struggle? Or Emancipatory Idea Whose Time Has Come?' (2011) 38(1) *The Journal of Peasant Studies* 81, 96.

[256] UNHRC, 'Promotion of the human rights of peasants and other people working in rural areas' (24 September 2012) UN Doc. A/HRC/21/L.23.

[257] UNHRC, 'Promotion of the human rights of peasants and other people working in rural areas' (11 October 2012) UN Doc. A/HRC/RES/21/19, para.1. The first session of the Working Group took place on 15-19 July 2013. UNHRC, 'Open-ended intergovernmental working group on a United Nations declaration on the rights of peasants and other people working in rural areas' www.ohchr.org/EN/HRBodies/HRC/RuralAreas/Pages/WGRuralAreasIndex.aspx accessed 31 October 2013.

[258] For a discussion of the negotiation process, see Geneva Academy of International Humanitarian Law and Human Rights, *Negotiation of a United Nations Declaration on the Rights of Peasants and Other People Working in Rural Areas* (2015) www.geneva-academy.ch/docs/publications/Briefings%20and%20In%20breifs/InBrief5_rightsofpeasants.pdf accessed 21 February 2016.

[259] Rose (n 23) 160.

first at the international level, where La Via Campesina's presence plays an educational role vis-à-vis government representatives, and at the grassroots level where rights talk 'plays a crucial role in efforts to form class consciousness and solidarity amongst peasants'.[260] As La Via Campesina activist Paul Nicholson puts it:

> The convention would specify the rights associated with food sovereignty ... it is also a mobilisation strategy. It is not simply beginning a very complicated administrative process in the United Nations, it is a process too, to identify our rights and to give meaning for national organisations in their negotiations and in their relationship with power. We want to legitimise, in a very visible way, what we are talking about ... The Convention on Peasant Rights would be a universal charter, but really it is a strategic vision for social mobilisations around these rights. It is clear that a formal declaration won't change anything ... it can be a reference at the international level, but it is more than that. It is a re-assertion of out mobilisations, and part of a global strategy.[261]

In other words, the very act of producing the declaration and negotiating it at the international level has an important communicative function of articulating alternative norms and an organisational function in helping to co-ordinate and lend legitimacy to mobilisations and forms of resistance against the predations of neo-liberal food governance. What these statements illustrate is the subtle and sophisticated understanding that La Via Campesina has with regard to their engagement with rights discourse. They are aware of the dangers of *trasformismo* and also of the limits of achieving formal declarations within the UN. Nevertheless, they also regard the pursuing of rights discourse in the international domain as providing a number of opportunities to promote and advance food sovereignty discourse. The following section will examine some of the benefits of evoking the right to food within this context from a counter-hegemonic perspective.

3.5 FOOD SOVEREIGNTY, THE RIGHT TO FOOD AND COUNTER-HEGEMONIC STRATEGY

It is apparent from the above discussion that despite the efforts of the international right to food and food sovereignty movements, and the

---

[260] ibid, 160–161.
[261] Paul Nicholson, quoted in Hannah Wittman, 'Interview: Paul Nicholson, La Via Campesina' (2009) 36(3) *The Journal of Peasant Studies* 676, 679.

serious undermining of the neo-liberal food regime's legitimacy as a result of the 2008 food crisis, global food governance is still largely guided by the ethico-political framework of neo-liberalised food security. Conceptions of the right to food advanced in various inter-governmental forums within the FAO are largely congruent with this module. The right to food is variously recast as a non-binding policy goal, a compensatory mechanism to address market shortcomings and the product of increased trade liberalisation and marketisation of agriculture. Moreover, it can be observed that La Via Campesina have turned their attention principally to the recognition of new 'peasants rights' rather than focusing on the right to food.

Does this mean that the turn to the right to food as a counter-hegemonic strategy can be regarded as a failure? I would argue that the adoption of the right to food by the food sovereignty movement has in fact been tactically useful in spite of such appropriation. Specifically, I think there are three identifiable advantages of using the right to food to advance the counter-hegemonic objectives of the food sovereignty movement. First, rights have been used to reframe the debate about what is just and unjust. Claiming food as a human right has allowed La Via Campesina and others to challenge the reductive economic approach of the neo-liberal model and illustrate that food must be treated differently from other commodities because it is essential to human life and has significant implications for the health of the planet's ecosystems. The right to food establishes obligations on states – and other actors in more radical interpretations – to ensure that every individual has physical and economic access to adequate food or to the means to procure such food. Against the moral outrage of widespread global hunger and malnutrition, these obligations provide a powerful yardstick to critique the failure of the current neo-liberal food regime.

Second, asserting that food is a fundamental human right has allowed the food sovereignty movement to frame its demands in universal terms. Food sovereignty is sometimes identified as paying insufficient attention to poor urban populations and unemployed workers, who make up a significant proportion of hungry people worldwide.[262] By arguing that the realisation of food sovereignty is a necessary prerequisite for the

---

[262] Tina Beuchelt and Detlef Virchow, 'Food Sovereignty or the Human Right to Adequate Food: Which Concept Serves Better as International Development Policy for Global Hunger and Poverty Reduction?' (2012) 29 *Agriculture and Human Values* 259, 265.

realisation of the right to food for all,[263] La Via Campesina has made the necessary counter-hegemonic step from promoting particular and sectorial needs to advancing an ideology framed in terms of universal benefit. Whilst it is true that much food sovereignty discourse focuses on the conditions of rural food producers, it also recognises consumers' rights to nutritious, safe and culturally appropriate food. This implies that the rights of rural food producers would be restricted insofar as they could possibly interfere with the rights of urban food consumers. Whilst there could be potential conflicts between the interests of urban and rural populations, these are not insurmountable. Indeed, there have been a number of urban food sovereignty projects,[264] and the fact that Venezuela, which is the most urbanised country in Latin America, has been able to integrate the concept of food sovereignty into its national constitution indicates that the principle can generate broad-based support.[265] As Walden Bello argues, food sovereignty is about striking a new balance between agriculture and industry and the countryside and city in ways that will benefit both the rural and urban poor. After all, the present subordination of agriculture to industry and urban elites has 'resulted in a blighted countryside and massive urban slums full of rural refugees'.[266] The human right to food can thus provide a universal goal that can unite the various subaltern classes that are excluded and under-served by the current neo-liberal food regime.

Finally, the use of the right to food has facilitated the integration of multiple ideologies, movements and institutions into an increasingly unified and coherent counter-hegemonic universal bloc. The recognition of the right to adequate food under international law in particular has enabled the food sovereignty movement to enter into alliances

---

[263] As La Via Campesina stated at the NGO/CSO forum of the World Food Summit in 1996 'Food is a basic human right. This right can only be realized in a system where food sovereignty is guaranteed' La Via Campesina, 'The Right to Produce and Access Land: A Future without Hunger' (1996) www.voiceoftheturtle.org/library/1996%20Declaration%20of%20Food%20Sovereignty.pdf accessed 2 November 2013.

[264] Daniel R Block, Noel Cha´vez, Erika Allen and Dinah Ramirez, 'Food Sovereignty, Urban Food Access, and Food Activism: Contemplating the Connections through Examples from Chicago' (2011) *Agriculture and Human Values*. Available at http://cis.uchicago.edu/outreach/summerinstitute/2012/documents/sti2012-block-sovereignty-access-activism-chicago.pdf accessed 1 December 2012.

[265] This is not to suggest that the food sovereignty experiment in Venezuela has been successful. For a critique see Rhoda E Howard-Hassmann, 'The Right to Food under Hugo Chavez' (2015) 37(4) *Human Rights Quarterly* 1024.

[266] Bello (n 78) 136.

with strategic partners within the UN human rights system as well as with specialist human rights NGOs with particular expertise and resources. Whilst such alliances often carry the danger of transforming 'the perceived grievances of oppressed, marginal, and exploited populations' into the language of 'hegemonic rational organization',[267] in the case of the alliances fostered by La Via Campesina it is clear that its efforts have in fact significantly shaped many understandings of the human right to food so that they reflect the concerns of grassroots movements. This is most apparent in the work of the UN special rapporteurs on the right to adequate food. The first rapporteur, Jean Ziegler, explicitly embraced food sovereignty as an alternative model in which the right to food could be realised. He concluded his 2008 report by stating:

> In the face of mounting evidence that the current world trading system is hurting the food security of the poorest and most marginalized, and generating ever-greater inequalities, the Special Rapporteur believes that it is now time to look at alternative means that could better ensure the right to food. The implementation of the concept of food sovereignty is a valuable solution.[268]

The subsequent special rapporteur, Olivier de Schutter, has also embraced food sovereignty as an essential component in the realisation of the right to food. In 2011 he criticised G20 leaders for their technocratic solutions to world hunger that have proceeded from 'the misdiagnosis of attributing global hunger to a simple lack of food (and) for years focused their efforts solely on increasing agricultural production by industrial methods alone'.[269] De Schutter argues that the structural inequalities of the present system need to be addressed by reorienting the food system to support the capacity of smallholder farmers to feed their communities and also by supporting 'the capacity of all countries to feed themselves based on the right to food'.[270] In his final report as special rapporteur, de Schutter called for a radical transformation of the

---

[267] Ronen Shamir, 'Corporate Social Responsibility: A Case of Hegemony and Counter-Hegemony' in Boaventura de Sousa Santos and Cesar A Rodriguez-Garivito (eds.) *Law and Globalization from Below: Towards a Cosmopolitan Legality* (Cambridge University Press, 2005) 113.

[268] UNHRC, 'Report of the Special Rapporteur on the Right to Food, Jean Ziegler' (10 January 2008) UN Doc. A/HRC/7/5, para.75.

[269] Olivier de Schutter, 'Food Crisis: Five Priorities for the G20' *Guardian* (Manchester, 16 June 2011).

[270] ibid.

world's food system, including a shift towards food democracy and agro-ecology.[271] The report concludes:

> Understood as a requirement for democracy in the food systems, which would imply the possibility for communities to choose which food systems to depend on and how to reshape those systems, food sovereignty is a condition for the full realization of the right to food.[272]

De Schutter's successor as special rapporteur – Hilal Elver – has continued to promote a similar message.[273] Whilst La Via Campesina focus their energy principally on peasants' rights as opposed to the right to food, these calls from the UN for radical transformation of the global food regime towards greater national self-sufficiency indicate the sustained and deepening synthesis of the right to food and food sovereignty discourses and hence the continued relevance of the right to food to counter-hegemonic praxis.

The continued counter-hegemonic expression of the right to food that can be found in the statements of various human rights representatives and NGOs as well as statements, draft declarations and codes of conduct adopted by global civil society at counter-summits and international events can be understood as a Gramscian 'war of position', whereby subaltern forces are creating alternative institutions and intellectual resources within the existing order whilst slowly building up the strength for the foundations of an alternative one.[274] They also indicate that La Via Campesina and its allies have largely avoided the dangers of *trasformismo* by adopting a twin-track approach of working within official global forums to act as points of visibility and pressure on governmental and inter-governmental institutions whilst also retaining their independence by organising and participating in an array of subaltern counterpublics, where they continue to formulate their own oppositional interpretations of rights.

---

[271] UNHRC, 'Report of the Special Rapporteur on the Right to Food, Oliver De Schutter. Final Report: The Transformation of the Food System' (24 January 2014) UN Doc. A/HRA/25/57.
[272] ibid, para.50.
[273] See, e.g., UNHRC, 'Interim Report of the Special Rapporteur on the Right to Food' (7 August 2014) UN Doc. A/69/275, para.63.
[274] See, for example, APM World Network, 'Food: People's Right to Produce, Feed Themselves and Exercise Their Food Sovereignty' in William F Fisher and Thomas Ponniah (eds.) *Another World Is Possible: Popular Alternatives to Globalization at the World Social Forum* (Zed Books, 2003) 161.

Food sovereignty and right to food-based approaches have informed subaltern social movements from the Brazilian Landless Peasants Movement (MST), which occupies lands and establishes agricultural production co-operatives, to the Indian Bija Satyagraha Movement, which struggles for farmers rights to share seeds in defiance of intellectual property rights (IPRs) laws protecting corporate monopolies.[275] Food Sovereignty has also been enshrined as a right in a number of constitutions of States committed to building alternative paths to the neo-liberal paradigm, including Venezuela, Nicaragua, Ecuador and Bolivia.[276] The continued power of human rights discourse to support subaltern struggles, even in the face of attempts to appropriate them by hegemonic interests, recalls the following observation by Upendra Baxi:

> It is unsurprising that human rights standards and norms, which are the product of diplomatic and civil service desire within the ever-expanding United Nations system, lend themselves to a whole variety of foreign policy and global corporate uses and abuses under the cover of "international consensus". The amazing aspect, however, is the resilient autonomy of human rights normativity that periodically interrogates such acts of expropriation of human rights in the pursuit of severely self-regarding national or regional interests.[277]

3.6 CONCLUSION

This chapter has assessed the relationship of the human right to food to the ethico-political framework of the neo-liberal food regime. It has been argued that since 1996, when the right to food began to gain prominence in international food governance discourse, there have in fact been two competing conceptions of the right to food. One conception, which is the outcome of a number of inter-governmental gatherings within the FAO, treats the right to food as a voluntary policy goal of domestic governments. It is often articulated in tandem with neo-liberal policy prescriptions such as agricultural trade liberalisation and technological innovations. This approach renders the right to food congruous with the hegemonic neo-liberal model of food security. Alternatively, subaltern forces within global civil society, alongside key allies within human rights institutions and organisations, have articulated a conception of the right to food that gives

---

[275] Schanbacher (n 43) 71; McMichael, 'The Land Grab' (n 33) 185.
[276] Beuchelt and Virchow (n 262) 264.   [277] Baxi (n 112) 9.

rise to binding obligations upon governments, civil society actors and inter-governmental organisations to respect, protect and fulfil the right to food. This conception of the right to food is closely allied to the counter-hegemonic conception of food sovereignty in that it supports the promotion of a policy space for all nations and communities to develop the capacity to feed themselves and determine their own policies to realise the right to food. Whilst the right to food movement has not yet been successful in creating internationally recognised binding obligations in relation to the right to food, it has used the right to food to raise consciousness, foster alliances and co-ordinate action internationally in ways that are helping to construct a global counter-hegemonic movement around the idea of food sovereignty.

4

INTELLECTUAL PROPERTY, THE RIGHT TO HEALTH AND THE GLOBAL ACCESS TO MEDICINES CAMPAIGN

'No other issue so clearly epitomizes the clash between human rights and intellectual property as access to patented medicines'.[1]

'If in the name of human rights we argue only over whether test data protection should be conferred for five or fifteen years, without perceiving the deeper effects of intellectual property policy on the logic underlying our world or equipping our activists with the tools to resist them, we will have lost even when we win'.[2]

4.1 INTRODUCTION

It is estimated that approximately two billion people, about a third of the world's population, lack access to life-saving and other essential medicine.[3] Unsurprisingly, it is those living in poverty who suffer from the lowest levels of access. In the poorest parts of Africa and Asia as much as half the population lack regular access to such medicines.[4] Although access to treatment for HIV/AIDS has improved since the turn of the

---

[1] Laurence R Helfer and Graeme W Austin, *Human Rights and Intellectual Property: Mapping the Global Interface* (Cambridge University Press, 2011) 90.
[2] Angelina Snodgrass Godoy, *Of Medicines and Markets: Intellectual Rights and Human Rights in the Free Trade Era* (Stanford University Press, 2013) 150.
[3] UNCHR, 'Report of the Special Rapporteur on the right of everyone to the enjoyment of the highest attainable standard of physical and mental health, Anand Grover' (31 March 2009) UN Doc. A/HRC/11/12, para.14. The United Nations Development Group defines 'access' in this context as 'having medicines continuously available and affordable at public or private health facilities or medical outlets that are within one hour's walk from the homes of the population'. MDG Gap Task Force, *Millennium Development Goal 8: Delivering on the Global Partnership for Achieving the Millennium Development Goals: MDG Gap Task Force Report 2008* (United Nations, 2008) 36.
[4] World Health Organization (hereafter 'WHO'), *WHO Medicines Strategy: Countries at the Core 2004–2007* (2004) 3 http://apps.who.int/medicinedocs/en/d/Js5571e/ accessed 11 November 2013.

century, in 2014 only 41 percent of adults living with HIV were accessing antiretroviral therapy.[5]

Numerous factors can contribute to lack of access to medicines, including inadequate health financing, unreliable medicine supply, poor quality medicines, inadequate consumer information, inadequate national health policies and unreliable health systems.[6] However, this chapter will focus on two specific factors in relation to lack of access: the high cost of medicines[7] and the lack of research and development (R&D) for new medicines, particularly in relation to diseases principally affecting poor populations in the Global South.[8]

These two latter barriers will be examined in the context of the international rules around patent protection in the pharmaceutical sector. It is widely recognised that drug prices are primarily determined by patents.[9] A patent is a form of intellectual property right (IPR) that grants the holder of the patent exclusive legal monopoly over any expression or implementation of a protected work for the period the patent is granted.[10] It also allows the patent holder to prevent anybody else from making, using or selling their invention.[11] Since 1995 all World Trade Organization (WTO) members have been required to protect these rights internationally under the Agreement on Trade Related Aspects of Intellectual Property Rights (TRIPS)[12] as well as under an array of regional and bilateral trade Free Trade Agreements (FTAs).[13]

---

[5] UNAIDS, 'Fact Sheet 2015' available at www.unaids.org last accessed 12 March 2016.
[6] Ellen t'Hoen, *The Global Politics of Pharmaceutical Monopoly Power: Drug patents, Access, Innovation and the Application of the WTO Doha Declaration on TRIPS and Public Health* (AMB, 2009) 3; Lisa Forman, 'Trade Rules, Intellectual Property and the Human Right to Health' (2007) 21(3) *Ethics & International Affairs* 337, 338; WHO, WHO Medicines Strategy (n 4) 25.
[7] Thomas Pogge, 'The Health Impact Fund: Enhancing Justice and Efficiency in Global Health' (2012) 13(4) *Journal of Human Development and Capacities* 537, 539–540.
[8] UN Millennium Project Task Force on HIV/AIDS, Malaria, TB and Access to Medicines, *Prescriptions for Healthy Development: Increasing Access to Medicines* (Earthscan, 2005) 31.
[9] Lisa Forman, 'From TRIPS-PLUS to Rights Plus? Exploring the Right to Health Impact Assessment of Trade-Related Intellectual Property Rights through the Thai Experience' (2012) 7(2) *Asian Journal of WTO and International Health Law and Policy* 347, 351.
[10] Adam D Moore, 'Introduction' in Adam D Moore (ed.), *Intellectual Property: Moral Legal and International Dimensions* (Rowman & Littlefield, 1997) 4. Other forms of IPR include copyright, trade-marks and trade secrets. See Moore, 2–7.
[11] ibid 5.
[12] Agreement on Trade Related Aspects of Intellectual Property, April 15, 1994, 33 I.L.M. 81 (hereafter TRIPs agreement).
[13] Forman, 'From TRIPS' (n 9) 351.

The dramatic expansion of IPRs over the last three decades has been described as a 'second enclosure movement'.[14] Intellectual Property is concerned with the creation of exclusive property rights in knowledge,[15] and therefore constitutes 'the enclosure of the intangible commons of the mind'.[16] It can be understood as an integral component of neo-liberalism's drive towards 'accumulation by dispossession' in the sense that it creates new frontiers of exclusion in areas that were formerly thought of as either common property or uncommodifiable.[17] Proponents of regimes of strengthened international IPR protection argue that they are necessary to incentivise R&D for new medicines through rewarding innovators in a costly, protracted and risk-laden area of investment.[18] Critics argue that, in practice, patents result in expensive medicines,[19] undermine health-oriented development strategies in the Global South[20] and provide inadequate incentive structures to promote R&D for the priority health concerns of developing countries.[21]

This chapter is concerned with the potential of the 'right to health' as a counter-frame to the protection of IPR under TRIPS and other international agreements. To what extent can engaging the 'right to health' contest the international IPR regime? The potential here lies in the ability of global justice movements to use the right to health to assert that affordable access to medicines and treatment for all must assume priority over the commercial property rights of the pharmaceutical industry.

---

[14] James Boyle, 'The Second Enclosure Movement and the Construction of the Public Domain' (2003) 66 *Law and Contemporary Problems* 33.

[15] Rajshree Chandra, *Knowledge as Property: Issues in the Moral Grounding of Intellectual Property Rights* (Oxford University Press, 2010) 14.

[16] Boyle (n 14) 37.

[17] David Harvey, *A Brief History of Neoliberalism* (Oxford University Press, 2005) 160.

[18] Commission on Intellectual Property Rights, *Integrating Intellectual Property Rights and Development* (2002) 14 www.iprcommission.org/papers/pdfs/final_report/ciprfullfinal.pdf accessed 17 April 2013.

[19] Oxfam, *Patent Injustice: How World Trade Rules Threaten the Health of Poor People* (Oxfam, 2001) 13 http://policy-practice.oxfam.org.uk/publications/patent-injustice-how-world-trade-rules-threaten-the-health-of-poor-people-114044 accessed 11 November 2013.

[20] UNDP, *UN Human Development Report 2000* (Oxford University Press, 2000) 84 ('It is estimated that industrialized countries hold 97% of all patents, and global corporations 90% of all technology and product patents. Developing countries have little to gain from the stronger patent protection from the TRIPS agreement because they have little research and development capacity').

[21] Liviu Oprea et al., 'Ethical Issues in Funding Research and Development of Drugs for Neglected Tropical Diseases' (2009) 35 *Journal of Medical Ethics* 310, 310.

The first substantive part of the chapter will begin by describing the development of the current global intellectual property regime and the political institutions and social forces that comprise it. A number of criticisms that have been levelled at these institutional arrangements will be identified. Following this, the ethico-political ideological framework that underpins these institutional arrangements will be set out, namely the legitimising frame of 'intellectual property rights (IPRs)', and then this will be contrasted with the alternative ethico-political framework of 'access to medicines' – a discourse based upon the twin values of open-knowledge and public health. The main part of the chapter is concerned with the ways in which the evocation of the 'right to health' has been used to bolster the access to medicines framework. It will examine this question in a number of contexts: first, how the right to health emerged as an oppositional frame to IPR within the global access to medicines campaign and the international human rights system; second, how the right to health was used in an attempt to reform the TRIPS regime itself; third, how the right to health has been mobilised to challenge 'TRIPS-Plus' agreements contained in bilateral and regional FTAs; and finally, how the right to health has been used to promote alternative R&D paradigms aimed at developing drugs for neglected diseases that affect poor populations in the Global South.

4.2 THE POLITICAL-INSTITUTIONAL CONTEXT: THE GLOBAL INTELLECTUAL PROPERTY REGIME

The TRIPS agreement was the first international treaty to establish a global regime of IPR protection.[22] Susan K. Sell argues that the emergence of this new universal IPR regime has to be understood in the context of broader structural shifts in the global economy and particularly with the increased mobility of capital.[23] The growth of the private R&D sector in the USA, Europe and Japan throughout the 1980s gave technology enhanced importance in international competition.[24] The reduction of trade barriers to the Global South increased commercial

---

[22] Peter Drahos, 'Thinking Strategically about Intellectual Property Rights' (1997) 21(3) *Telegraph Communications Policy* 201, 201.
[23] Susan K Sell, *Private Power, Public Law: The Globalization of Intellectual Property Rights* (Cambridge University Press, 2003) 17.
[24] Carlos M Correa, *Intellectual Property Rights, the WTO and Developing Countries: The TRIPS Agreement and Policy Options* (Zed, 2000) 3.

opportunities to export the products of R&D to new markets. However, there was a perception within the R&D industry that their products could be replicated and distributed by other producers in countries that lacked robust IPR protection. This perception was 'astutely and effectively promoted by industrial lobbies', which persuaded the US government to link trade to IPRs in order to secure returns for R&D and to prevent imitation.[25] Lobbying organisations such as the Pharmaceutical Research and Manufacturers of America (PhRMA) were subsequently successful in winning international support from the USA, the European Community, Switzerland and Japan to include IPR protection within the Uruguay round of trade negotiations leading up to the establishment of the WTO.[26] In turn, these countries, and particularly the USA, played a critical role during the Uruguay negotiations in securing the agreement of Southern States to TRIPS. This was achieved through a combination of threats of economic sanctions for states that did not sign up,[27] the promise of expanded market access for the agricultural and textile products of countries that did,[28] and by relying on technical and legal expertise provided by national industrial associations such as PhRMA to undermine counter-proposals to TRIPS from India and Brazil.[29]

The TRIPS agreement established universal minimum standards of IPR protection for all WTO member states.[30] This chapter is concerned with the protection of patent rights because these pertain directly to the question of access to medicines and the right to health. Prior to TRIPS, patent protection varied significantly across the countries of the Global South, and in general these countries enforced less stringent IPR regimes than those in operation in Europe and North America.[31] In the field of

---

[25] ibid, 4.
[26] Susan K Sell, 'Access to Medicines in the Developing World: International Facilitation or Hindrance?' (2002) 20 *Wisconsin International Law Journal* 481, 482–494.
[27] Sell, 'Private Power' (n 23) 109–110.
[28] Duncan Matthews, *Globalising Intellectual Property Rights: The TRIPS Agreement* (Routledge, 2002) 44; Sell, 'Private Power' (n 24) 110; Daniel Gervais, 'International Intellectual Property and Development: A Roadmap to Balance?' (2005) 2(4) *Journal of Generic Medicines* 327, 328.
[29] Duncan Matthews, 'Is History Repeating Itself? The Outcome of the Negotiations on Access to Medicines, The HIV/AIDS Pandemic and Intellectual Property Rights in the World Trade Organisation' (2004) *Law, Social Justice & Global Development* 1, 4.
[30] TRIPS (n 12) article 1(1).
[31] United Nations Development Project, *United Nations Human Development Report 1999* (Oxford University Press, 1999) 69.

pharmaceutical products, many countries did not issue patents at all.[32] Some countries, such as India, that did provide IPR protection in the field of pharmaceutical research only granted patents in relation to pharmaceutical *processes* and not *products*.[33] This permitted the reverse engineering and re-manufacture of patented drugs through different processes. In turn, this enabled the manufacture and distribution of identical, non-patented 'generic medicines', usually at a considerably lower price.[34] Article 27(1) of TRIPS requires that WTO Members grant patents for pharmaceutical products, thereby rendering reverse engineering a *prima facie* breach of TRIPS obligations unless it is within certain limited circumstances (to be discussed below). The TRIPS regime also required a minimum of a 20-year term of protection from the filing date of the patent: nearly four times the period that was granted in some countries.[35] Furthermore, TRIPS has dramatically reduced the previously robust trade in generic versions of patented medicines between middle and low-income countries, particularly those that had insufficient or no manufacturing capacities in the pharmaceutical sector.[36]

In addition to TRIPS, IPR standards have been ratcheted up through a 'complex and coordinated set of strategies pursued simultaneously in multiple fora'.[37] For example, the USA has often imposed stringent IPR standards unilaterally by placing countries that it deems to have serious defects in their IPR protection on 'watch lists', exposing them to the risk of trade sanctions or actions in the WTO (so called 'Section 301' procedures).[38] IPR standards have been incorporated into regional and bilateral FTAs that the USA and the European Union have negotiated with low and middle-income countries.[39]

In summary, the Global IPR regime is constituted by an array of multilateral, regional and bilateral agreements – as well as unilateral measures by Northern States – promoted primarily by the branch of

---

[32] ibid.   [33] Indian Patents Act 1970, Section (5)(1).
[34] For example, prior to India relaxing its patent laws in 1970, the domestic price of pharmaceuticals was amongst the highest in the world. See generally, Janice Mueller, 'The Tiger Awakens: The Tumultuous Transformation of India's Patent System and the Rise of Indian Pharmaceutical Innovation' (2007) 68 *University of Pittsburgh Law Review* 491, 504–509.
[35] TRIPS (n 12) Article 33; UNCHR, 'Report of the Special Rapporteur' (n 13) para.76.
[36] Vanessa Bradford Kerry and Kelley Lee, 'TRIPS, the Doha Declaration and paragraph 6 decision: what are the remaining steps for protecting access to medicines?' (2007) *Globalization and Health* http://researchonline.lshtm.ac.uk/9769/1/1744–8603-3-3.pdf accessed 13 April 2013.
[37] Godoy (n 2) 44.   [38] ibid.   [39] See Section 2.4.3 for further discussion.

the Transnational Capitalist Class (TCC) involved in R&D and backed by the governments of northern states, as well as the WTO itself. It comprises an integral component of neo-liberal globalisation to the extent that it constitutes a new enclosure movement involved in the suppression of the rights of the commons in the area of knowledge.

4.2.1 CRITICISMS OF THE GLOBAL IPR REGIME IN RELATION TO ACCESS TO MEDICINES: 'A TRAGIC DOUBLE JEOPARDY'

Enhanced global IPR protection has been controversial from the beginning, particularly from a public health perspective.[40] A central concern of public health advocates is that increased patent protection leads to higher drug prices.[41] Increased prices will constitute a significant barrier to access for poor people, particularly in developing countries that lack comprehensive public healthcare systems. As a result, patents widen the access gap between rich and poor.[42] Furthermore, the enforcement of WTO and other IPR rules will negatively impact upon the local manufacturing capacity of poorer countries to produce generic, innovative, quality drugs priced more appropriately for their population's purchasing power.[43] Global IPR protection is therefore also likely to widen health inequalities between the Global North and Global South.

Many have called into question the extent to which enhanced regimes of IP protection are actually effective in promoting innovative advancements in medicine. It has been pointed out that increased patent protection can in fact undermine medical innovation in a number of ways. For example, research by the FDA suggests that patent protection has led to an increase in the production of so-called "me too" drugs (i.e. minor modifications to existing therapeutic drugs) and a parallel decrease in the development of new drugs more likely to be truly innovative.[44] Increased

---

[40] Helfer and Austin (n 1) 90.
[41] See, e.g., UNDP, *UN Human Development Report 2000* (n 20) 84. For detailed microeconomic and empirical inquiry see Holger Hestermeyer, *Human Rights and the WTO: The Case of Patents and Access to Medicines* (Oxford University Press, 2008) 138–153.
[42] Kavaljit Singh, 'Patents vs Patients: AIDS, TNCs and Drug Price Wars' (Public Interest Research Centre, 2001) www.madhyam.org.in/admin/tender/Patents%20vs.%20Patients, %20%20AIDS,%20TNCs%20and%20Drug%20Price%20Wars.pdf accessed 11 November 2013.
[43] Ellen 't Hoen, 'TRIPS, Pharmaceutical Patents, and Access to Medicines: A Long Way from Seattle to Doha' (2002) 3(1) *Chicago Journal of International Law* 39, 42.
[44] Cited in Claudia Chamas, Ben Prickril and Joshua D Sarnoff 'Intellectual Property and Medicine: Towards Global Health Equity' in Tzen Wong and Graham Dutfield (eds),

patents have also been linked to the practice of 'ever-greening': the process where manufacturers extend the life of their patent monopolies by filing new applications on minor modifications to the invention disclosed in the original patent. These new 'upgraded' drugs often offer little if any therapeutic advantage over the first drug.[45] Most significantly, IPR protection does not encourage adequate R&D in developing countries for diseases such as malaria and tuberculosis because poor countries often do not provide sufficient profit potential to motivate R&D investment by the pharmaceutical industry.[46] Therefore, while patents – at least as they are currently calibrated – clearly promote certain forms of innovation, these tend to be commercially driven rather than driven by the greatest medical need.

The combination of the high drugs costs resulting from monopoly pricing and the inherent flaws of the patent system's incentive structure has led Angelina Godoy to describe the present global IPR regime as embodying 'a tragic double jeopardy', wherein drugs are not made available for diseases that disproportionately affect the world's poor, and where drugs are available for diseases that afflict the wealthy and poor alike, they are priced out of the reach of the poor.[47]

4.3 THE POLITICS OF KNOWLEDGE: OWNERSHIP VS. ACCESS

It is clear that there are disputes surrounding the overall effectiveness of patents as a form of R&D innovation for medicines and medical technologies. However, the contours of the dispute do not simply revolve around competing sets of empirical claims. Deeper contestations surrounding counter-posed value systems are at stake as well. Proponents of strong IPR protection argue variously that: (1) IPRs should be protected as natural rights, based on creators being entitled to the fruits of their intellectual labour; (2) that IPRs are necessary to incentivise innovation and promote free trade, and; (3) that there are sufficient flexibilities and limitations within the present system to offset any potential negative

---

Intellectual Property and Human Development: Current Trends and Future Scenarios (Cambridge University Press, 2011) 66.

[45] ibid 67.

[46] WHO, *Public Health Innovation and Intellectual Property Rights, A Report of the Commission on Intellectual Property Rights, Innovation and Public Health* (2006) 85. www.who.int/intellectualproperty/documents/thereport/CIPIHReport23032006.pdf accessed 11 November 2013.

[47] Godoy (n 2) 42.

consequences that may arise from enhanced patent protection. What these various strands of argument share in common is a conception (or at any rate, an acceptance) of medicine (and more dramatically, *medical knowledge*) – as a 'private commodity'.[48] In contrast, the opponents of enhanced IPR protection argue that both public health and the medical knowledge upon which it depends are goods of a fundamentally public nature.

4.3.1 THE NEO-LIBERAL ETHICO-POLITICAL FRAMEWORK: INTELLECTUAL PROPERTY RIGHTS

A number of commentators have compellingly argued that an integral aspect of the R&D industry lobbyists' campaign for achieving strengthened international IP protection was to deploy 'framing' techniques.[49] Sell suggests that the framing of IP protection as a 'right' was a particularly significant aspect of this strategy. Historically, patents and other IPRs were considered 'grants of privilege' rather than rights. Whereas a grant favours the grantor of the privilege, 'rights talk' favours the person claiming the right. As Sell puts it '[b]y wrapping themselves in the mantel of "property rights", [the IP lobbyists] suggested that the rights they were claiming were somehow natural, unassailable and automatically deserved'.[50]

The moral grounding of IPRs has historically rested on two distinct legitimising frameworks: natural rights and utilitarianism.[51] The natural

---

[48] Benjamin Mason Meier, 'Employing Health Rights for Global Justice: The Promise of Public Health in Response to the Insalubrious Ramifications of Globalization' (2006) 39 *Cornell International Law Journal* 711, 729.

[49] John S Odell and Susan K Sell, 'Reframing the Issue: The WTO Coalition on Intellectual Property and Public Health, 2001' in John S Odell (ed.), *Negotiating Trade: Developing Countries in the WTO and NAFTA* (Cambridge University Press, 2006) 87; Sell, 'Access to Medicines' (n 26) 489. Framing serves the purpose of diagnosing certain solutions as problematic, offering solutions and calling for action. See RD Benford and David Snow, 'Ideology, Frame Resonance, and Participant Mobilization' (1988) 26 *Annual Review of Sociology* 197, 199.

[50] Sell, 'Access to Medicines' (n 26) 490. A second discursive frame deployed was to cast IP rights as essential to 'free trade'. In the past, IP rights were regarded as State-backed 'monopolies' that actually threatened to undermine free market competition. The counter-frame to this conception advanced by the IP lobbyists is that if companies could not guarantee that their products would not be protected against 'piracy' in countries with weak IP protection, there would be little incentive for them to trade with such countries.

[51] Chandra (n 15) 27 ('Two principles currently dominate the theoretical literature on IRPs – self ownership ... and utility'). Based on theories deriving from John Locke, the 'self-ownership' of individuals forms the basis of their natural rights. A third justificatory

rights argument derives from the classical Lockean conception of property as a fundamental right. According to this viewpoint, it is the labour that an individual has invested into the creation of a thing that confers the rights of ownership in that thing.[52] Advocates of IPR extend this logic to the finished products of intellectual labour, including newly synthesised drugs.[53] Whilst TRIPS identifies IPRs as 'private rights' rather than fundamental natural rights,[54] much of the justificatory discourse surrounding the need for enhanced global IPR protection draws upon the strong moral language associated with natural rights discourse.[55] For example, a number of commentators have noted how TNCs and northern governments frame IPRs using the emotive language of protection against 'piracy' and 'theft'.[56] By presenting the production and distribution of generic drugs as illegal and immoral, the natural rights argument forecloses policy consideration about the potential benefits that can derive from these practices.[57]

However, the strongest and most widely used justification for IPRs is the utilitarian argument that IPRs provide incentives for innovators to create valuable intellectual works.[58] It is argued that in the absence of IPRs there would be no incentive for innovators to spend vast amounts of time, money and resources if they were not guaranteed protections against 'free-riders' copying and selling their end product.[59] This argument is premised on 'the tragedy of the commons': the hypothesis that, in the absence of property rights, common resources will be over-exploited or depleted by utility-maximising individuals.[60] Even though the latter utilitarian argument is the most prevalent justification for IPR protection,

---

argument for the grounding of IPR is sometimes identified in 'personality theory', which describes property as an expression of the self. This chapter will not consider this line of justification as it has not played a prominent role in the context of controversies surrounding TRIPS. See Justin Hughes, 'The Philosophy of Intellectual Property' (1988) 77 *Georgetown Law Journal* 287, 288.

[52] Aurora Plomer, 'The Human Rights Paradox: Intellectual Property Rights and Rights of Access to Science' (2013) 35 *Human Rights Quarterly* 143, 158.
[53] Robert Nozick, *Anarchy, State and Utopia* (first published 1974, Blackwell, 2001) 182; Adam D Moore, 'Towards a Lockean Theory of Intellectual Property' in Moore (ed.) (n 10) 81–103; Robert P Merges, *Justifying Intellectual Property* (Harvard University Press, 2011) 31–67.
[54] TRIPS (n 12) preamble.   [55] Pogge (n 7) 540.   [56] Odell and Sell (n 49) 87.
[57] Thomas Owen, 'From "Pirates" to "Heroes": New Discourse Change, and the Contested Legitimacy of Generic HIV/AIDS Medicines' (2013) 8(3) *The International Journal of Press/Politics* 259, 263.
[58] Edwin C Hettinger, 'Justifying Intellectual Property' in Moore (ed.) (n 10) 20.
[59] ibid.   [60] See Garett Hardin, 'Tragedy of the Commons' (1968) 162 *Science* 1243–1248.

it is often used in tandem with the natural rights argument. The following quote from the U.S. Chamber of Commerce's Global Intellectual Property Centre demonstrates the presence of both legitimising frameworks for IPR:

> Even as some nations are taking steps to improve IP rights, a range of anti-IP NGOs ... are working ... to undermine respect of IP. These groups ... believe that no matter how much is invested in research and development and individual risk and hard work, the world should have virtually free and open access to new drugs ... Should these activists prevail, there would be fewer resources available for medical and technological advances that save lives ... Their agenda is to take property away from those who create and own it without just compensation or rule of reasonable law.[61]

As demonstrated by this quote, both the natural rights and utilitarian frames share a number of assumptions that have historically been associated with the justifications for the enclosure of common property. These are: (1) *commodification* – medicines are most appropriately to be regarded as private commodities rather than public goods;[62] (2) *economic rationalism* – profit is the incentive that will spur on investment and innovation in medicine;[63] and (3) *individualism* – medical research is understood as the product of the investment and innovation by *individuals* or *individual legal entities* such as corporations.[64]

---

[61] Global Intellectual Property Centre, *Intellectual Property: Creating Jobs, Saving Lives, Improving the World* quoted in Godoy (n 2) 42.

[62] Regina Hertzlinger, *Market-Driven Health Care: Who Wins, Who Loses, in the Transformation of America's Largest Service Industry* (Addison-Wesley, 1997) ('health care should be treated like any other commodity, i.e. its cost, price, availability and distribution should be left up to the workings of a free marketplace constrained by a minimum of government regulation') Quoted in Timothy P Doty, 'Healthcare as a Commodity: The Consequences of Letting Business Run Healthcare' (March 2008) www.ucalgary.ca/familymedicine/system/files/Resident+Research+Review+Report.pdf accessed 2 December 2013.

[63] See, for example, the U.S. Supreme Court ruling in *Eldred v. Ashcroft, 537 U.S. 186 (2003)* ('The profit motive is the engine that ensures the progress of science'), per Justice Ruth Bader Ginsburg writing for the majority.

[64] Ranjit Shahani, Vice Chairman and Managing Director of Novartis India Limited, argues that 'We strongly believe that original innovation should be recognized in patents to encourage investment in medical innovation'. Novartis, 'Supreme Court Denial of Glivec Patent Clarifies Limited Intellectual Property Protection and Discourages Future Innovation in India' (2013), Media Release April 2013. www.novartis.com/newsroom/media-releases/en/2013/1689290.shtml accessed 7 May 2013. The idea of knowledge production as 'original innovation' is premised on an individualism that ignores the social and historical dimensions of the development of knowledge.

In addition to these justificatory frameworks, the IPR regime contains a number of ethico-political compromises intended to limit the scope of, and provide exemptions from, IPR protection. These include the so-called TRIPS flexibilities, which allow states to depart from strict IPR protection in a number of limited circumstances. Such flexibilities involve, *inter alia*, the potential for states to engage in the process of 'parallel importing' of patented drugs – that is to say, the importing of a patented product from a low-pricing country to a higher-pricing country without the consent of the patent holder.[65] TRIPS also allows for the limited use of 'compulsory licencing',[66] which involves a government authorising a third party to make, use or sell a patented invention without the consent of the patent holder.[67] In principle, this would allow governments to authorise companies to produce and distribute cheap generic versions of medicines provided that the legal requirements of article 31 of TRIPS are satisfied.[68] One of the most problematic requirements is in article 31(f), which states that compulsory licencing must be limited to usage 'predominantly for the supply of the domestic market'. In practice, this means that compulsory licencing cannot be used for the export of generic drugs to countries with little or no local manufacturing capacities. It is extremely difficult therefore for countries that lack manufacturing capacities to make effective use of TRIPS flexibilities.

Southern states trying to make use of compulsory licenses and other TRIPS flexibilities have run into all manner of difficulties more broadly due to the complex and burdensome national legal provisions required for implementation as well as their lack of administrative capacity.[69] To that extent, the flexibilities can be understood as hegemonic compromises of an ethico-political kind that in reality are very limited exceptions to the IPR paradigm. Nevertheless, the flexibilities have, as this chapter will discuss, enabled access to medicines activists to contest the restrictive application of IPRs under TRIPS.

---

[65] Parallel importing in the field of pharmaceutical products can be a useful strategy when the prices of those products are high domestically compared to the price in other countries. The TRIPS agreement is silent on the legality of parallel importing but states that 'nothing in this Agreement shall be used to address the issue of the exhaustion of intellectual property rights' (Art. 6). Carlos Correa suggests that this allows Member States to adopt the principle of international exhaustion of rights and therefore to admit parallel imports. Carlos M Correa, *Trade Related Aspects of Intellectual Property Rights: A Commentary on the TRIPS Agreement* (Oxford University Press, 2007) 16.
[66] ibid, article 31.   [67] ibid.   [68] See TRIPS, article 31 (a)–(l).
[69] Duncan Matthews, *Intellectual Property, Human Rights and Development* (EE, 2011) 25.

4.3.2 THE ALTERNATIVE ETHICO-POLITICAL FRAMEWORK: ACCESS TO MEDICINES

Over the last twenty years numerous groups have emerged to contest the recent expansion of intellectual property. Disparate campaigning organisations for farmers' rights, access to medicines, free and open source computer software and alternative models for scientific research have begun to forge alliances and coalesce around the idea of 'access to knowledge' (A2K) as a counter-frame to IPR.[70] This movement contains an array of positions ranging from understandings that IPR 'restrictions on access (to knowledge) ought to be the exception, not the other way around'[71] to more radical commitments 'to reject all monopolistic appropriation of knowledge'[72] and even for the recognition of a 'natural right to imitation ... to learn from each other and to elaborate and develop each other's ideas'.[73] The A2K movement incorporates a spectrum of perspectives, but its general thrust can be understood as counter-hegemonic to the extent that it seeks to challenge neo-liberalism's drive towards accumulation by dispossession and the suppression of the rights of the commons in the area of knowledge.

The access to medicines movement can be understood as one of the most important components of the A2K milieu.[74] Shortly after the establishment of TRIPS, a transnational coalition of civil society organisations and local groups began to articulate access to medicines as a public health issue. A central component of this movement's mission has been to reframe 'the property-orientated, technical and economically instrumental language' of the IPR into a 'more public interest-orientated and less parochial discourse' explicitly connected to public goods such as global public health.[75] The global access to medicines movement has

---

[70] Jack M Balkin and Reva B Seigel, 'Principles, Practices and Social Movements' (2006) 154 *University of Pennsylvania Law Review* 926, 948–949; Amy Kapczynski, 'The Access to Knowledge Mobilization and the New Politics of Intellectual Property (2008) 117 *Yale Law Journal* 804, 825–839.

[71] Geneva Declaration on the Future of the World Intellectual Property Organization (2004) www.futureofwipo.org/futureofwipodeclaration.pdf accessed 2 December 2004.

[72] François Houtart, 'Knowledge, Copyright and Patents (ii): Conference Synthesis' in William F Fisher and Thomas Ponniah (eds.) *Another World Is Possible: Popular Alternatives to Globalization at the World Social Forum* (Zed, 2003) 147.

[73] Usha Memnon, quoted in Volker Heins, 'Human Rights, Intellectual Property and Struggles for Recognition' (2008) 9 *Human Rights Review* 213, 219.

[74] Hoen, 'TRIPS, Pharmaceutical Patents' (n 42) 39.

[75] Marget Chon, 'A Review of Intellectual Property, Human Rights and Development: The Role of NGOs and Social Movements by Duncan Matthews' (2012) 2(2) *The IP Law Book Review* 63, 66–67.

emerged as a diverse and amorphous campaign involving direct action protest, lobbying, publicising activities and the provision of technical assistance to governments. It has sought to highlight the negative ramifications of TRIPS for access to medicines, particularly in the context of the HIV/AIDS crisis.[76] In 1999 the Consumer Project on Technology (CPTech), Health Action International (HAI) and Médecins Sans Frontières (MSF) produced the Amsterdam Statement, which provided a public health framework for the interpretation of key features of the WTO agreement.[77] The Amsterdam Statement informed the adoption of the Revised Drug Strategy (RDS) by the WHO's World Health Assembly (WHA) in May 1999.[78] The new resolution urged WHO Member States 'to ensure that public health rather than commercial interests have primacy in pharmaceutical and health policies and to review their options under the TRIPS Agreement to safeguard access to essential drugs'.[79]

Informed by the values of public health and open-knowledge, the access to medicines framework challenges the assumptions of neo-liberal common sense in a number of ways: (1) given that the supply and provision of essential medicines plays such an important role in the protection of public health it should be regarded as a *fundamental public good* that cannot be left to the whims of market outcomes; (2) the economic rationalist argument that profit generated through monopoly ownership rights is the only, or best, incentive for medical research is deficient. There are many other incentives that could drive individuals to create new medical products, including 'fees, awards, acknowledgement, gratitude, praise, security, power, status and public financial support';[80] (3) the claim that knowledge creation is the product of individuals obscures the fact that knowledge is *socially produced*. As Rajshree Chandra puts it, 'Far from being the achievement of a few men of genius, scientific knowledge is the result of a long, complex and irregular social process'.[81] Knowledge does not emerge in a vacuum; it builds upon extant ideas in ways that make the task of identifying an *individual* knowledge creator of a particular product far from a

---

[76] Ruth Mayne, 'The Global Campaign on Patents and Access to Medicines: An Oxfam Perspective' in Peter Drahos and Ruth Mayne (eds.), *Global Intellectual Property Rights: Knowledge, Access, Development* (Palgrave Macmillan, 2002) 244.
[77] Amsterdam Statement to WTO Member States on Access to Medicine (1999). www.cptech.org/ip/health/amsterdamstatement.html accessed 2 December 2013.
[78] Sell, 'Access to Medicines' (n 26) 505–506.
[79] Quoted in Matthews, *Intellectual Property, Human Rights and Development* (n 69) 29.
[80] Hettinger (n 58) 25.   [81] Chandra (n 15) 68–69.

straightforward task.[82] This understanding of knowledge has particular salience in the context of so-called private medical R&D, which is heavily parasitic upon public research as well as often in receipt of significant sums of public funding.[83] The granting of monopoly ownership rights in such instances – and the consequent increase in prices – looks more like a transfer of wealth from the public to the private sphere, rather than private investment procuring a public good.[84]

The natural rights justificatory framework of IPR is undermined to the extent that the right to benefit from the fruits of one's intellectual labour in the form of profits is subordinated to the right of individuals to essential drugs. The former is a private commercial interest or, at best, a right that has to be limited to the extent that it undermines public health concerns. The latter, by contrast, is necessary for the basic wellbeing and indeed survival of entire populations.[85] Access to medicines discourse also undermines the utilitarian framework in so far as its proponents can highlight the deficiencies in the distributive outcomes and incentive structures created by the IPR regime, as outlined above in part 2(2). From a public health perspective, the inherent flaw of the present regime is its commercial rather than needs-orientation.

Highlighting these deficiencies alone is not enough for a counter-hegemonic position, however. A global counter-hegemonic position would be one in which the interests of the subaltern classes are elaborated as an alternative ideology 'expressed in universal terms'.[86] In the context of the global IPR debate, this would entail the articulation and pursuit of concrete, plausible alternative mechanisms to achieve R&D aimed at stimulating medical innovation that would better serve humanity (such as for neglected 'tropical diseases' predominantly affecting the Global South) as well as allowing for greater access once those drugs are brought into the market. Actors within the global access movement have been attempting to articulate such alternatives since 1999.[87]

---

[82] Hettinger (n 58) 22–23.
[83] Bhaven Sampat and Frank Lichtenberg, 'What Are the Respective Roles of the Public and Private Sectors in Pharmaceutical Innovation?' (2011) 30(2) *Health Affairs* 332.
[84] See, e.g., Ha-Joon Chang, 'Company Profits Depend on the "Welfare Payments" They Get from Society' *Guardian*, 5 April 2013; George Monbiot, 'Property, Theft and How We Must Breach this Sacred Line' *Guardian*, 25 March 2013.
[85] Chandra (n 15) 341.
[86] Robert Cox, 'The Way Ahead: Toward a New Ontology of World Order', in Catherine Eschle and Brice Maiguashea (eds.) *Critical Theory and World Politics* (Boulder, 2001) 58.
[87] See discussion in Section 4.4.4.

Table 4.1 *Intellectual property rights and access to medicine paradigms compared*

	Intellectual property rights paradigm	Access to medicines paradigm
Best mechanisms for advancing scientific knowledge	Competitive Private Exclusive property rights over inventions	Collaborative Open/Shared knowledge A variety of mechanisms, possibly in conjunction with qualified IPRs.
Key advocates	WTO, R&D industry, US and EU governments	A2K movements, public health NGOs, HIV/AIDs advocates, Some southern governments
Rights status of intellectual property	Natural/Fundamental	Derivative/Instrumental
Agents of knowledge advancement	Natural and legal individuals	Collectives and social processes
Incentives for knowledge production	Commercial profit	Multifarious

4.4 THE RIGHT TO HEALTH AND ACCESS TO MEDICINES CAMPAIGN

As the previous section argued, the negative impact of the Global IPR Regime on access to medicines has led to the emergence of a global access to medicines campaign that has mobilised around the potentially counter-hegemonic oppositional frame of access to medicines. An interrelated discourse that has been evoked by this movement has been the 'right to health'.[88] This is not simply a moral discourse, but also one that has a basis under international law. The Universal Declaration of Human Rights (UDHR) proclaims that '(e)veryone has the right to a standard of living adequate for health and well-being ... including medical care and necessary social services'.[89] The Preamble to the 1946 World Health Organization (WHO) Constitution declares that 'the enjoyment of the highest attainable standard

---

[88] Matthews, *Intellectual Property Rights, Human Rights and Development* (n 69) 44–45.
[89] Article 25(1), Universal Declaration of Human Rights 1948 GA Res 217A(III), 10 December 1948, A/810.

of health is one of the fundamental rights of every human being' and that 'governments have a responsibility for the health of their peoples which can be fulfilled only by the provision of adequate health and social measures'.⁹⁰ The International Covenant on Economic, Social and Cultural Rights (ICESCR) made the aspirations set out in the UDHR and WHO Constitution into a legally binding international obligation expressed as the 'right of everyone to the enjoyment of the highest attainable standard of physical and mental health'.⁹¹ To fulfil their duties in relation to this right, states parties are required to take steps to progressively realise the reduction of infant mortality and the healthy development of the child;⁹² prevent, treat and control diseases;⁹³ and assure medical service and attention to all in the event of sickness.⁹⁴ These provisions of international law are increasingly finding expression at the national level, with at least ninety-five national constitutions including rights to health facilities, goods and services.⁹⁵

It has been argued that the concept of the right to health is anathema to the neo-liberal approach to healthcare.⁹⁶ Whereas the neo-liberal approach views healthcare as a commodity or service like any other, the 'right to health' promotes the equal enjoyment of all to receive medical care on the basis of need.⁹⁷ Within this understanding, health services are reframed as important public goods 'to be universally accessible without conditions' and hence conceptually distinct from commercial goods or services purchased in the market.⁹⁸ This constitutes an 'important paradigm shift, helping to roll back the dominance of the market approach to health care'.⁹⁹

---

⁹⁰ Preamble, Constitution of the World Health Organization 1946, 14 UNTS 185.
⁹¹ Article 12, The International Covenant on Economic, Social and Cultural Rights (ICESCR), 993 UNTS 3 entered into force 3 January 1976.
⁹² ibid, article 12(a).   ⁹³ ibid, article 12(c).   ⁹⁴ ibid, article 12(d).
⁹⁵ S Katrina Perehudoff, 'Health, Essential Medicines, Human Rights and National Constitutions' (2008) WHO, xvii available at http://apps.who.int/medicinedocs/documents/s17421e/s17421e.pdf accessed 2 April 2013.
⁹⁶ Tony Evans, 'The Human Right to Health?' (2002) 201 23(2) *Third World Quarterly* 197, 201; Paul O'Connell, 'The Human Rights to Health in an Age of Market Hegemony' in John Harrington and Maria Stuttaford (eds.) *Global Health and Human Rights: Legal and Philosophical Perspectives* (Routledge, 2010) 191.
⁹⁷ ibid, 199–200.
⁹⁸ People's Health Movement, 'Health for All Now: Using the Right to Health while Developing It' 2 (n/d phmovement.org). www.phmovement.org/sites/www.phmovement.org/files/Conceptual,%20introductory%20section%20of%20right%20to%20health_0.pdf accessed 19 November 2013.
⁹⁹ ibid.

Furthermore, the right to health constitutes a moral and legal 'trump' in relation to the ownership rights in intangible assets.[100] Whereas the former is necessary for all humanity to ensure their wellbeing and dignity, the latter is 'a legally protected interest of a lower order than a human right'.[101]

However, the right to health frame is not without its own potential pitfalls and risks as a counter-hegemonic discourse. Benjamin Mason Meire has argued that the right to health under international law has been framed as an *individual* right that focuses 'upon individual access to health services at the expense of collective health promotion and disease programs through public health systems'.[102] Such 'limited and atomised' readings of the right to health are argued to bolster neo-liberal conceptions of healthcare as a private good[103] and may, in some instances, even deepen existing health inequalities within public health systems.[104] Furthermore, whilst public health advocates may assert that the right to health assumes primacy over patent rights, the reality of the international legal order is that the TRIPS agreement contains detailed rules and is linked to a hard-edged enforcement regime backed up by the threat of retaliatory measures.[105] By contrast, the provisions contained within the ICESCR are relatively vague and lack the comprehensive enforcement mechanisms that the WTO possesses. Defenders of strong international IPR protection can argue that the right to health is a mere 'soft law' or an aspirational norm that may inform the application of the TRIPS agreement, but cannot overrule or abrogate it.[106] In that sense the right to health can be seen as a weak legal discourse to challenge TRIPS.

Taking note of these potential strengths and weaknesses, this section will examine how right to health discourse has interpolated the global conflict surrounding IPRs and access to medicines.

---

[100] Heins (n 73) 228.
[101] Stephen Marks and Adriana Lee Benedict, 'Access to Products, Vaccines and Medical Technologies' in José M Zuniga, Stephen P Marks and Lawrence O Gostin (eds.), *Advancing the Human Right to Health* (Oxford University Press, 2013) 305, 305.
[102] Benjamin Mason Meire, 'Employing Human Rights for Global Justice: The Promise of Public Health' (2006) 39 *Cornell International Law Journal* 711, 738.
[103] Aeyal Gross, 'Is There a Human Right to Private Health Care?' (2013) 41(1) *The Journal of Law, Medicine and Ethics* 138–144.
[104] Octavio Luiz Motta Ferraz, 'The Right to Health in the Courts of Brazil: Worsening Health Inequities?' (2009) 11(2) *Health and Human Rights* 33–45.
[105] Laurence R Helfer, 'Regime Shifting: The TRIPs Agreement and New Dynamics of International Intellectual Property Lawmaking' (2004) 29(1) *Yale Journal of International Law* 1, 45.
[106] Evans (n 96) 201.

4.4.1 THE RIGHT TO HEALTH AS OPPOSITIONAL FRAME

4.4.1.1 THE RIGHT TO HEALTH AND THE GLOBAL ACCESS TO MEDICINES MOVEMENT

The global access to medicines movement arguably has its origins in the struggles of national public health campaigns in sub-Saharan Africa, Latin America and Asia for access to affordable generic drugs. In particular, the international HIV/AIDS pandemic, affecting tens of millions of people in the Global South, generated demands for access to anti-retroviral (ARV) treatments and medications patented by northern-based pharmaceutical TNCs.[107] In response to the prohibitively high costs of these and other life-saving medicines, a number of states – including South Africa, Brazil, India and Thailand – introduced legislation aimed at circumventing strict IPR protection through the use of compulsory licencing and parallel importing.[108] In turn, these countries faced threats of trade sanctions and corporate litigation from both the pharmaceutical industry and northern governments.[109] In response to these attacks, a number of public health movements and NGOs in these countries sought recourse to internationally and constitutionally enshrined commitments to the 'right to health', and other human rights principles, in support of their government's efforts to make affordable drugs available.[110]

Of all of these national movements, it is perhaps the South African example that did the most to catalyse the global access to medicines movement.[111] In response to legislation introduced by the South African Government in 1997 to enable the generic substitution of on-patent medicines through compulsory licences and the use of parallel importation,[112] forty-one pharmaceutical companies initiated a legal challenge against the South African government on the basis that, *inter alia*, the

---

[107] Matthews *Intellectual Property, Human Rights and Development* (n 69) 26.
[108] Sell, 'TRIPS and Access to Medicines' (n 26) 496, 500–502.
[109] Lisa Forman, 'Trade Rules, Intellectual Property, and the Right to Health' (2007) 21(3) *Ethics and International Affairs* 337, 342.
[110] Matthews *Intellectual Property, Human Rights and Development* (n 69) 92–201.
[111] Erika George, 'The Human Right to Health and HIV/AIDS: South Africa and South-South Cooperation to Reframe Global Intellectual Property Principles and Promote Access to Essential Medicines' (2011) 18(1) *Indiana Journal of Global Legal Studies* 176, 187.
[112] The Medicines and Related Substances Control Amendment Control Act, No. 90 of 1997 was passed by the national assembly on 31 October 1997 and signed by then-President Mandela on 25 November 1997, sections 15(c)(b) and 22F(1)(a).

legislation was contrary to South Africa's obligations under TRIPS.[113] In June 1998 the US government joined the dispute, threatening South Africa with trade sanctions and the withdrawal of aid if it failed to repeal the provisions in its legislation that it regarded as contrary to TRIPS obligations.[114] South Africa's Treatment Action Campaign (TAC) – a grassroots NGO that campaigns for access to healthcare as a human right – intervened in the ensuing court case as *amicus curie* for the government. As Mark Heywood has noted, the inclusion of the TAC transformed the case from a 'dry legal contest into a matter about human lives'.[115] TAC brought human rights arguments from international and constitutional law, arguing that the right to health provided authority for the legislation to be prioritised over corporate property rights. TAC also provided 'extensive empirical evidence to undercut corporate claims about the cost of R&D, and its link to innovation, as well as the personal testimony of poor people unable to buy medicines, to illustrate the costs of the litigation'.[116]

Simultaneously, outside of the court the TAC organised demonstrations, mass protests and acts of disobedience such as importing thousands of generic drugs in defiance of patent laws.[117] In the process it enlisted the support of over 250 NGOs worldwide.[118] The international pressure that was created by this coalition had a dramatic impact, forcing the pharmaceutical coalition to drop their case[119] and the US government to abandon the threat of sanctions against the South African government.[120] Similar transnational alliances and victories occurred in the context of India and Brazil.[121] In all three countries NGOs mobilised against the TRIPS agreement's insistence on patent

---

[113] *Pharmaceutical Manufacturers' Association of South Africa v. President of the Republic of South Africa*, Case no. 4283/98, High Court of South Africa (Transvaal Provincial Division).

[114] Heinz Klug, 'Campaigning for Life: Building a New Transnational Solidarity in the Face of HIV/AIDS and TRIPS' in Boaventura De Sousa Santos and Cesar A Rodriguez-Garavito (eds.) *Law and Globalization from Below: Towards a Cosmopolitan Legality* (Cambridge University Press, 2005) 127.

[115] Mark Heywood, 'Debunking "Conglomo-Talk": A Case Study of the Amicus Curie as an Instrument for Advocacy, Investigation and Mobilisation' (2002) 6 *Law Democracy and Development* 12.

[116] Lisa Forman, '"Rights" and "Wrongs": What Utility for the Right to Health in Reforming Trade Rules on Medicines?' (2008) 10(2) *Health and Human Rights* 37.

[117] Klug (n 114) 131.   [118] ibid, 132.   [119] George (n 111) 186.

[120] Klug (n 114) 133.

[121] Matthews *Intellectual Property, Human Rights and Development* (n 69) 125–201.

protection for pharmaceuticals by using the right to health recognised by their respective constitutions.[122] In these cases the right to health was used both as tool of litigation but also as a political discourse 'to mobilize, galvanize, educate and inform'.[123]

Following the efforts of national NGOs to frame access to medicines as a fundamental human rights issue, the right to health also became an important discursive frame in the arena of global civil society. In the midst of the contestation between the PMA and the South African government, the XIII International AIDS Conference in Durban hosted the first global demonstration for access to HIV/AIDS treatment. The conference issued a declaration beginning: 'We are united with a single purpose, to ensure that everyone with HIV and AIDS has access to fundamental rights of healthcare and access to life-sustaining medicines.'[124] Duncan Matthews has also documented the critical role that international NGOs involved in the access to medicines campaign played in framing intellectual property and access to medicines as human rights issues.[125] Such NGOs combined evidence-based critique of TRIPS as a barrier to access to medicines in the Global South with rhetoric drawn from human rights discourse.[126] Intellectual property was framed as 'not only or primarily a trade issue, but also as one relevant to health and human rights'.[127] In engaging this discourse, these NGOs helped to undercut both the utilitarian justification for IPR, by highlighting their negative impact on access, and the natural rights justification by reframing the upholding of IPRs as being capable of violating higher order human rights, such as the right to health, which assumes priority over rights to ownership of intangible property.

Heinz Klug argues that the evocation of the right to health frame to contest TRIPS had a number of important effects on the international intellectual property regime.[128] First, it promoted the acceptance of a human rights perspective that the health impact of any trade rule should be given consideration in assessing the validity of any particular claim of

---

[122] ibid. In the Indian context the right to health was 'read in' to the constitutionally enshrined right to life.
[123] Katherine G Young, 'Securing Health through Rights' in Thomas Pogge, Matthew Rimmer and Kim Rubenstein (eds.), *Incentives for Global Health: Patent Law and Access to Essential Medicines* (Cambridge University Press, 2010) 363.
[124] XIII International AIDS Conference, Global Manifesto (9 July, 2000). www.actupny.org/reports/durban-access.html accessed 21 November 2013. See also Mark Schoofs, 'AIDS: The Agony of Africa: South Africa Acts Up' (1999) 44 *The Village Voice* 2, 23.
[125] Matthews, *Intellectual Property, Human Rights and Development* (n 69) 43.
[126] ibid.   [127] ibid, 44.   [128] Klug (n 114) 137.

intellectual property.[129] Second, it established awareness of health and human rights considerations as essential correctives to potential injustices that arise from enhanced IPR protection.[130] Third, it provided developing countries with a legal means to justify decisions 'privileging policies securing access to medicines over concerns about their international trade commitments'.[131] Klug notes that this alternative perspective could form the basis for an alternative approach to medical provision:

> Instead of relying on thin strands of flexibility, NGOs, international organizations, and governments attempting to address the global HIV/AIDS pandemic might look to the building of this new, multifaceted, and complex solidarity to assert a human-rights based interpretation to the international trade regime and intellectual property regime – one which places public health ahead of property claims.[132]

The counter-hegemonic potential of the right to health lies in its ability to support a discourse that favours an international institutional arrangement in which public health assumes primacy over private profit. The campaigns and victories in the South African, Brazilian and Indian contexts indicate such potentiality.

4.4.1.2 ACCESS TO MEDICINES AND THE UN HUMAN RIGHTS SYSTEM

The global controversy generated by the conflict between TRIPS and access to medicines also attracted the attention of the UN human rights system. In 1998 UNAIDS and the Office of the High Commissioner for Human Rights published international guidelines requiring states to, *inter alia*, provide access to ARVs at affordable prices and to 'incorporate to the fullest extent any safeguards and flexibilities' in international IP treaties to ensure that they satisfy their domestic and international commitments to human rights.[133] On 11 May 2000 the Committee on Economic, Social and Cultural Rights (CESCR) published General

---

[129] ibid.   [130] ibid, 137–138.   [131] ibid, 138.   [132] ibid.
[133] Office of the United Nations High Commissioner for Human Rights and the Joint United Nations Programme on HIV/AIDS, 'International Guidelines on HIV/AIDS and Human Rights 2006 Consolidated Version' available at http://data.unaids.org/Publications/IRC-pub07/jc1252-internguidelines_en.pdf (updated 2006) accessed 14 April 2013. See also UN Declaration of Commitment on HIV/AIDS, G.A. Res. 33/2001, para. 15, UN. GAOR, 26th Special Sess. (25–27 June 2001) (recognising 'that access to medication in the context of pandemics such as HIV/AIDS is one of the fundamental elements to achieve progressively the full realization of everyone to the enjoyment of the highest standard of physical and mental health').

Comment 14 on the right to health under article 12 of the ICESCR.[134] The comment established that the provision of essential drugs is a core component of the right to health. States must therefore ensure that such drugs – whether privately or publicly provided – are available in sufficient quantity within a state party, are 'affordable for all' and that everyone has equal and timely access to their provision.[135] As part of this obligation, States Parties must refrain from denying or limiting equal access to medicines;[136] ensure that essential medicines provided by third parties, such as pharmaceutical companies, are available on an equal basis for all;[137] and promote medical research, in particular with respect to certain categories of diseases including HIV/AIDS.[138]

The General Comment also establishes a number of international obligations under the right to health.[139] States Parties have to respect the enjoyment of the right to health in other countries, and to prevent third parties from violating the right in other countries, if they are able to influence these third parties by way of legal or political means.[140] Specifically, States Parties should recognise the essential role of international co-operation and comply with their commitment to take joint and separate action to achieve the full realisation of the right to health, taking into account the existing gross inequality in the health status between people in developed and developing countries.[141] States Parties should ensure that the right to health is given due attention in international agreements and should take steps to ensure that these instruments do not adversely impact upon the right to health.[142] Finally, States Parties also have an obligation to ensure that their actions as members of international organisations take due account of the right to health.[143]

These international obligations suggest, at the very least, that States Parties to the ICESCR should: (1) not take measures, such as filing complaints through the WTO Dispute Settlement Body or using economic, political or diplomatic pressure and other unilateral measures, that could negatively impact on access to medicines in other countries; (2) give due consideration to the impact on the right to health when negotiating IPRs in pharmaceuticals in the WTO or in regional or bilateral trade agreements and; (3) ensure that such international agreements do not

---

[134] CESCR 'General Comment No. 14: The Right to the Highest Attainable Standard of Health (Art. 12 of the Covenant)' (11 August 2000) UN Doc. E/C.12/2000/4.
[135] ibid, paras.12(a), 12(b) (iii) and 17.   [136] ibid, para.34.   [137] ibid, para.35.
[138] ibid, para.36.   [139] ibid, paras.38–42.   [140] ibid, para.39.   [141] ibid, para.38.
[142] ibid, para.39.   [143] ibid.

adversely impact on the right to health in other countries, for example by making the cost of essential medicines beyond the means of the poor.

The delineation of the normative content of the right to health and its relation to access to medicines in General Comment 14 generated further possibilities for contesting the TRIPS agreement by recourse to international human rights law. In August 2000 the UN human rights system first confronted TRIPS head-on when the UN Sub Commission on the Protection and Promotion of Human Rights adopted Resolution 2000/7 on 'Intellectual Property Rights and Human Rights'.[144] The resolution begins by noting 'actual or potential conflicts exist between the implementation of the TRIPS Agreement and the realisation of economic, social and cultural rights in relation to, inter alia ... restrictions on access to patented pharmaceuticals and the implications for the enjoyment of the right to health'.[145] The TRIPS regime does not adequately reflect the 'fundamental nature and indivisibility of all human rights, including ... the right to health'.[146] Where conflicts arise, the Sub-Commission stressed that governments should be reminded of 'the primacy of human rights obligations over economic policies and agreements'.[147]

The Sub-Commission's resolution constituted the beginning of 'an ambitious new agenda for a review of intellectual property issues within the U.N human rights system, an agenda animated by the principle of human rights primacy'.[148] The subsequent response of the UN human rights bodies to clarifying the relationship between intellectual property and the right to health was 'nothing short of overwhelming'.[149] Resolution 2000/7 requested the Secretary-General to submit a report on the question of IP rights and human rights at its fifty-first session. To this end, letters were sent to international organisations and NGOs on the 6 March 2001 requesting information that would be relevant to the report.[150] The WTO was one of the organisations to respond to this request.[151] The WTO reply suggested that the TRIPS agreement was consistent with human rights principles and the right to health. The balance between

---

[144] Sub-Commission on Human Rights, 'Intellectual property rights and human rights', ESCOR, Commission on Human Rights, Sub-Commission on the Promotion and Protection of Human Rights, 52nd Session, 25th meeting, UN Doc. E/CN.4/Sub.2/Res/2000/7.
[145] ibid, preamble.   [146] ibid, preamble.   [147] ibid, para.3.
[148] Helfer and Austin (n 1) 53.   [149] ibid.
[150] UNCHR, 'Intellectual Property Rights and Human Right: Report of the Secretary-General' (14 June 2001) UN Doc. E/CN.4/Sub.2/2001/12.
[151] ibid, 7–9.

the equitable treatment of inventors and the public interest is adequately protected by article 7 of TRIPS, which requires that IPRs should be enforced in a manner that is to 'the mutual advantage of producers and users of technological knowledge' and is 'conducive to social and economic welfare'.[152] The WTO suggested that article 7 mirrors the balance of interests contained within article 15(1) of the ICESCR, which recognises the right of everyone: (a) to take part in cultural life; (b) to enjoy the benefits of scientific progress and its applications; and (c) to benefit from the protection of the moral and material interests resulting from any scientific, literary or artistic production of which he or she is the author.[153] The WTO went on to argue that TRIPS creates the incentives required to develop the medicines needed to realise the right to health and any conflict that may arise between creators' rights and health users' rights can be reconciled through utilising existing TRIPS flexibilities.[154]

The WTO's defence of the TRIPS agreement's human rights compatibility represents a shift in the terms under which international IP protection is justified. In the run up to the adoption of TRIPS the justifications were based upon the private commercial rights of innovators and utilitarian justifications about promoting free trade and innovation. When faced with a challenge to the legitimacy of TRIPS from a human rights perspective, the WTO chose not to simply dismiss the relevance of such critiques but to address them on their own terms. In doing so, the WTO articulated a reading of the right to health, one that attempted to justify the TRIPS status quo by reference to human rights standards. The argument can be summarised as follows: (1) IPRs are themselves human rights; (2) IPRs are instrumental in protecting rights like the right to health by encouraging innovation to develop essential medicines that would not otherwise be forthcoming; and (3) the flexibilities contained within TRIPS allow for an optimal balance between the rights of innovators on the one hand and the health rights of those in need of medicines on the other hand.[155] All three of IPR's legitimising frameworks identified in Section 4.3 are present in the WTO's response: IPRs as a natural right, IPRs justified

---

[152] ibid, 7.   [153] ibid.   [154] ibid, 8.
[155] For neo-liberal readings of IPR as a human right see Ernst-Ulrich Petersmann, 'The WTO Constitution and Human Rights' (2000) 3(1) *Journal of International Economic Law* 19, 19 and 21; Robert D Anderson and Hannu Wager, 'Human Rights, Development, and the WTO: The Cases of Intellectual Property and Competition Policy' (2006) 9(3) *Journal of International Economic Law* 707, 721–730.

on utilitarian grounds and the ethico-political compromises within TRIPS as sufficient to balance creators' rights with health service users' rights.

Despite the WTO's attempt to justify TRIPS in human rights terms, the UN human rights system continued to produce a number of highly critical documents relating to the TRIPS regime. In June 2001 the UN High Commissioner for Human Rights produced a detailed analysis of the impact of TRIPS on human rights. Whilst the report recognised that TRIPS flexibilities could facilitate a human rights compatible reading, there were nonetheless a number of features of TRIPS that raised questions about its human rights consistency.[156] First, the overall thrust of the TRIPS agreement is the promotion of innovation through creating commercial incentives. The various links to the subject matter of human rights, such as public health, are expressed in terms of exceptions to the rule rather than as guiding principles. A human rights approach, by contrast, would explicitly place the right to health and other rights at the heart of intellectual property protection.[157] Second, whilst TRIPS sets out the content of IPRs in detail, it only alludes to the responsibilities of IP holders in general terms without establishing what these obligations are or how they should be implemented. A human rights-based approach would require a more concrete delineation of the obligations of IP holders in relation to human rights protection.[158] And third, the TRIPS requirement that WTO members provide patent protection to cover all pharmaceuticals could undermine human rights-orientated development strategies which require states to have the right and duty to formulate national development strategies to improve the wellbeing of their populations.[159] The TRIPS agreement could be inappropriate for this end in the Global South because it is based on forms of protection that have developed in the northern states.[160]

In relation to the right to health, the High Commissioner's report identifies two core problems with the TRIPS regime that public health advocates had been highlighting for many years. The first was that the effect of patents on affordability is significant and drug prices fall

---

[156] UNCHR, 'The Impact of the Agreement on Trade-Related Aspects of Intellectual Property Rights on Human Rights, Report of the High Commissioner' (27 June 2001) UN Doc. E/CN.4/Sub.2/2001/13, para.21. See also UN Sub-Commission on Human Rights 'Intellectual Property and Human Rights', Res. 2001/21 (16 Aug. 2001), UN Doc. E/CN.4/Sub.2/2001/L.11/Add.2.
[157] UNCHR, ibid, para.22.   [158] ibid, para.23.   [159] ibid, para.24.
[160] ibid, para.25.

sharply when generic substitutes enter a market to compete with drugs upon patent expiry.[161] The second problem related to medical research. The report notes that IPRs are limited commercial rights geared towards economic rewards. Human rights are at best a secondary consideration. The report highlights research from the WHO indicating that less than 1 percent of new chemical entities developed between 1975 and 1996 were for the treatment of tropical diseases. The High Commissioner suggests that this 'could mean that *alternative mechanisms to patents* might need to be considered by states in implementing articles 12 and 15 of ICESCR'.[162] In relation to the question of the counter-hegemonic potential of the right to health it is particularly noteworthy that the High Commissioner indicates that it might be necessary to create institutional frameworks for innovation outside of the patents system to comply with human rights obligations under the ICESCR.[163]

Later that year the CESCR released a statement explicitly addressing the link between human rights and intellectual property.[164] The statement sought to establish clear distinctions between IPRs and human rights:

> human rights are fundamental as they derive from the human person as such, whereas intellectual property rights derived from intellectual property systems are instrumental, in that they are a means by which States seek to provide incentives for inventiveness and creativity from which society benefits ... In contrast to human rights, intellectual property rights are generally temporary in nature and can be revoked ... while ... human rights are timeless expressions of fundamental entitlements ... any intellectual property regime that makes it more difficult for a State Party to comply with its core obligations in relation to health ... is inconsistent with the legally binding obligations of the State Party.[165]

---

[161] ibid, para.44.   [162] ibid, para.38 (my emphasis).
[163] On this point see J Olkoa-Onyango and Deepika Udagama, 'Globalization and its impact on the full enjoyment of human rights' (2 July 2001) UN Doc. E/CN.4/Sub.2/2001/10, para.74 (arguing for a *sui generis* regime of protection in the area of pharmaceuticals that moves away from a situation of monopoly rights in such a crucial area of human existence).
[164] CESCR, 'Human Rights and Intellectual Property' (14 December 2001) UN Doc. E/C.12/2001/15.
[165] Paras.6 and 12. In 2005 the CESCR published General Comment 17 on creators' rights under article 15(1)(c) of the ICESCR. The comment established that: (1) IPRs and human rights are conceptually distinct; (2) corporations are not entitled to protection under the ICESCR and; (3) in any event, in a conflict between article 15(1)(c) rights and the right to access medical treatment the latter trumps the former. See CESCR, 'General Comment 17: The Right of Everyone to Benefit from the Protection of the Moral and

The right to health discourse that emerged in the UN human rights system further assisted the access to medicines movement by reframing the existing legal discourses concerning IPRs and medicines. Actors within the UN human rights system were able to draw upon the experiences of countries that had contested strict TRIPS enforcement in part through utilising right to health discourse – as well as empirical research by international NGOs and UN agencies like WHO and UNAIDS – to reveal the negative impact of TRIPS enforcement on access to medicines and hence the right to health.

The UN human rights discourse helped to shift the focus of the analysis in relation to the legal obligations of states with regard to access to medicines in a number of ways. First, human rights and IPR are conceptually distinct categories: the former are of an intrinsic and inalienable nature, whereas the latter are instrumental and temporary. Second, where there is a conflict between human rights and IPR, the former have primacy. Third, the TRIPS regime, as it is currently constituted, is not consistent with human rights standards, including the right to health. Fourth, there may be the need to restructure medical R&D towards the treatment of neglected diseases and the health needs of the world's poor. The UN human rights discourse does not regard IPR protection for pharmaceuticals in itself as *inherently* incompatible with the right to health, but it does regard the latter as a higher order legal norm and that the former needs to be both reformed and supplemented in order for states to be able to fulfil their right to health obligations.

In adopting this position the UN human rights system contested the IPR regime's natural rights framework by undermining the claim that IPRs are the same as creators rights under article 15(1)(c) of the ICESCR. It also challenged the utilitarian justifications for IPR by highlighting the negative consequences for public health resulting from TRIPS and suggested that the TRIPS flexibilities, as they were then (under)applied, may not be appropriate to address right to health obligations. However, all the positions adopted within the UN human rights system were of a non-binding nature. Whilst the right to health does not have strong international enforcement machinery, IPRs under TRIPS can be enforced through the panel and Appellate Body reports of the WTO. The question that remained, therefore, was

Material Interests Resulting from Any Scientific, Literary or Artistic Production of Which He Is the Author (Art. 15(1)(c))' (12 January 2006) UN Doc E/C.12/GC/17, paras.1, 2, 3 and 7.

whether the human rights and right to health-based critiques of TRIPS could be incorporated into the TRIPS agreement itself.

4.4.2 THE RIGHT TO HEALTH AND THE REFORM OF TRIPS

The social movement activism orientated around the constitutionally grounded right to health in South Africa, Brazil and India, combined with the activities of public health-orientated transnational advocacy networks and UN human rights agencies, is widely regarded to have been a significant causal factor leading to the TRIPS Council organising a special session in Doha in November 2001 concerning the relationship between TRIPS and public health. The South African case, in particular, which generated significant international civil society mobilisation and global media attention, is regarded as playing a catalytic role that led to the Doha negotiations.[166] Katherine Young suggests that while the rhetorical force of 'the right to health' is not exactly traceable in relation to the Doha negotiations, it nonetheless 'undoubtedly created important legal and institutional channels'.[167] During the negotiations, international NGOs such as HAI, CPTech, MSF, Oxfam and the Third World Network (TWN) supported countries in the Global South through the provision of technical advice and by helping to garner global public support for their position.[168] On the eve of the TRIPS Council Meeting, Oxfam, MSF and TWN issued the following statement:

> Governments need a permanent guarantee that they can put public health and welfare of their citizens before patent rights, without having to face legal pressures or threat of trade sanction experienced by South Africa and Brazil ... People all over the world will be watching whether WTO member countries meet the challenge of tackling the global health crisis and demonstrate their commitment to the prevention of further unnecessary deaths.[169]

During the negotiations, a group of eighty countries, led by the Africa group as well as India and Brazil, were successful in their efforts to secure a declaration affirming an interpretation of TRIPS that would permit

---

[166] Erica George suggests that, in particular, the TAC legal and moral arguments around the right to health 'catalysed an international movement across the Global South which culminated in the adoption of the Doha Declaration'. George (n 111) 197.
[167] Young (n 123) 371.
[168] Matthews *Intellectual Property, Human Rights and Development* (n 69) 37.
[169] Quoted in Odell and Sell (n 49) 100.

them to pursue policies affording access to essential medicines without fear of retribution from other WTO members, despite the opposition to such proposals by the USA and the pharmaceutical industry.[170] In that sense, the Doha Declaration represented a 'real success' of the access to medicines movement and its campaign to reframe intellectual property as a public health and human rights issue.[171]

The Doha Declaration confirmed that 'the TRIPS Agreement does not and should not prevent Members from taking measures to protect public health'.[172] Furthermore, in language 'redolent of human rights and the right to health',[173] the declaration acknowledges 'the right to protect public health and, in particular, promote access to medicines for all'[174] and 'the right' to do so using TRIPS flexibilities and parallel imports[175] as well as the right of Member States to determine 'what constitutes a national health emergency or other circumstances of extreme emergency'.[176] Whilst these principles confer rights on *states* as opposed to *individuals* it can be argued that they comport with the notion of the extraterritorial obligation of the WTO and its members to *respect* the right to health, that is, the obligation to refrain from enforcing decisions that interfere with the right to health in other countries.[177]

A point unresolved by the Doha Declaration was the problem posed by the TRIPS agreement, which requires compulsory licencing to be limited to usage 'predominantly for the supply of the domestic market'.[178] This had the practical effect of preventing exports of generic drugs to countries with little or no local manufacturing capacities. Paragraph 6 of the Declaration acknowledged that such countries faced difficulties in making effective use of compulsory licensing under TRIPS and set a deadline of the end of 2002 to find an expeditious solution to the problem. The paragraph 6 negotiations were held within the TRIPS

---

[170] ibid, 107; Sell, 'TRIPS and Access to Medicines' (n 26) 515.
[171] Matthews, *Intellectual Property, Human Rights and Development* (n 69) 37.
[172] World Trade Organization, Declaration on the TRIPS Agreement and Public Health, WT/MIN(01)/DEC2 (20 November 2001), (hereafter 'Doha Declaration') para.4 www.wto.org/english/thewto_e/minist_e/min01_e/mindecl_trips_e.htm accessed 26 November 2013.
[173] Forman, '"Rights" and "Wrongs"' (n 116) 44.   [174] Doha Declaration (n 172) Para 4.
[175] ibid, paras.5(b) and 5(d).   [176] ibid, para.5 (c).
[177] In General Comment 14 (n 134) the CESCR state that 'To comply with their international obligations in relation to article 12, States parties have to respect the enjoyment of the right to health in other countries ... In relation to the conclusion of other international agreements, States parties should take steps to ensure that these instruments do not adversely impact upon the right to health', para.39.
[178] TRIPS (n 12) art.31(f).

council from early 2002 onwards, culminating in the WTO decision at the end of August 2003, which was formally incorporated into TRIPS as Article 31 *bis* in December 2005. Whereas the original TRIPS negotiations were characterised mainly by support of northern states by national industry associations and other business interests, the paragraph 6 negotiations involved a far higher degree of involvement by NGOs such as Oxfam, MSF and CPTech, assisting the southern countries.[179] After heated negotiation, the final decision provided a temporary waiver of Members' obligations under Article 31(f) until such time as that article is amended. Countries eligible to import under the agreement include any 'least- developed' WTO Member in the case of a national emergency or other circumstances of extreme urgency.[180]

The decision also sets out a number of strict conditions which permit least developed and other countries to import generic medicines made under compulsory license. These requirements have been described as 'cumbersome and complex procedural impediments'[181] that require countries wanting to import generic brands to 'jump through multiple hoops to prove they are truly in need, unable to afford patented drugs and incapable of producing the medicines domestically'.[182] Despite the significant role played by the NGOs and developing countries in redefining the rules in international IP law, Duncan Matthews concludes that the final outcome of the paragraph 6 decisions 'was characterised by the dominance of the US and EC as key institutional actors, assisted by industry groups'.[183]

In the run-up to the Doha Declaration, international NGOs co-ordinated their activities effectively and collaborated with developing countries on the access to medicines issue. However, during the Doha negotiations the coalition began to fragment, with some international NGOs prioritising other trade issues and focusing less specifically on TRIPS. Matthews

---

[179] Duncan Matthews, 'Is History Repeating Itself? The Outcome of the Negotiations on Access to Medicines, The HIV/AIDS Pandemic and Intellectual Property Rights in the World Trade Organisation' (2004) *Law, Social Justice & Global Development* 1, 6.
[180] Implementation of paragraph 6 of the Doha Declaration on the TRIPS Agreement and public health, available at www.wto.org/english/tratop_e/trips_e/implem_para6_e.htm IP/C/W/405, 30 August 2003, para.1(b). The WTO General Council formalised the 2003 decision in a resolution amending TRIPS and incorporating article 31 *bis*, WT/L/641 (6 December 2005).
[181] George (n 111) 193.
[182] Naomi Klein, 'Bush's AIDS Test', *The Nation* (27 October 2003) www.thenation.com/article/bushs-aids-test# accessed 28 November 2013.
[183] Matthews, Is History Repeating Itself?' (n 179) 13. See also Correa (n 65) 341.

suggested that it remained impossible to keep a broader NGO coalition together after the Doha Declaration given the specialist nature of the TRIPS negotiations. Whilst international NGOs continued to support developing countries in the negotiations, 'the increasingly technical nature of the proposals being formulated and differences in the viewpoints of NGOs and developing countries on what could be achieved altered the dynamics of the access to medicines campaign'.[184] The Post-Doha negotiations became increasingly bogged down in technical IPR issues to the extent that the human rights content of the original challenge to TRIPS was progressively suppressed.

This technicalisation of the access to medicines campaign has implications for its counter-hegemonic potential. Part of the potential of the right to health lies in the fact that it is an easily comprehended rhetorical demand, which can be used to mobilise large numbers of people into a campaign for public health. This comports with Gramsci's vision of a counter-hegemonic project as entailing the construction of 'an intellectual-moral bloc which can make politically possible the intellectual progress of the mass and not only small intellectual groups'.[185] The critical energy that characterised the mass mobilisations in South Africa, Brazil and elsewhere around the 'right to health' was being progressively surpassed with an exclusionary and technical discourse accessible only to international lawyers with expertise in IP law. The discourse of the Doha negotiations is reflective of what Andrew Lang calls the neo-liberal 'formal-technical turn' in the global trade regime, that is, the specialisation and legalisation of trade politics.[186] Lisa Duggan has argued that central to neo-liberal discourse is the presentation of economic policy as a matter of neutral, technical expertise 'separate from politics and culture, and not properly subject to specifically political accountability or cultural critique'.[187] Given the presence of such discourse in the post-Doha negotiations, we may share Ronen Shamir's concern that transnational campaigns may often result in the 'perceived grievances of oppressed, marginal, and exploited populations ...' being transformed into the

---

[184] Matthews, *Intellectual Property, Human Rights and Development* (n 69) 40.
[185] Antonio Gramsci, *Selections from the Prison Notebooks of Antonio Gramsci* (Quintin Hoare and Geoffrey Nowell Smith eds. and trans.) (Lawrence & Wishart, 1971) 332–333.
[186] Andrew Lang, *World Trade Law after Neoliberalism: Re-Imagining the Global Economic Order* (Oxford University Press, 2011) 103 and 221.
[187] Lisa Duggan, *The Twilight on Equality* (Beacon Press, 2003) xiv.

language of 'hegemonic rational organizational and managerial systems characteristic of contemporary capitalism'.[188]

Whilst the Doha Declaration may have at least part of its origins in right to health discourse and may contain some residues of that discourse, public-interest-orientated human rights discourse was sidelined in favour of the property-oriented, technical and economically instrumental language of IPRs advanced by the hegemonic bloc. If the Doha Declaration does embody any right to health standard it is one that is closer to the WTO's right to health discourse than the more substantive right to health obligations articulated by the UN human rights bodies. The burdensome requirements for states to effectively utilise TRIPS flexibilities indicate that IPR protection remains the dominant concern within TRIPS, and public health considerations are at best secondary. The counter-hegemonic right to health discourse that subverts IPR to higher order human rights norms has been replaced with an IP regime in which human rights concerns are at best *post hoc* considerations and exceptions. In that sense, whilst the Doha Declaration opens up further possibilities for contestation of the IPR regime, it can also be understood as an instrument of *trasformismo*, that is, in the sense that it co-opts subaltern movements to work within the neoliberal discursive framework.

4.4.3 AFTER DOHA: THE RIGHT TO HEALTH IN THE 'TRIPS PLUS' ERA

The Post-Doha landscape opened up opportunities and new challenges for access to affordable medicines in the face of renewed efforts by northern states and the pharmaceutical industry to limit the scope of southern states' use of the Declaration and TRIPS flexibilities. Despite the limitations of the Doha Declaration identified above, it did nonetheless create openings for southern states to contest IPRs protection.[189] Beginning in 2002 a number of developing and middle-income countries began to issue compulsory licences for HIV/AIDS drugs, including Brazil, Cameroon, Ghana, Indonesia, Malaysia, Mozambique, Rwanda, Thailand, Zambia and Zimbabwe. As a consequence, the

---

[188] Ronen Shamir, 'Corporate Social Responsibility: A Case of Hegemony and Counter-hegemony, in Boaventura De Sousa Santos and Cesar A Rodriguez-Garavito (eds.) (n 114) 113.
[189] George (n 111) 191–192.

price of antiretroviral medicines has fallen in these countries.[190] The Indian Supreme Court has also cited the Doha Declaration, in conjunction with the TRIPS flexibilities, to affirm a rejection of a product patent application by Novartis for a modified version of a drug used to treat leukaemia.[191]

The Doha Declaration has provided an important moral and legal anchor for NGOs in their advocacy regarding access to medicines and has also been used by national courts to allow for more flexible interpretations of TRIPS. For example, national governments have also used the threat of compulsory licencing to negotiate with pharmaceutical companies to lower the price of products.[192] Furthermore, it is also noteworthy that there have been no complaints against a WTO Member filed under the Dispute Settlement Understanding on matters relating to IPRs and public health since the adoption of the declaration.[193]

In spite of these successes, a 2011 report on the impact of the Doha Declaration suggests that on the whole the Doha Declaration has not triggered a widespread incorporation and use of TRIPS flexibilities to increase access to medicines.[194] This relates in part to the declaration's complex and cumbersome procedural requirements identified in the previous section, particularly for countries with little or no domestic pharmaceutical manufacturing ability.[195] However, an even more potent factor undermining the potential of TRIPS flexibilities has been the proliferation of enhanced IPRs standards in regional and bilateral FTAs.[196] These agreements are referred to as 'TRIPS Plus' because they contain IPRs protection standards more rigorous than those contained within TRIPS, oblige poorer countries to implement TRIPS fully before the end of their transition periods, or require countries to accede to other

---

[190] Cynthia M Ho, A New World Order for Addressing Patent Rights and Public Health (2007) 82 *Chicago Law Review* 1469 cited in Laurence R Helfer, 'Pharmaceutical Patents and the Human Right to Health: The Contested Evolution of the Transnational Legal Order on Access to Medicines' in Terence C Haliday and Gregory Shaffer (eds.) *Transnational Legal Orders* (Cambridge University Press, 2014) 330.

[191] *Novartis AG v. Union of India & Others* (2013) Civil Appeal Nos. 2706–2716 of 2013, paras.64, 65, 76, 77 and 101.

[192] Brazil and Thailand are two examples of countries that have used the threat of compulsory licences to negotiate lower prices for drugs. See Helfer and Austin (n 1) 127–134.

[193] Carlos Correa and Duncan Matthews, *The Doha Declaration Ten Years on and Its Impact on Access to Medicines and the Right to Health: Discussion Paper* (UNDP, 2011) 19.

[194] ibid, 20.   [195] Helfer and Austin (n 1) 132–134.

[196] Correa and Matthews (n 193) 20; UNCHR, 'Report of the Special Rapporteur' (n 13) para.94.

multilateral IPR agreements.[197] The USA itself has signed bilateral and regional trade agreements containing TRIPS-plus standards with over sixty countries, 'many of which are developing countries with extremely high disease burdens, including HIV/AIDS'.[198] Enhanced IPR standards have also been incorporated into regional trade agreements such as the Central American Free Trade Agreement (CAFTA), the Free Trade Agreement of the Americas (FTAA) and the European Union (EU).[199] Currently, negotiations are underway on two FTAs – the Trans-Pacific Partnership Agreement (TTPA) and the Transatlantic Trade and Investment Partnership (TTIP) – which could further deepen IPR protections at the international level.[200]

Despite the challenges posed by the mobilisation of 'TRIPS Plus', the right to health frame has been utilised by the UN Human Rights System and public health activists to critique and mobilise against these measures. United Nations Special Rapporteurs and treaty bodies like the CESCR have published an array of reports and concluding observations expressing concern about the potential and actual negative impact that TRIPS plus agreements have had on the availability of affordable medicines in developing countries.[201] Activist networks and southern countries have evoked the right to health to resist the imposition of TRIPS plus conditions. In 2006 the Ministers of Health of ten South American countries issued a joint declaration on intellectual property, committing themselves to avoid 'TRIPS plus' provisions in bilateral and regional trade agreements, to facilitate the use of compulsory licensing and parallel importing and to avoid broadening the scope of patentability and the extension of

---

[197] ibid, 40. [198] Forman, 'Trade Rules' (n 110) 342.
[199] Forman, 'From TRIPS-Plus' (n 9) 352.
[200] HAI Europe and Oxfam, *Trading Away Access to Medicines – Revisited: How the European Trade Agenda Continues to Undermine Access to Medicines* (2014) www.oxfam.org/sites/www.oxfam.org/files/file_attachments/bp-trading-away-access-medicines-290914-en.pdf accessed 12 March 2016; Fran Quigley, 'The Trans-Pacific Partnership and Access to Medicines' (*Health and Human Rights Journal*, 18 June 2015) www.hhrjournal.org/2015/06/the-trans-pacific-partnership-and-access-to-medicines/ accessed 12 March 2016.
[201] See, for example, CESCR, 'Concluding Observations of the CESCR: Ecuador' (7 June 2004) UN Doc. E/C.12/1/Add.100, para.30; CESCR, 'Concluding Observations of the CESCR: Costa Rica' (4 January 2008) UN Doc. E/C.12/IMD/CO/4, para.27; UNCHR, 'Report of the Special Rapporteur' (n 13) para.108; UNCHR, 'Report of the Special Rapporteur on the right of everyone to the enjoyment of the highest attainable standard of physical and mental health, Anand Grover: Mission to Guatemala' (16 March 2011) UN Doc. A/HRC/17/25/Add.2, para.82; UNCHR, 'Report of the Special Rapporteur on the right of everyone to the enjoyment of the highest attainable standard of physical and mental health, Anand Grover: Mission to Vietnam' (4 June 2012) UN Doc. A/HRC/20/15/Add.2, paras.35–41.

patentable areas.²⁰² The Declaration stated that 'access to medicines and critical raw materials is an integral part of the right to health, which is a basic human right of every individual and a fundamental prerequisite that governments have a duty to ensure.'²⁰³

Many of these countries are part of the Bolivarian Alliance for the Peoples of Our Americas (ALBA).²⁰⁴ ALBA emerged as an alternative to the United States Promoted FTAA and other FTAs which contained significant 'TRIPS Plus' components. ALBA is opposed to the enhanced protection of intellectual property which the northern countries imposed in these treaties.²⁰⁵ Principle 18 of ALBA's Fundamental Principles of the People's Trade Agreement prioritises 'the right to development and health over the intellectual property of multinationals'.²⁰⁶ The right to health has also served as an important discursive tool for campaigning networks such as the coalition of Medicines Transparency Alliance (MeTA) Philippines, the Coalition for Health Advocacy and Transparency and the EU-ASEAN FTA Campaign Network, which mobilised together to oppose TRIPS Plus provisions within the EU-ASEAN FTA.²⁰⁷ Lisa Forman has documented the critical role performed by a Right to Health Impact Assessment conducted by the National Human Rights Commission of Thailand in mobilising Thai society against the stringent IPR conditions being negotiated as part of a FTA with the USA.²⁰⁸

It is clear from the above examples that the right to health continues to operate as an important counter-frame to IPRs provisions in TRIPS and TRIPS Plus agreements. However, it cannot be suggested that all right to health-based opposition to intellectual property is inherently

---

²⁰² Martin Khor, 'South American Ministers Vow to Avoid TRIPS-plus Measures' (*Third World Network*, 1 June 2006) www.twnside.org.sg/title2/twninfo414.htm accessed 2 December 2013.
²⁰³ ibid.   ²⁰⁴ See Chapter 1.
²⁰⁵ Venezuelan Bank of External Commerce (Bancoex), 'What Is the Bolivarian Alternative for Latin America and the Caribbean?' *Venezuela Analysis* (5 February 2004) available at http://venezuelanalysis.com/analysis/344 accessed 3 May 2013.
²⁰⁶ ALBA, 'Fundamental Principles of the Peoples Trade Treaty' (Cochabamba, Bolivia – 17 October 2009) www.alba-tcp.org/en/contenido/fundamental-principles-tcp accessed 2 December 2013.
²⁰⁷ See Medicines Transparency Alliance (MeTA) Philippines, the Coalition for Health Advocacy and Transparency and the EU-ASEAN FTA Campaign Network 'Defend the Right to Health and Access to Affordable Medicines! No to TRIPS Plus Provisions in Bilateral FTAs!' (2011) http://focusweb.org/philippines/fop-articles/statementsdeclarations/568-defend-right-to-health-and-access-to-affordable-medicines accessed 26 November 2013.
²⁰⁸ Forman, 'From TRIPS-Plus' (n 9) 356–362.

counter-hegemonic. Angelina Snodgrass Godoy has published a thought-provoking monograph on human rights-based strategies to challenge IPR protection in relation to CAFTA.[209] Whilst Godoy acknowledges the international access movement's argument that intellectual property can drive up prices by reducing competition, she suggests that the over-emphasis on competition as a key factor in securing affordable access is myopic in its focus and has diverted attention away from the proactive measures required of states to ensure access to affordable medicines.[210] Using the pharmaceutical markets in a number of Central American economies as her case study, Godoy illustrates that these markets are not truly innovative in the first place, and are characterised by the production of aggressively marketed 'me too' type medicines without significant clinical advantage.[211] Godoy also suggests that drug markets, especially in small countries like those of Central America, are not particularly competitive even without the introduction of IP.[212] Whilst the introduction of generic competitors can reduce the cost of medicines, this is by no means an automatic process. A strong regulatory state is also required to, *inter alia*, rigorously monitor the drug supply, ensure the safety of all drugs for human consumption, and establish an equitable and transparent bidding process for the government purchase of drugs.[213] Furthermore, Godoy suggests that IPRs were not a primary focus of concern for health activists in the Central American context. Instead, activism was orientated around ensuring access to primary health care for all or promoting improved outcomes through community-based, participatory interventions like the delivery of potable water or improved sanitation systems. The disconnection between these local concerns and international focus on IPRs led to a lack of uptake of struggles against IPRs by local activists.[214]

Of course, there is not necessarily a dichotomy between adopting a human rights-based strategy that both opposes stringent intellectual property protection *and* demands that the state adopts a plethora of appropriate measures to ensure the availability, accessibility, acceptability and quality of medicines.[215] Nor is there a contradiction in raising these demands within the broader context of campaigning to improve the 'underlying determinants of health'.[216] However, it is worth recalling here the neo-Gramscian insight that the domain of global civil society is 'a

---

[209] See generally, Godoy (n 2).   [210] ibid, 15.   [211] ibid, 66–67.   [212] ibid, 67–71.
[213] ibid, 71–77.   [214] ibid.
[215] CESCR, 'General Comment 14' (n 134) paras.12 and 33.   [216] ibid, para.11.

discursive space, which helps to reproduce global hegemony'.[217] Social movements must therefore be aware that 'they are positioned within the hegemonic constellation, and ... that there are structural and discursive forces at play, of which the very framework of global civil society is itself a part, and which social movements themselves may actually be actively reproducing, rather than challenging'.[218] What is relevant for the purposes of this book is that organisations like MSF and Oxfam's overemphasis on competition as the key to affordable medicines may inadvertently promote the neo-liberal faith in the 'free' market and divert attention from the critical task of building state- and community-led public health systems. The excessive focus by international NGOs on IPRs to the detriment of other public health issues may also constitute a stumbling block to creating effective and truly representative international alliances that could constitute a truly counter-hegemonic movement.[219]

The above underscores a recurrent theme in this chapter: that the market, at least as it is currently constituted, has been shown to be an abject failure as an adjudicator of health rights, particularly in relation to its inability to stimulate R&D for neglected diseases affecting the peoples of the Global South. As Benjamin Mason Meire argues

> ... even assuming that a developing state were permitted to engage in compulsory licensing or parallel importation ... this would not be enough. From a public health perspective, developing states need research in those diseases that most affect them – not just medicines for the diseases endemic to developed states. Only through research mechanisms can public health create appropriate life-saving medications, incentivising research for medicines necessary to treat "tropical disease" and making these medications physically and economically accessible to all.[220]

The next section will explore the extent to which the right to health has been able to frame and advance demands for such alternative research mechanisms.

4.4.4 BEYOND TRIPS: THE RIGHT TO HEALTH AND ALTERNATIVES TO IP

Even to the extent that southern states can make use of parallel importing and compulsory licences in the aftermath of Doha, such strategies do not

---

[217] Lucy Ford, 'Challenging Global Environmental Governance: Social Movement Agency and Global Civil Society' (2003) 3(2) *Global Environmental Politics* 120, 129.
[218] ibid.   [219] Godoy (n 2) 100–106.   [220] Meire (n 102) 730.

address the other concern that has heavily featured in the right to health critiques of extant IP protection: the lack of R&D for diseases that afflict the global poor. This 'global drug gap' has sometimes been expressed as the '10/90 disequilibrium', which denotes that only 10 percent of global health research is committed to conditions that account for 90 percent of the global disease burden.[221] TRIPS flexibilities, even when successfully utilised, cannot address the global drug gap. As the 2001 Report of the High Commissioner made clear, it may be necessary to supplement the patent system with alternative institutional arrangements to stimulate medical research so that states can comply with their obligations to realise the right to health.

Mere recourse to compulsory licencing and parallel importing without sustained public investment in medical provision and R&D can easily look like forms of trade protectionism that accept the basic contours of the current medical model. In the context of access to medicines, this requires not only the adoption of *post hoc* measures to mitigate the potential negative effects of monopoly pricing but also a challenge to the very idea that monopoly rights on medical knowledge offers the best solution to global medical needs.

Efforts towards this paradigm shift at the global level have been underway since 1999 when MSF set up the Neglected Diseases Group (NDG) to address the diseases that disproportionately affect poor populations. The NDG began to discuss global legal reforms aimed at transforming the medical R&D paradigm.[222] Further opportunities were opened up in 2003 when the WHA passed a resolution that led to the creation of the Commission on Intellectual Property Rights and Public Health (CIPIH) to collect data and proposals on incentives for the creation of health technologies for developing countries.[223] In 2005, a coalition of more than 150 NGOs including MSF, Oxfam, Health Action

---

[221] Médecins sans Frontières, Access to Essential Medicines Campaign & Drugs for Neglected Diseases Working Group, 'Fatal Imbalance: The Crisis in Research and Development for Drugs for Neglected Diseases' (2001), 10 available at www.doctorswithoutborders.org/publications/reports/2001/fatal_imbalance_short.pdf accessed 10 May 2013.

[222] Rachel Kiddell-Monroe, Johanne Helene Iversen and Unni Gopinathan, 'Medical R&D Convention Derailed: Implications for the Global Health System' (2013) *Journal of Health Diplomacy* 1 www.hsph.harvard.edu/global-health-and-population/files/2013/06/Medical-R-and-D-Convention-Derailed-Implications-For-The-Global-Health-System-3.pdf accessed 1 December 2013.

[223] WHA, 'Intellectual Property Rights, Innovation and Public Health' (28 May 2003) WHA56.27.

International and Knowledge Ecology International called on the WHO to consider a proposal for the Medical Research and Development Treaty (MRDT).[224]

The treaty's advocates argued that the global framework for supporting medical R&D had a number of profound flaws, including problems of access to medicine, costly and excessive marketing of products, barriers to follow-on research, incentives to research products that offer little therapeutic benefit and scant investment in treatments for the poor, basic research or public goods.[225] The treaty proponents argued that a new approach was needed to support medical R&D to 'reconcile different policy objectives, including the promotion of both innovation and access, consistent with human rights and the promotion of science in the public interest'.[226]

The treaty's core goal was to advance the promotion and sharing of medical research and investment in response to the greatest global health needs.[227] It aimed to do so by setting out minimum obligations to finance R&D based on internationally agreed-upon proportions of nations' Gross Domestic Product[228] and through creating a system of tradable 'credits' for investments in priority R&D, traditional medical knowledge and open public goods.[229] In relation to intellectual property, the proposed treaty required that Member States forgo certain WTO TRIPS dispute resolution cases as well as bilateral or regional trade sanctions in areas where compliance with the terms of the MRDT provides a superior framework for supporting innovation.[230] Members are also encouraged to adopt 'minimum exceptions to patents rights for research purposes' within five years of ratification and forego patent applications for inventions based upon data from certain open or 'public goods databases'.[231]

The MRDT arguably provided a rudimentary basis for an international counter-hegemonic framework for medical innovation. Rather than relying predominantly on profit-orientated private sector investment, the treaty placed considerable emphasis on the role of public sector investment in medical research orientated towards global health needs rather

---

[224] Letter to Ask World Health Organization to Evaluate New Treaty Framework for Medical Research and Development (24 February 2005) available at www.cptech.org/workingdrafts/rndsignonletter.html accessed 3 May 2013.
[225] ibid. [226] ibid.
[227] Article 2.1 Medical Research and Development Treaty (MRDT) (draft 4, 7 February 2005) www.cptech.org/workingdrafts/rndtreaty4.pdf accessed 3 May 2013.
[228] ibid, article 4.2. [229] ibid, article 12. [230] ibid, articles 2.3 and 16(d).
[231] ibid, articles 14.1 and 14.2.

than simply on the most profitable diseases. The treaty expressly required human rights and the sharing of scientific benefits to take legal precedence over IPRs, allowing the latter in medical research only under limited conditions. The MRDT thus inverted the TRIPS regime, in which IP protection in medicines is regarded as *prima facie* justified with some exceptions for public health considerations. By contrast, the MRDT has human rights and public health at the centre of its objectives and only allows for IP protection in medical research when it can be justified as the best strategy for the realisation of the treaty's objectives.

In 2006 the WHA established an Intergovernmental Working Group on Public Health, Innovation and Intellectual Property to consider and implement the CIPIH's recommendations[232] and in 2008 adopted a Global Strategy and Plan of Action on Public Health, Innovation and Intellectual Property.[233] The Global Strategy acknowledges that intellectual property is 'an important incentive for the development of new health-care products' but also that the incentive 'alone does not meet the need for the development of new products to fight diseases where the potential paying market is small or uncertain'.[234] The strategy therefore acknowledges the shortcomings of the IPRs regime's commercial orientation as identified by right to health advocates and voices within the UN human rights system. The strategy also quotes provisions of the UDHR on sharing scientific benefits and protection of moral and material interest resulting from scientific production[235] and states that 'the enjoyment of the highest attainable standard of health is one of the fundamental rights of every human being'.[236] Reflecting a key demand of the right to health activists, the strategy calls for states to develop proposals for 'health-needs driven research', including 'where appropriate, addressing the de-linkage of the costs of research and development and the price of health products'.[237]

The text of the strategy indicates the success that the global public health movement has had in framing access to medicines as a human rights issue on the international stage. However, the strategy only constitutes a starting point. Its lack of concrete targets or proposals for financing

---

[232] WHA, 'Public Health, Innovation, Essential Health Research and Intellectual Property Rights: Towards a Global Strategy and Plan of Action' (27 May 2006) WHA59.24.
[233] WHA, 'Global Strategy and Plan of Action on Public Health, Innovation and Intellectual Property' (24 May 2008) WHA61.21.
[234] ibid, annex, para.7.   [235] ibid, annex, para.10.   [236] ibid, annex, para.16.
[237] ibid, annex, para.4.

mechanisms has prevented effective utilisation or adoption of accountability measures.[238] On 26 May 2012 the WHA adopted a resolution calling for an inter-governmental meeting to examine the Consultative Expert Working Group on Research and Development's (CEWG) proposals to start multilateral negotiations for the possible adoption of a binding convention on health R&D.[239] However, efforts towards this goal have been derailed by a combination of pharmaceutical lobbying, opposition from northern states (particularly the USA) and the exclusion of civil society from policy and decision-making in the inter-governmental processes.[240]

In November 2012, twenty-five WHO Member States, led by the USA, drafted a resolution calling for the postponing of discussions on the treaty until 2016,[241] although at the 2013 WHA the USA signalled a willingness to reopen debate on a treaty before the three years were up.[242] The future of the global R&D treaty therefore remains open. However, given that the pharmaceutical companies and their allies have consistently opposed attempts to de-link medical prices from R&D – on the basis that they might undermine patents and corporate profits – and the USA has opposed attempts to impose binding obligations upon states to fund research, it is likely that the final content of any R&D treaty is likely to be far removed from what right to health campaigners have argued for. It is also unlikely that it will have the effect of displacing IPRs as the dominant framework for medical R&D. Whilst human rights and public health-based campaigns aimed at establishing an alternative paradigm for medical research on the global stage are likely to be forestalled in inter-governmental settings, there may be more fruitful possibilities for expanding forms of 'south-south' collaboration in medical research capacity building.[243] For example, Erika

---

[238] Marks and Benedict (n 102) 322.
[239] WHA, 'Follow-up of the Report of the Consultative Expert Working Group on Research and Development: Financing and Coordination' (26 May 2012) WHA65.22.
[240] Monroe et al. (n 222) 2–3.
[241] James Love, 'WHO Negotiators Propose Putting off R&D Treaty Discussions until 2016' (*Knowledge Ecology International*, 28 November 2012) http://keionline.org/node/1612 accessed 1 December 2013.
[242] Rachel Marusak Hermann, 'World Health Assembly: Members Debate US Proposed Advisory Meeting on Health R&D' (IP Watch, 24 May 2013) www.ip-watch.org/2013/05/24/world-health-assembly-members-debate-us-proposed-advisory-meeting-on-health-rd/ accessed 1 December 2013.
[243] Sachin Chaturvedi and Halla Thorsteinsdóttir, 'BRICS and South-South Cooperation in Medicine: Emerging Trends in Research and Entrepreneurial Collaborations' (RIS Discussion Paper no. 177, March 2012) www.ris.org.in/images/RIS_images/pdf/dp177_pap.pdf accessed 1 December 2013.

George notes that countries across the Global South are 'cooperating in order to enhance access by increasing their capabilities to conduct research, develop, produce, and distribute medicines through the Technological Network on HIV/AIDS'.[244] Furthermore, as HAI and Oxfam note, there are a number of projects underway aimed at advancing technologies tailored to the public health needs of the Global South:

> New approaches to biomedical innovation are based on sharing knowledge and data, rather than shrouding in secrecy and IP protection. Increasingly, public and private R&D initiatives engage in collaborative and open forms of innovation that allow for open access to research results ... New product development partnerships ... financing mechanisms ... medical patent pools, open data pools and prize funds have been created or conceived. These could generate and ensure access to technologies to meet the public health needs of LMICs [low and middle income countries].[245]

The transformative capacity of the right to health as a component of the global justice struggle for access to medicines will depend in part on its ability to spur on and lend authority and legitimacy towards these new ventures.

4.5 CONCLUSION

The way in which the international protection of intellectual property is framed and understood has undergone dramatic transformation since the entry into force of TRIPS in 1995. In the negotiations leading up to TRIPS the pharmaceutical industry was successful in narrowly framing international IP protection as an issue concerning commercial property interests and free trade. Health-orientated national social movements, international NGOs and actors within the UN human rights system have been successful in reframing IPRs as an issue that has serious human rights implications, particularly by drawing attention to the adverse consequences that stringent protection has had on marginalised, disadvantaged and poor populations in the Global South. Despite being 'one of the most extensive and complex human rights in the international lexicon',[246] the right to health has functioned as an easily comprehended rhetorical demand that can be grasped and acted upon by anybody, and hence has been used to broaden and strengthen a global public health

---

[244] George (n 111) 192.   [245] HAI and Oxfam (n 200) 10.
[246] Paul Hunt, cited in O'Connell (n 96) 191.

movement and develop its mass character.[247] When an individual or community suffering from HIV is denied access to available treatment due to its prohibitively high cost, the claim that such an individual or community has had their 'right to health' violated is one that has intuitive moral resonance, regardless of the potentially complex legal conception of that right. The UN human rights system has also lent legitimacy and moral weight to the global access campaign by producing various statements and decisions stipulating that physical and affordable access to medicines is a key component to the human right to health. It has also stipulated the distinctions between human rights and IPRs, and argued that the former must be afforded primacy where they come into conflict with the latter, thus promoting public health needs over private commercial interests.

Despite the right to health being developed as a counter-frame to IPRs within the domain of global civil society, attempts to institutionalise its primacy over the international protection of IPRs have been less successful. Despite the openings made available by the Doha Declaration, the text has significant shortcomings as a means to realise the right to health – particularly for countries with limited domestic drug manufacturing capacities – and has been further undermined by the adoption of 'TRIPS-Plus' provisions in FTAs. Furthermore, as Godoy argues, some international NGOs have actually often framed their opposition to IPRs in a way that re-inscribes the neo-liberal faith in the 'free market', despite the fact that drug markets, even without IPRs, are often uncompetitive, expensive and not particularly innovative. The presence of such discourse within the access to medicines movement highlights some of the critical differences that may exist between professional development NGOs and grassroots human rights and social movements, thereby undermining the capacity to build a genuine transnational movement around health rights. Such discordance undermines the possibility for the access movement to form a viable counter-hegemony, spanning south and north, capable of drawing together 'subaltern social forces around an alternative ethico-political conception of the world'.[248]

In order to promote a genuine alternative to the neo-liberal private market model, elements within the access to medicines movement have attempted to articulate new mechanisms for the creation and distribution

---

[247] People's Health Movement (n 98) 1.
[248] William K Carroll, 'Hegemony, Counter-Hegemony, Anti-Hegemony' (2006) 2(2) *Socialist Studies* 9, 21.

of medicines that are congruent with a human rights framework. Such attempts have been made within the context of the WHO, but efforts towards a new global R&D treaty have been derailed by northern states, most notably the USA. Given the prospects for the undermining and *trasformismo* of right to health discourse within multilateral forums such as the WTO and WHA, it is arguable that new forms of south-south co-operation and open research projects will be required to achieve a medical funding model more attuned to the demands of the right to health.

5

A COMMODITY OR A RIGHT? EVOKING THE HUMAN RIGHT TO WATER TO CHALLENGE NEO-LIBERAL WATER GOVERNANCE

'Water is of course the most important raw material we have in the world today. It's a question of whether we should privatize the normal water supply for the population and there are two different opinions on this matter. The one opinion, which I think is extreme, is represented by the NGOs who bang on about declaring water a public right. That means that as a human being you should have a right to water. That's an extreme solution. And the other view says that water is a foodstuff like any other and like any other foodstuff it should have a market value'.[1]

'They wanted to take away our right to water. But they could not because they could never take away our thirst'.[2]

5.1 INTRODUCTION

Clean, safe water is the most basic necessity of life, yet it is estimated that 884 million people worldwide lack access to it.[3] More than 3.4 million people die each year from water- and sanitation-related causes, the vast majority of these in the developing world.[4] Lack of access to water is

---

[1] Peter Brabeck, chairman and former CEO of Nestle, the largest distributor of bottled water in the world. Quoted in Alison K Grass, 'How to Crash a Nestlé Waters Press Conference' (*foodandwaterwatch.org*, 5 April 2013) www.foodandwaterwatch.org/blogs/how-to-crash-a-nestle-waters-press-conference/ accessed 27 May 2013 (click on the third hyperlink for a video containing the full quote).

[2] Eduardo Hughes Galeano, Uruguayan Journalist and Activist. Quoted in Susan Spronk, 'International Solidarity for the Struggle for Water Justice in El Alto, Bolivia' (*Znet*, 10 May 2005). www.zcommunications.org/international-solidarity-for-the-struggle-for-water-justice-in-el-alto-bolivia-by-susan-spronk.html accessed 10 December 2013.

[3] WHO/UNICEF Joint Monitoring Programme for Water Supply and Sanitation, *Progress on Sanitation and Drinking Water 2010 Update* (WHO, 2010) 7.

[4] WHO, Safer Water, Better Health: Costs, Benefits, and Sustainability of Interventions to Protect and Promote Health (WHO, 2008) 12.

extremely uneven, both in and between nations. It is estimated that the water used on a typical United States golf course in a day could satisfy the daily water needs of 30,000 Africans.[5] Even where people have physical access to clean, safe water, the cost of water services may be prohibitively high. In 2014 tens of thousands of poor Detroit households had their water supplies cut off after they were unable to pay their water bills.[6] Many estimate that water shortages will affect more people in the future. A recent report suggests that within the space of two generations the majority of the 9 billion people on Earth will live with severe pressure on fresh water supplies as the result of climate change, pollution and the over-use of resources.[7]

It is clear that whilst water scarcity and unequal access to water are partly related to climatic and ecological factors,[8] they also reflect 'power relations and policy choices at national and international levels'.[9] According to neo-liberal discourse, the lack of access to water and the depletion of water resources can be explained primarily by governmental inefficiency in the management of water services and the failure to adequately price water. The solution is increased private sector participation in the water sector.[10] For activists within what will be called 'the water justice movement', the neo-liberal prescriptions for the global water crisis have worsened problems associated with access to water. The profit-driven approach is argued to have exacerbated inequalities in

---

[5] Richard P Hiskes, 'Missing the Green: Golf Course Ecology, Environmental Justice, and Local "Fulfillment" of the Human Right to Water' (2010) 32(2) *Human Rights Quarterly* 326, 326.

[6] UN News, 'In Detroit, City-Backed Water Shut-Offs "Contrary to Human Rights," say UN Experts' (UN News, 20 October 2014). www.un.org/apps/news/story.asp?NewsID= 49127#.VYvl0Vz4tlI accessed 26 June 2015. See further, Georgetown Law Human Rights Institute, *Tapped Out: Threats to the Human Right to Water in the Urban United States* (Georgetown Law, 2013).

[7] Fiona Harvey, 'Global Majority Faces Water Shortages "within Two Generations"' *Guardian* (24 May 2013).

[8] These factors should not, however, be thought of simply as 'natural' but rather as often created and exacerbated by human actions. As Vandana Shiva notes 'Deforestation and mining have destroyed the ability of water catchments to retain water. Monoculture agriculture and forestry have sucked ecosystems dry. The growing use of fossil fuels has led to atmospheric pollution and climate change, responsible for recurrent floods, cyclones, and droughts'. Vandana Shiva, *Water Wars: Privatization, Pollution and Profit* (South End Press, 2002) 2.

[9] Kevin Murray, 'Whose Right to Water?' (2003) *Dollars & Sense* (November/December edition) 23, 23.

[10] See *infra*, Part 2.

access, led to more expensive services, and undermined participation, accountability and democracy.[11]

This chapter will explore the ways in which 'right to water' discourse has intervened in the debate around the provision of water services. It will begin by identifying the political-institutional framework of global water governance – termed 'the neo-liberal water regime' – and will then identify a number of criticisms that have been levelled against this regime's impact on the provision of affordable water services. Following this, the neo-liberal water regime's framing of 'water as commodity' will be contrasted to the emerging counter-hegemonic discourse based around the idea of 'water as commons'. Having established the political contours of this dispute, the chapter will then consider the various manifestations of the 'right to water' within the debates around global water governance, including by anti-privatisation movements, United Nations bodies and neo-liberal actors such as the World Bank and the World Water Council (WWC). A case study from South Africa will then be used to illustrate some of the problems associated with the neo-liberal interpretation of the right to water. Finally, the chapter will conclude by critically evaluating the counter-hegemonic potential of the right to water.

5.2 THE POLITICAL-INSTITUTIONAL FRAMEWORK: THE NEO-LIBERAL WATER REGIME

In the 1980s privatised water systems were extremely rare, and international funding was directed exclusively to public entities until 1990.[12] The 1990s witnessed a global retreat from the subsidised public provision of water and concomitantly saw growing private sector provision of water services on a for-profit basis.[13] By the end of 2000 at least ninety-three countries had private sector involvement in their piped water services,[14] a transformation that affected over 2000 water and sewage projects[15]

---

[11] ibid.
[12] James Winpenny (World Panel on Financing and Infrastructure), *Financing Water for All* (2003) 7 www.unwater.org/downloads/FinPanRep_MainRep.pdf accessed 28 January 2014.
[13] Michael Goldman, 'How "Water for All!" Policy Became Hegemonic: The Power of the World Bank and Its Transnational Policy Networks' (2007) 38 *Geoforum* 786–800.
[14] Colin Kirkpatrick and David Parker, 'Domestic Regulation and the WTO: The Case of Water Services in Developing Countries' (2005) 28(10) *World Economy* 1491, 1495.
[15] ibid.

and made more than 460 million people dependent upon global firms for their water supply.[16]

This major shift in water policy was principally pursued by Transnational Corporations (TNCs) involved in energy and water service delivery. It was also actively promoted by the World Bank through its Water Policy Building Program and through loan conditionalities and structural adjustment programmes.[17] Through these strategies, the World Bank was able to recruit parliamentarians, policymakers, technical specialists, development agencies, think-tanks, civil society leaders, third world elites and NGOs into a 'transnational water policy network'.[18] For the purpose of this chapter this hegemonic bloc will be described as the 'the neo-liberal water regime'.

One particularly important node within this global network has been the WWC. Established in 1996, the WWC is an international water policy think-tank which provides decision makers with advice and assistance on global water issues.[19] It is the main organiser of the triennial global World Water Forum events and is 'intimately linked to the for-profit water services industry through its leadership and members'.[20] Hilal Elver notes that the development of private international think tanks such as the WWC – and others such as the Global Water Partnership and the Stockholm Water Symposium – in the 1990s coincided with the relative decline in the leadership of the UN system and heralded the beginning of the *de facto* privatisation of global water policy.[21] What unites the various

---

[16] Goldman (n 13) 790. See also Steven Shrybman, 'Thirst for Control: New Rules in the Global Water Grab' (2002) Prepared for the Council of Canadians. www.canadians.org/water/documents/campaigns-tfc.pdf accessed 30 May 2013.

[17] A study of World Bank loans between 1996 and 1999 reported that 58 percent had privatisation of water services as a condition. See Centre for Public Integrity, *Promoting Privatization* (*Centre for Public Integrity*, 4 February 2002) www.publicintegrity.org/2003/02/03/5708/promoting-privatization accessed 6 June 2013. See also Sara Grusky, 'The IMF, the World Bank and the Global Water Companies: A Shared Agenda' (2001) available at www.internetfreespeech.org/documents/sharedagenda.pdf accessed 6 June 2013.

[18] Goldman (n 13) 790.

[19] David Hall, Emanuele Lobina, Philipp Terhorst and Emma Lui, 'Controlling the Agenda at WWF – the Multinationals' Network' (2009) www.waterjustice.org/uploads/attachments/wwf5-controlling-the-agenda-at-wwf.pdf accessed 30 March 2013.

[20] Polaris Institute, 'The Corporate Stranglehold over the United Nations: How Big Business Already Wields Significant Power over the UN Water Agenda' (2009), 6 available at www.polarisinstitute.org/files/UNreport.pdf accessed 30 March 2013.

[21] Hilal Elver, 'The Emerging Global Freshwater Crisis and the Privatization of Global Leadership' in Stephen Gill (ed.), *Global Crises and the Crisis of Global Leadership* (Cambridge University Press, 2012) 117.

actors within the global water regime is the belief that the state alone is unable to provide the infrastructure and management required for effective and equitable water and sanitation services due to its inefficiency, corruption and lack of capital.[22] In place of state monopoly control of water these new actors insisted upon private sector participation in the water sector and the principle of cost recovery to ensure the profitability, and thus the sustainability, of the market-led model.[23]

5.2.1 CRITICISMS OF THE NEO-LIBERAL WATER REGIME

The introduction of private sector involvement in water services in the 1990s was justified on the basis that it would attract the finance required to build and expand infrastructure, particularly in developing countries. It was also argued that public utilities had largely failed to provide water to the poor due to corruption, inefficiency and lack of investment.[24]

Whilst the public provision of water can often leave a lot to be desired, the experience of the 1990s seriously called into question whether privatisation was the solution to the world's water problems.[25] First, the claim that private sector involvement would generate new sources of capital and relieve pressure from limited public budgets did not appear to materialise, on the whole. The main funding for services with private sector involvement came from the collection of service fees, direct subsidies to private companies and borrowing rather than private capital.[26] Second, privatisation measures in the 1990s usually produced exorbitant price increases

---

[22] Shiney Varghese, 'Transnational Led Privatization and the New Regime for the Global Governance of Water' (2003) 9/10 (1/2) *Water Nepal: Journal of Water Resources Development* 77, 80.

[23] Bronwen Morgan, 'The Regulatory Face of the Human Right to Water' (2004) 15 *Journal of Water Law* 179, 180.

[24] See discussion in Section 5.3.1.

[25] Experiences of water privatisation are, of course, not uniform. For an analysis of the various factors that may affect the success of a privatisation see Tanya Kapoor, 'Is Successful Water Privatization a Pipe Dream?: An Analysis of Three Global Case Studies' (2015) (40) *Yale Journal of International Law* 157.

[26] David Hall and Emanuele Lobina, *Pipe Dreams: The Failure of the Private Sector to Invest in Developing Countries* (World Development Movement, 2006) http://gala.gre.ac.uk/3601/1/PSIRU_9618_%2D_2006%2D03%2DW%2Dinvestment.pdf accessed 11 December 2013: Jose Esteban Castro, 'Neoliberal Water and Sanitation Policies as a Failed Strategy: Lessons from Developing Countries' (2008) 8(1) *Progress in Development Studies* 63, 68. See also Rebecca Brown, 'Unequal Burden: Water Privatization and Women's Human Rights in Tanzania' (2010) 18(1) *Gender and Development* 59, 62.

that resulted in water services becoming even more out of reach for the poorest sectors of society.²⁷ This created 'water poverty' for many poor citizens due to having to pay a much higher proportion of their household incomes towards water bills. Third, there is little evidence to indicate that privatised water services were any more efficient that public-sector controlled ones. Privatisation actually often resulted in shoddy maintenance of infrastructure, deterioration in the quality of water supplied and unsustainable price rises for services.²⁸ Fourth, privatisation reduced public control and accountability, with monopoly TNCs being accountable neither to their customers nor to civil society organisations in the countries where they operate.²⁹ This accountability deficit is exacerbated by the power disparity between states in the Global South and powerful TNCs.³⁰ While the latter often do not fulfil the terms of the concessions, they may seek compensation for the termination or modification of contracts which can be prohibitively expensive for many poorer countries.³¹

5.3 DUELLING VISIONS: 'WATER AS COMMODITY' VS. 'WATER AS COMMONS'

5.3.1 THE NEO-LIBERAL DISCURSIVE FRAMEWORK: WATER AS A COMMODITY

The move to privatised water provision in the 1990s was justified within the context of two discursive shifts. The first was the growing recognition that the earth's water is a scarce resource that needs to be efficiently managed in such a way that it would be available in sufficient quantities

---

[27] Castro, ibid., 70–75. See also Sarah L Hale, 'Water Privatization in the Philippines: The Need to Implement the Human Right to Water' (2006) 15 *Pacific Rim & Policy Journal* 765, 772 (showing that the 1997 privatisation of the Metropolitan Waterworks and Sewage System in Manila, Philippines, led to rate increases of 500 percent to 700 percent nine years after privatisation).

[28] Varghese (n 22) 84.

[29] ibid, 85. See also Miloon Kothari, 'Privatizing Human Rights – The Impact of Globalization on Access to Adequate Housing, Water and Sanitation' (*Social Watch*, 2003). http://unpan1.un.org/intradoc/groups/public/documents/apcity/unpan010131.pdf accessed 30 June 2013.

[30] Violeta Petrova, 'At the Frontiers of the Rush for the Blue Gold: Water Privatization and the Human Right to Water' (2005–2006) 31 *Brooklyn Journal of International Law* 577, 590.

[31] ibid. See also *Suez Sociedad General de Aguas de Barcelona S.A. and Vivideni S.A v. Argentina* ICSID Case No. ARB/03/19 (ruling that a price freeze on water services made by Argentina in the context of an economic crisis violated investor rights).

for future generations.³² This idea, which came to be known as 'sustainable development', would inform a new approach to water governance that stressed the need to incorporate economic analysis into environmental management.³³ Whereas water was regarded in the past as a free or highly subsidised good, there was now growing opinion that this was neither a sustainable nor an efficient means to distribute water.³⁴ It was further argued that the failure by governments to appropriately price water meant that they lacked resources to invest in expensive piped water supplies and sewage disposal systems.³⁵

The other trend in the 1980s, associated with neo-liberal discourse, was the ideological shift towards the promotion of withdrawal of the state from the provision of social services and its progressive replacement with the for-profit private sector.³⁶ In this period the World Bank argued that governments should work to reduce barriers to the involvement of the private sector in the provision of water services.³⁷ In 1989 the privatisation of water utilities in the United Kingdom marked the start of a new era in which water utilities would become potentially lucrative markets for private sector providers.³⁸ The ideas of sustainable development, commodification and privatisation converged into a development paradigm that Karen Bakker has termed 'market environmentalism'.³⁹ The discourse of market environmentalism offers to combine economic growth, efficiency and environmental preservation through the establishment of property rights, the deployment of markets as allocation mechanisms and the use of pricing to limit environmental externalities.⁴⁰ At the heart of the neo-liberal conception of water governance is the view of water as a commodity. From this perspective, water is an economic good rather than a public good and users are individual consumers rather than

---

[32] See, e.g., World Commission on Environment and Development, *Report of the World Commission on Environment and Development: Our Common Future* (1987) www.un-documents.net/our-common-future.pdf accessed 17 June 2013.

[33] ibid; see generally, chapter 2.   [34] ibid, chapter 8, para.53.

[35] ibid, chapter 9, para.55.

[36] Maude Barlow, *Blue Covenant: The Global Water Crisis and the Coming Battle for the Right to Water* (The New Press, 2007) 36–38.

[37] John Briscoe and David de Ferranti, *Water for Rural Communities: Helping People Help Themselves* (World Bank, 1988) 2.

[38] Barlow (n 36) 37.

[39] Karen Bakker, 'The "Commons" versus the "Commodity": Alter-Globalization, Anti-Privatization and the Human Right to Water in the Global South' (2007) 39(3) *Antipode* 430, 421.

[40] ibid, 432.

a collective of citizens. Giving water a market value is argued to be the most effective means to incentivise conservation.[41]

In order to attempt to maintain its hegemonic status, neo-liberal water governance discourse has also adapted and changed in the face of problems posed by national economic crises, fierce public opposition and difficulties in extracting profits from poor consumers.[42] In between 2000 and 2003 high profile cancellations of water supply concession contracts took place in many cities, including Atlanta, Buenos Aires, Jakarta, La Paz and Manila.[43] The World Bank and other IFIs began to acknowledge that the privatisation of water services may have negative consequences, particularly for the poorest.[44] A number of adjusted 'pro-poor' policy prescriptions were implemented throughout Latin America, Africa and Asia as a result.[45] Such reforms included: (1) the promotion of differentiated (cheaper) services for poor households, (2) participation to determine users' willingness to pay (and thus the appropriate level of service), and (3) government subsidy to meet service needs for the poor.[46] The new 'social turn'[47] in water governance, whilst signifying a softening of earlier approaches, did not constitute a break with the fundamental tenants of neo-liberalism. Despite the fact that the overall level of investment by TNCs in water services in the Global South has declined since the 1990s, World Bank water services and sanitation projects have continued to promote various forms of privatisation and cost recovery principles, often in the form of 'public-private partnerships'.[48]

---

[41] ibid, 441.
[42] Peter R Robbins, 'Transnational Corporations and the Discourse of Water Privatization' (2003) 15 *Journal of International Development* 1073, 1073.
[43] Bakker (n 39) 440; The mood of private investors in that period was reflected in the rhetorical question posed by a water business expert in 2004: 'Can anyone imagine investing hard currency in water projects in countries like the Philippines, Argentina and Bolivia now?' Cited in Castro (n 26) 66.
[44] Elver (n 21) 110–111.
[45] Kathryn Furlong, 'Neoliberal Water Management: Trends, Limitations, Reformulations' (2010) 1 *Environment and Society: Advances in Research* 46, 61.
[46] ibid, 62.
[47] Andrienne Roberts, 'Privatizing Social Reproduction: The Primitive Accumulation of Water in an Era on Neoliberalism' (2008) 40(4) *Antipode* 535, 549–553.
[48] Food and Water Watch, *Dried Up, Sold Out: How the World Bank's Push for Private Water Harms the Poor* (Food and Water Watch, 2009) 17 http://documents.foodandwaterwatch.org/doc/worldBank.pdf accessed 29 June 2013; Philippe Marin, *Public-Private Partnerships for Urban Water Utilities: A Review of Experiences in Developing Countries* (World Bank, 2009) 1–2; Shiva (n 8) 88–92.

The 'social turn' in neo-liberal water governance can therefore be understood as reflecting the hegemonic shift from the Washington Consensus to the Post-Washington Consensus (PWC) in neo-liberal governance. The challenges of high risks, low profitability and social protest associated with water privatisation have forced neo-liberals to engage with 'a range of unsavoury challenges of intervention, amelioration, and reregulation'.[49] Part of this strategy has involved what Arne Rucket has termed 'accumulation by subsidization': state transfer payments to the poor who cannot become 'normal customers' in recently privatised markets.[50] Such an approach is not committed to the decommodification of important aspects of life through the granting of social citizenship rights but rather focuses on the subsidisation of impoverished consumers to (re)produce stable markets in highly sensitive areas, such as water services.[51] Private sector participation, commodification and commercialisation are still encouraged, but regulatory and supplementary measures by the state are required as safety valves for capital investment.[52] In short, both the economic and ethico-political compromises of the 'social turn' in neo-liberal water governance satisfy Gramsci's formula for ruling class hegemony in the sense that they actually strengthen and reproduce the economic interests of the Transnational Capitalist Class (TCC).[53]

5.3.2 THE COUNTER-HEGEMONIC DISCURSIVE FRAMEWORK: WATER AS 'COMMONS'

The privatisation of water services in the 1990s was met with popular resistance in a number of countries, including Bolivia, Panama, Argentina, Uruguay, Brazil, Peru, Indonesia, Pakistan, India, Ghana, South Africa, Poland, Hungary, the USA and Canada.[54] At the same time, campaigns

---

[49] Jamie Peck, *Constructions of Neoliberal Reason* (Oxford University Press, 2010) 23–24.
[50] Arne Ruckert, 'Towards an Inclusive-Neoliberal Regime of Development: From the Washington to the Post-Washington Consensus' (2006) 39(1) *Labour, Capital and Society* 35, 63.
[51] ibid.
[52] Toby Carroll, 'Auctioning off Manila's Water Services: Market Extension, the World Bank and Socio-Institutional Neoliberalism' (2006) Asia Research Centre Working Paper No.138 19. wwwarc.murdoch.edu.au/publications/wp/wp138.pdf accessed 9 June 2013.
[53] Antonio Gramsci, *Selections from the Prison Notebooks of Antonio Gramsci* (Quintin Hoare and Geoffrey Nowell Smith eds. and trans.) (Lawrence & Wishart, 1971) 161.
[54] William Finnegan, 'Letter from Bolivia: Leasing the Rain' *New Yorker* (New York, 8 April 2002) 43, 53; Maude Barlow and Tony Clarke, *Blue Gold: The Battle against the Corporate Theft of the World's Water* (Earthscan, 2002) 188–191.

against water privatisation became increasingly internationalised, reflecting that the 'globalized nature of the water industry and the market itself require that community-based campaigns take on international dimensions'.[55] In 2000 the Council of Canadians launched the Blue Planet Project to challenge the dominant neo-liberal vision of water governance. Subsequently, a host of movements, including the Inter-American Vigilance for the Defence and Right to Water Network, the Centre on Housing Rights and Evictions (COHRE), the Heinrich Boil Foundation and Food and Water Watch have organised internationally to contest privatisation and argue that states have to ensure the right to water for all of their citizens.[56] Without wishing to oversimplify the diverse array of perspectives within this network of national movements and international NGOs, it is possible to identify the emergence of a 'water justice movement' united in its shared opposition to the privatisation of water services and common commitment to an alternative model of water governance based around the ideal of the 'commons'.[57]

In contrast to the view of water as an economic good, the conception of water as commons stresses that water is so essential for life and ecosystem health that it should be regarded as the common heritage of the earth.[58] Within this conceptual framework, common ownership of water services is preferable to private commercial ownership because, as documented above, water supply is subject to multiple market failings. Water service provision is a classic case of a 'natural monopoly', that is, an industry where, by virtue of inherent technical characteristics, it is most economically efficient for a single service provider to produce the entire industry output.[59] As such, once a contract or concession is granted by a government to a company there will be no competition to regulate the behaviour of the monopoly in the interests of society.[60] The negative consequences of private monopoly ownership are particularly undesirable

---

[55] Barlow and Clarke (n 54) 203.
[56] Adam Davidson-Harden, Anil Naidoo and Andi Harden, 'The Geopolitics of the Water Justice Movement' (2007) 11 *Peace, Conflict and Development* 1, 30.
[57] Bakker (n 39) 440–447.   [58] ibid, 441.   [59] Varghese (n 22) 83.
[60] In theory, there can be competition at the bidding stage where a government awards the contract or concession. Furthermore, the state can in theory determine the length of the contract and terminate it where its conditions have not been fulfilled. However, given the power discrepancies that often exist between TNCs and governments, the lack of transparency and accountability that often characterise such bidding processes, and the complex web of international legal rules in favour of private investors, it is unlikely that fair competition will often be assured. See Varghese (n 22) 84.

Table 5.1 *The commons vs. the commodity debate*[67]

	Commons	Commodity
Definition	Public good	Economic good
Pricing	Free or 'lifeline'	Full-cost pricing
Regulation	Command and control	Market based
Goals	Social equity and livelihood	Efficiency and water security
Manager	Community	Market

in relation to water, given its necessity for protecting ecological and public health as well as its cultural and, often, spiritual dimensions.[61]

Contrary to the neo-liberal insistence on commodification as the basis for environmental protection, advocates of the commons argue 'that conservation is more effectively incentivised through an environmental, collectivist ethic of solidarity, which will encourage users to refrain from wasteful behaviour'.[62] The world's water crisis does not principally stem from the poor not paying enough for water services but rather from practices and inequalities associated with neo-liberal capitalism: 'megadams, inappropriate irrigation, fish destocking, bulk water diversions, bottled water, abuse of water by golf courses and extractive firms like Coca-Cola and Nestle'.[63] In other words, the water crisis is a socially produced scarcity arising from the 'short-term logic of economic growth, twinned with the rise of corporate power'.[64]

Whilst the commodity can be understood, following Marx, as the 'cell-form' of capitalism,[65] the commons can be understood as the basis for an alternative political economy, connecting 'multifarious struggles against capital's ... on-going primitive accumulation'.[66] The table below gives an indication of the alternative ethico-political foundations of a commons approach to water governance.[67]

---

[61] Bakker (n 39) 441. [62] ibid.
[63] Patrick Bond and Jackie Dugard, 'Water, Human Rights and Social Conflict: South African Experiences' (2008) *Law, Social Justice & Global Development (An Electronic Law Journal)* 1, 12.
[64] Bakker (n 39) 441.
[65] Karl Marx, *Capital: A Critical Analysis of Capitalist Production Volume 1* (Lawrence & Wishart, 1961) 8.
[66] Nick Dyer-Witherford and Greig de Peuter, 'Commons and Cooperatives' (2010) 4(1) *Affinities: A Journal of Radical Theory, Culture and Action* 30, 31.
[67] Adapted from Bakker (n 39) 441

Implementing water governance based on the idea of water as commons has been attempted using diverse forms of public water management under a variety of socioeconomic, cultural and political circumstances.[68] These have included co-operatives, public utility companies and community-managed water.[69] Whilst opposed to private provision of water services, many defenders of water as commons are also dissatisfied with traditional models of centralised, bureaucratic state provision.[70] As Greig de Peuter and Nick Dyer-Witherford argue:

> Beyond resistance to enclosures, commons politics involve renewed attention on the forms of association through which communities can govern shared resources ... Commons discourse resumes older discussions about "public goods", but breaks new ground, both in the range of ecological, biogenetic, and cultural domains it addresses, and in its interest in the possibilities for the organization of resources from below, rather than according to the models of command economies or bureaucratic welfare states.[71]

However, commons models of water governance do not usually reject all involvement for the state. Indeed, a prominent 2005 world report on public models of water provision stressed the 'crucial role of national governments' in providing finance and support for water utilities.[72] In surveying a range of public water models, Bakker concludes that the most progressive solutions involve reforming state governance while 'fostering and sharing alternative local models of resource management'.[73] In opposition to PWC neo-liberal calls for 'public-private partnerships', advocates of the commons call for 'public-public partnerships'. These are collaborations between two or more public authorities or organisations, based on solidarity, to improve the capacity and effectiveness of one partner in providing public water or sanitation services, with neither partner seeking to make a commercial profit.[74]

---

[68] Belén Balanyá, Brid Brennan, Olivier Hoedeman, Satoko Kishimoto and Philipp Terhorst (eds.), *Reclaiming Public Water: Achievements, Struggles and Visions for around the World* (Transnational Institute, 2005) 247; Mike Gonzalez and Marianella Yanes, *The Last Drop: The Politics of Water* (Pluto Press, 2015) 34.
[69] ibid.   [70] Bakker (n 39) 442.   [71] Dyer-Witherford and de Peuter (n 66) 31.
[72] Balanyá et al. (eds.) (n 68) 263.   [73] Bakker (n 39) 446.
[74] David Hall et al., *Public-Public Partnerships (PUPs) in Water* (PSI-TNI-PSIRU, 2009) 2 www.tni.org/sites/www.tni.org/files/download/pupinwater.pdf accessed 10 December 2013.

5.4 THE RIGHT TO WATER

Having sketched the contours of hegemonic and counter-hegemonic visions of water governance, this section will now critically evaluate the role the human right to water has played in these contestations and particularly around the issue of water privatisation. Whilst, as will be discussed below, many within the water justice movement have articulated demands for the right to water as a means to oppose privatisation and promote the idea of water as commons, some are sceptical about the discursive counter-hegemonic potential of the right to water. In particular, Bakker argues that:

> Human rights are individualistic, anthropocentric, state-centric, and compatible with private sector provision of water supply ... "rights talk" offers us an unimaginative language for thinking about new community economies, not least because pursuit of a campaign to establish water as a human right risks reinforcing the public/private binary upon which this confrontation is predicated, occluding possibilities for collective action beyond corporatist models of service provision.[75]

Given that the view of water as commons is often related to collectivist values, non-anthropocentric environmentalism and grass roots water management, these perceived biases within human rights discourse could undermine its potential to reflect counter-hegemonic values. Moreover, in so far as human rights discourse under international law is formally neutral with regard to different economic and political systems, it is conceivable that the privatisation of water services could be compatible with human rights obligations. Therefore, the manner in which the right to water is articulated, and its interaction with other discursive elements, is of critical importance in assessing its counter-hegemonic potential. This section will evaluate the ways in which the right to water – a norm that has emerged in international law of the last thirty years – has been articulated by various actors supporting and contesting global water governance.

5.4.1 THE RIGHT TO WATER IN THE TWENTIETH CENTURY

The right to water was not explicitly recognised in the Universal Declaration of Human Rights (UDHR) or either of the two human rights Covenants of 1966 (hereafter collectively 'the International Bill of

---

[75] Bakker (n 39) 447.

Human Rights').[76] However, the International Bill of Human Rights does contain a number of provisions that suggest the drafters considered water to be both a fundamental right and 'derivative' of the rights explicated. Most important among these rights are the right to life,[77] the right to an adequate standard of living[78] and the right to the highest attainable standard of health.[79]

Whilst a right to water might be implicit in the International Bill of Human Rights, the exact nature of the right was largely unknown in the twentieth century. In the 1970s, a concern with basic needs from international development organisations and a growing interest regarding environmental issues resulted in the emergence of water access as a concern in global governance.[80] Reflecting this context, the 1977 UN Water Conference in Mar del Plata adopted a series of resolutions and a plan of action that placed a strong emphasis on equitable water access with a particular emphasis on 'developing' countries.[81] The 1977 conference also explicitly recognised the right to water for the first time, declaring that 'All peoples, whatever their stage of development and social and economic conditions, have the right to have access to drinking water in quantities and of a quality equal to their basic needs.'[82] The right to water in this context was wedded to a state-led development paradigm with supporting bilateral and multilateral assistance.[83] In that period the principle source of funding for water infrastructure came from international development agencies, international financial institutions and government agencies such as the U.S. Agency for International Development.[84]

---

[76] The Universal Declaration of Human Rights (UDHR) G.A. res 217A (III), UN Doc. A/810 at 71 1948; The International Covenant on Economic, Social and Cultural Rights (ICESCR), 993 UNTS 3 entered into force 3 January 1976; International Covenant on Civil and Political Rights (ICCPR), adopted 16 Dec. 1966, entered into force 23 March 1976, 999 UNTS 171, reprinted in 6 ILM – 368 (1967).

[77] UDHR, article 3; ICCPR, article 6.   [78] UDHR, article 25; ICESCR, article 11(1).

[79] UDHR, article 25, ICESCR, article 12. See further, Pierre Thielbörger, 'Re-Conceptualizing the Human Right to Water: A Pledge for a Hybrid Approach' (2015) 15 (2) *Human Rights Law Review* 225, 228–232.

[80] Oriol Mirosa and Leila M Harris, 'Human Right to Water: Contemporary Challenges and Contours of a Global Debate' (2012) 44(3) *Antipode* 932, 934.

[81] United Nations Water Conference, Mar Del Plata, 14–15 March 1977 (United Nations publication, Sales No. E.77II.A.12) www.ielrc.org/content/e7701.pdf accessed 28 May 2013

[82] ibid, para.II(a).

[83] ibid, see 'plan of action'. See also Terence Lee, 'Water Management since the Mar Del Plata Action Plan: Lessons for the 1990s' (1992) 16(3) *Natural Resources Forum* 202–211.

[84] Thomas M Kerr, 'Supplying Water Infrastructure to Developing Countries via Private Sector Project Financing' (1995) 8 *Georgetown international Environmental Law Review* 91, 91.

Though the normative content of the right to water was not clarified at Mar del Plata, the conference was nevertheless a catalyst for more explicit inclusion of the right in international human rights law. Both the subsequent 1979 Convention on the Elimination of All Forms of Discrimination against Women (CEDAW) and the 1989 Convention on the Rights of the Child (CRC) contain provisions recognising state obligations in relation to the right to water.[85] The widespread ratification of these treaties[86] is indicative of a growing consensus that a free-standing right to water exists, even if the content of that right and the nature of its obligations had not been clearly defined.

5.4.2 THE NEO-LIBERALISATION OF THE RIGHT TO WATER

The solidification of neo-liberal global governance and ideology in the 1980s and 1990s had a discernible impact upon the ways in which the right to water was understood. This was particularly identifiable in the 'Dublin Principles' that arose out of the 1992 International Conference on Water and Environment. Principle 4 identified water both as an 'economic good' and a 'basic right of all human beings'.[87] In full it stated:

> Water has an economic value in all its competing uses and should be recognised as an economic good. Within this principle, it is vital to recognize first the basic right of all human beings to have access to clean water and sanitation at an affordable price. Past failure to recognize the economic value of water has led to wasteful and environmentally damaging uses of the resource. Managing water as an economic good is an important way of achieving efficient and equitable use, and of encouraging conservation and protection of water resources.[88]

The Dublin Principles can be understood not only as the crystallisation and condensation of the emergent hegemonic reframing of water as a commodity but also as tying this conception to the principles of human rights and sustainable development. However, Principle 4 goes further

---

[85] The Convention on the Elimination of All Forms of Discrimination against Women ("CEDAW"), 1979, UN Doc. A/34/46, article 14(2)(h); The Convention on the Rights of the Child ("CRC"), 1989, UN Doc. A/44/49, article 24(2). See also: African Charter on the Rights and Welfare of the Child art. 14(2), 11 July 1990, OAU Doc. CAB/LEG/153 (entered into force 29 November 1999).

[86] 187 Member States are party to CEDAW and 193 are party to CRC.

[87] International Conference on Water and the Environment, 26–31 January 1992, The Dublin Statement on Water and Sustainable Development, principle 4, UN Doc. A/CONF.151/PC/112 (12 March 1992).

[88] ibid.

than merely regarding the commodification of water as being compatible with the human right to water. Indeed, it regards the treatment of water as a commodity to be a *necessary prerequisite* for the realisation of the human right to water. Without the proper valuation and efficient rationing of water resources through pricing there will not be sufficient supplies of water available for the world's population.[89]

Whilst Principle 4 does not explicitly mandate the privatisation of water services, it can be interpreted as prefigurative of privatisation in the sense that it discursively recasts water as a commodity rather than a public good and therefore also implicitly rescripts users as individual consumers rather than a collective of citizens. Indeed, Bakker argues that neo-liberal water governance can be understood as resting upon three analytically distinct (if overlapping) concepts: privatisation, commercialisation and commodification.[90] *Privatisation* involves an organisational change in the ownership or management of water and water services; *commercialisation* entails the introduction of commercial principles, methods and objectives into the management of water services; and *commodification* involves converting water into an economic good through the application of market pricing mechanisms.[91] Privatisation will be used in this chapter in the broad sense to describe 'private sector participation' in the provision of water and sanitation services.[92]

---

[89] In June of 1992 the United Nations Conference on Environment and Development in Rio adopted an agenda in which the economic approach to water was less accentuated than in the Dublin Principles. Principle 18.8 of agenda 21 declared that water is 'an integral part of the ecosystem, a natural resource and a social and economic good'. It states that 'priority has to be given to the satisfaction of basic needs and the safeguarding of ecosystems' but '[b]eyond these requirements, however, water users should be charged appropriately'.

[90] Karen Bakker, 'Neoliberalizing Nature? Market Environmentalism in Water Supply in England and Wales' (2005) 95(3) *Annals of the Association of American Geographers* 542, 544.

[91] ibid.

[92] This includes a wide array of private commercial interests ranging from transfer of property rights over assets (full disclosure) through to concessions and service contracts. See Castro (n 26) 65. The sale of water utilities in Britain in 1989 represents a form of 'fully fledged privatisation' which has not on the whole been followed by other countries. Instead the dominant model has been the 'public-private' arrangement, in which various assets or/and operations that were previously under the ownership and control of public bodies are transferred to the private sector. Whilst formal ownership of water may still be with the public sector, the term 'privatisation' will still be deployed to connote the tendency towards treating water as a private economic good as a result of the commercial orientation of for-profit water service providers. See Shiva (n 8) 89–92, Castro (n 26) 66.

However, the significance of the right to water in the neo-liberal global water governance regime in the 1990s should not be overstated. In fact, in this period the idea of the human right to water became increasingly muted within international water discourse. Particularly noteworthy in this regard is the absence of any reference to the right to water in the Declarations of the first two World Water Forums held in Marrakech in 1997 and The Hague in 2000, both of which recognised water as a human *need* rather than a human *right*.[93] Nevertheless, the Declarations were committed to putting into practice the Mar del Plata and Dublin Principles, both of which recognised the right to water.[94] Furthermore, in language redolent of the Dublin Principles, the Hague Declaration stated that the pricing of water services 'should take account of the need for equity and the basic needs of the poor and the vulnerable'.[95] Arundhati Roy has also noted that there was much 'pious talk' of having access to drinking water as a basic human right during the Hague Conference.[96] Within the neo-liberal discursive framework it followed that to realise the right to water it was necessary to commodify water and to remove barriers to private sector involvement in water and sanitation services. However, the right to water was not regarded as an international binding legal obligation on governments but rather as an aspiration or goal to guide progress in the reform of water governance.

5.4.3 THE EMERGENCE OF THE RIGHT TO WATER AS OPPOSITIONAL FRAME

The failure of the WWC to explicitly recognise the right to water drew significant criticism from a number of NGOs that had attended the Hague Conference. A statement issued by fifty-five NGOs and trade unions rejected the Ministerial Declaration for promoting a 'technocratic'

---

[93] World Water Council, 'The Declaration of Marrakech' First World Water Forum, Marrakech 21–22 March 1997 www.worldwatercouncil.org/fileadmin/wwc/Library/Official_Declarations/Marrakech_Declaration.pdf accessed 17 June 2013; Ministerial Declaration of The Hague on Water Security in the 21st Century, The Hague Wednesday 22 March 2000, (hereafter 'the Hague Declaration'). www.idhc.org/esp/documents/Agua/Second_World_Water_Forum%5B1%5D.pdf accessed 17 June 2013.
[94] The Declaration of Marrakech (ibid) states 'The Forum calls on governments, international organizations, NGOs and the peoples of the World to work together in a renewed partnership to put into practice the Mar del Plata and Dublin Principles'. See also The Hague Declaration (ibid), principle 2.
[95] The Hague Declaration, principle 3.
[96] Arundhati Roy, *Power Politics* (South End Press, 2001) 40.

and 'top down' 'corporate vision of privatisation'.[97] The statement also challenged the legitimacy of the WWC due to its unaccountability[98] and affirmed access to basic water and sanitation as universal rights that cannot be negotiated as commodities.[99] For water justice activists critical of privatisation, this failure to recognise water as a binding human right represented a deliberate strategy on the part of the event's corporate sponsors:

> [the conveners of the World Water Forum] wanted water to be officially designated as a 'need' so that the private sector, through the market, would have the right and responsibility to provide this vital resource on a *for-profit* basis. If, on the other hand water was officially recognised as a universal human right, then governments would be responsible for ensuring that all people would have equal access on a *non-profit* basis.[100]

In this critique, the right to water is actually *counter-posed* to the privatisation of water. Whereas the neo-liberal discursive formation regarded water privatisation as a requirement for the realisation of the right to water, the alternative formation suggests that the provision of water services for profit runs contrary to a human rights-based approach. For this incipient global water justice movement the gathering at the Hague forum provided a springboard to build a counter-hegemonic movement against the neo-liberal global water regime.[101] However, it would be the outcome of the unfolding 'water war' against privatisation in Cochabamaba, Bolivia, that would provide the impetus and narrative for a new global water justice movement against privatisation as well as for the development of a counter-hegemonic conception of the right to water.

The often regressive impact of the privatisation of water services led to large protests, non-payment campaigns and other forms of civil disobedience in many of the countries where such policies were introduced.[102] The most famous instance of resistance was in Cochabamba, Bolivia, following the World Bank-mandated privatisation of water services in 1999.[103] The socioeconomic consequences of the water privatisation were

---

[97] NGO Major Group Statement to the Ministerial Conference (21 March 2000, The Hague) 1.
[98] ibid, 2.   [99] ibid, 3.   [100] Barlow and Clarke (n 54) 80 (emphasis in original).
[101] Barlow (n 35) 124–125.   [102] Petrova (n 30) 578–579.
[103] For an account of the details of the privatisation see Melina Williams, 'Privatization and the Human Right to Water: Challenges for the New Century' (2007) 28 *Michigan Journal of International Law* 469, 496.

severe, with the cost of water bills rising by an average of 35 percent.[104] In one of the poorest countries in the Western Hemisphere – where approximately 70 percent of the population live in poverty – this had a devastating impact on low-income households. For the poorest, water bills came to represent 22 percent of the household salary.[105] These intolerable conditions provoked a fight back. In January 2000, a citizens' alliance comprising labour, environmental, human rights and community activists formed to reverse the privatisation law. After months of civil disobedience and protest, the government was forced to revoke the concession contract and repeal the law allowing for the privatisation of water.[106]

The victory against water privatisation in Cochabamba would quickly take on an international dimension. Just days after the contract was cancelled one of the leading activists in the protests, Oscar Olivera, attended a forum organised by the International Forum of Globalization (IFG) in Washington DC as part of a series of protests against World Bank meetings.[107] In December 2000 the IFG, alongside other international activists, were invited to an international conference in Cochabamba to forge a partnership between the citizens of Cochabamba and the international movement against corporate globalisation.[108] At the conference the IFG were joined by workers, students, lawyers, farmers and others in drafting a 'Declaration of Cochabamba', which stated amongst other things that water is a basic human right.[109] This Declaration can be understood as the catalyst for the global water justice movement. It begins:

> We, citizens of Bolivia, Canada, United States, India, Brazil – farmers, workers, indigenous people, students, professionals, environmentalists, educators, nongovernmental organizations, retired people, gather together today in solidarity to combine forces in the defense of the vital right to water.

---

[104] Jennifer Naegele, 'What Is Wrong with Fully Fledged Privatization?' (2004) 6 *Journal of Law and Social Challenges* 99, 125.
[105] Carlos Crespo, Nina Laurie, and Carmen Ledo, 'Cochabamba-Bolivia Case Study Report' (2003) cited in Castro (n 26) 71.
[106] Finnegan (n 53) 55.
[107] Vandan Shiva, 'Water Democracy' (Foreword) in Oscar Olivera, *Cochabamba! Water War in Bolivia* (South End Press, 2004), ix .
[108] Antonia Juhasz, 'Cochabamba Water War Presents Globalization Alternative to the World' (IFG Newsletter, February 2001) www.tyrannyofoil.org/article.php?id=90 accessed 5 June 2013.
[109] ibid.

> Here, in this city which has been an inspiration to the world for its' retaking of that right through civil action, courage and sacrifice standing as heroes and heroines against corporate, institutional and governmental abuse, and trade agreements which destroy that right ...[110]

What is particularly striking about the Declaration is that it directly *counter-poses* the commodification and privatisation of water with the idea of water as a fundamental human right: 'Water is a fundamental human right and a public trust to be guarded by all levels of government, therefore, it should not be commodified, privatised or traded for commercial purposes.'[111] This is in stark contrast to the Dublin principles, which sought to subsume the right to water within the conception of water as a commodity.[112] It is also noteworthy that the Declaration also calls for an internationally binding treaty imposing obligations upon national governments in relation to the right to water.

Furthermore, whilst the Declaration foregrounds the right to water, it is not individualist, anthropocentric or state-centric, contrary to Bakker's concerns outlined in s.5.4. This conception of the right to water is embedded within a *non-anthropocentric* environmentalist paradigm, that is to say, it does not reduce the value of water solely to the benefits that it confers on human beings but also recognises the inherent or intrinsic value of water: 'Water belongs to the earth and all species and is sacred to life, therefore, the world's water must be conserved, reclaimed and protected for all future generations and its natural patterns respected.'[113] This understanding clearly also frames water as a collective entitlement rather than simply as an individual right. The Declaration further rejects bureaucratic and top-down state-centric approaches to water governance. Instead, it argues that water should be 'protected by local communities and citizens who must be respected as equal partners with governments in the protection and regulation of water'.[114] Hence the Declaration recognises both the limitations of government provision of water services as well as the important role for governments to play within the context of a 'public-public partnership' with local communities.

---

[110] Cited in Jeff Malpas, 'The Forms of Water: In the Land and in the Soul' (2006) 1(2) *Transforming Cultures eJournal*, fn13.
[111] ibid, para.2.
[112] Recall that in the Dublin Principles it is only 'within' the principle that water is an economic good that human beings have a right to affordable water.
[113] Cited in Malpas (n 110) fn 13, para.1.   [114] ibid, para.3.

The Cochabamba 'Water War' not only resulted in a local victory against privatisation, but it also gave birth to a new international movement for water justice and the right to water. The human right to water subsequently became a core discursive element in a number of national and international struggles against water services privatisation. For example, in June 2001 the Ghana National Coalition against the Privatisation of Water produced the Accra Declaration on the Right to Water, which recognised water as a 'fundamental human right'[115] that 'should not be a common commodity to be bought and sold'.[116] At the same time, increasingly internationalised campaigns against water privatisation also began to evoke the right to water. On 8 July 2001 the Water for People and Nature Summit held in Vancouver, Canada, adopted 'The Treaty Initiative to Share and Protect the Global Water Commons'.[117] The Initiative, drafted by Blue Planet Project activists Maude Barlow and Jeremy Rifkin and unanimously endorsed by 800 delegates from 35 countries, echoed the Cochabamba Declaration in recognising water as a fundamental human right and public trust that belongs 'to the earth and all species, and therefore must not be treated as a private commodity to be bought, sold and traded for profit'.[118] It also called for a global treaty to oblige national governments to declare waters in their territories to be a public trust and to enact strong regulatory structures to protect them.[119] Thus, national movements and coalitions resisting privatisation became accompanied by a global movement challenging the legitimacy of neo-liberal water governance policies and posed alternative approaches based on human rights.

5.5 THE UNITED NATIONS AND THE RIGHT TO WATER

The new millennium witnessed two significant shifts in the discourse of global water governance. The first, as touched upon in Section 5.3.1, was the neo-liberal 'social turn' towards a PWC model of water governance,

---

[115] The Accra Declaration on the Right to Water (adopted 19 May 2001), www.liberationafrique.org/IMG/article_PDF/article_73.pdf accessed 6 June 2013.
[116] ibid.
[117] The Treaty Initiative to Share and Protect the Global Water Commons (Vancouver, Canada, Endorsed 8 July 2001) cited in 54) xvii–xviii.
[118] ibid, xviii.
[119] ibid, xviii. See also 'The Porto Alegre Water Declaration in the Spirit of Cochabamba (Bolivia), Narmada (India) and Ghana and Other Fights' (Porto Alegre, 1–5 February 2002).

which involved, in some respects, the softening of past neo-liberal policy prescriptions. The second was increased attention towards the normative content of the right to water within the UN human rights system. This section will pay attention to this later development, and consider the ways in which it interacted with the former.

5.5.1 THE UNITED NATIONS AND THE RIGHT TO WATER: BACKGROUND

Despite a number of declarations and statements recognising the right to water at the international level, scant attention had been paid to the normative content and scope of the right under international law in the twentieth century. This began to change at the end of 2001. The United Nations General Assembly adopted a resolution on the 'Status of Preparations for the International Year of Freshwater 2003' in December 2001.[120] Member States, the United Nations system and other groups were encouraged to work towards raising awareness of the essential importance of freshwater resources and for promoting action at the local, national, regional and international levels in preparation for the Third World Water Forum to be held in Japan in March 2003.[121] As part of this, the Committee on Economic, Social and Cultural Rights (CESCR) was asked to prepare a General Comment clarifying the issue of the right to water.[122] Committee member Eide Riedel was appointed as Rapporteur on the right to water with the task of drafting the Comment. In April 2002 the Human Rights Commission approved the appointment of Mr. El-Hadji Guissé as Special Rapporteur to conduct a detailed study on the relationship between the enjoyment of economic, social and cultural rights and the promotion of the realisation of the right to drinking water supply and sanitation.[123]

5.5.2 THE PRELIMINARY DISCUSSIONS ON THE GENERAL COMMENT ON THE RIGHT TO WATER

In November 2002 the CESCR hosted a preliminary discussion of a draft general comment on the right to water prepared in July of that

---

[120] UNGA, 'Status of preparations for the International Year of Freshwater, 2003: Resolution adopted by the General Assembly' (7 February 2002) UN Doc. A/RES/56/192.
[121] ibid, para.2.
[122] CESCR, 'Twenty-ninth session, Summary Record of the First Part (Public) of the 50th Meeting' (26 November 2002) UN Doc. E/C.12/2002/SR.50, para.2.
[123] Commission on Human Rights Decision 2002/105 (22 April 2002) (Forty- ninth meeting).

year.¹²⁴ Taking part in the debate were Eide Reidel, El Hadj Guissé, Jean Ziegler (Special Rapporteur on Food) and Miloon Kothari (Special Rapporteur on Adequate Housing). Also participating were representatives of the WHO, the World Bank, a number of NGOs and SUEZ – a French-based TNC with operations primarily in water.¹²⁵ It is noteworthy that no individuals from grassroots movements against water privatisation were present in the discussions. However, despite such exclusions, the high profile social conflicts that had arisen as a result of water privatisation were clearly an important influence on the discussion.

Much controversy revolved around the role of the private sector in water services provision as well as the related issue of the status of water. Amongst the Special Rapporteurs, Committee members and NGO participants there was a widespread consensus that privatisation had negatively affected the right to water, making it especially inaccessible to the poor.¹²⁶ It was felt that the General Comment should stress that human rights-based ethics, and not profitability, must guide the community design and planning of water supply systems.¹²⁷ Additionally, it was argued that the General Comment should recognise water as a natural resource and a public good rather than a commodity or a private good.¹²⁸

However, the World Bank and SUEZ representatives were keen to stress that privatisation had a role to play in ensuring the right to water. The World Bank representative argued that discussions of privatisation should not be clouded by a focus on certain bad examples.¹²⁹ He suggested that governments lacked the capacity to provide the world's population with reasonable access to water. Private companies therefore had an important role to play in the provision and rational use of water resources.¹³⁰ Furthermore, whilst water must be considered a public good it is *also* an economic good and a commodity¹³¹ and users of water services should be expected to contribute to the provision of water if they receive value in return.¹³² Similarly, the SUEZ representative welcomed 'a universal right of access to water' but argued that the General Comment should be neutral in respect of public or private

---

¹²⁴ CESCR, 'Twenty-ninth session, Summary Record 46th Meeting' (22 November 2002) UN Doc. E/C.12/2002/SR.46 (hereafter, Twenty-ninth session).
¹²⁵ CESCR, 'Committee on Economic, Social and Cultural Rights holds discussion on right to water' (29th Session) (11 November 2013) UN Doc. HR/4630.
¹²⁶ Twenty-ninth session (n 124) see, e.g., paras.17, 24, 39, 47, 53 and 57.
¹²⁷ ibid, para.20.   ¹²⁸ ibid, paras.24, 39 and 41.   ¹²⁹ ibid, para.26.   ¹³⁰ ibid.
¹³¹ ibid, para.59.   ¹³² ibid, para.26.

methods of delivery.[133] The private sector should not be presumed bad and could even be regarded as 'one of the weapons of mass salvation' for the poor.[134]

The preliminary discussion hosted by the CESCR reveals a shift in the contested terrain of global water governance. Whereas counter-hegemonic forces challenged forums like the WWC for treating water as a commodity and failing to recognise it as a human right, in the preliminary discussions a divergence developed around the *meaning* of the human right to water. On the one side, NGOs and human rights officials regarded the human right to water as being threatened by privatisation, whereas the World Bank and SUEZ officials argued that privatisation could play an important role in the realisation of the right to water. This latter argument can be viewed as an attempt to (re)incorporate the right to water into the hegemonic discursive formation of neo-liberal water governance.

5.5.3 GENERAL COMMENT 15

The CESCR subsequently released General Comment 15 (GC15) on the right to water.[135] The Comment suggests that the ICESCR provides a legal basis for the right to water in articles 11(1) (the right to an adequate standard of living) and 12(1) (the right to the highest attainable standard of health).[136] The right to water entitles everyone to sufficient, safe, acceptable, physically accessible and affordable water for personal and domestic uses.[137] While GC15 deals in depth with multifarious aspects of the right to water and the obligations of States Parties in relation to it, this section will focus on the aspects of the Comment that are most pertinent towards the right to water in the context of privatisation.

The first important point to consider is how water is conceptualised in GC15. As already highlighted, the neo-liberal attempt to commodify water services in the 1990s produced a hegemonic discourse in which water was primarily framed as an economic good. In contrast, in the early 2000s an array of subaltern social forces reconceptualised water as a human right and a public good that *must not* be treated as an economic good. GC15 adopted an approach that rests somewhere in between these

---

[133] ibid.   [134] ibid.
[135] CESCR, General Comment No. 15: The Right to Water' ('GC15') (20 January 2003) UN Doc. E/C.12/2002/11.
[136] ibid, para.3.   [137] ibid, para.2.

two conceptions. Water is identified as 'a limited natural resource',[138] a 'public good',[139] 'a social and cultural good, and not primarily an economic good'.[140] Hence, GC15 recognises that water *can* be an economic good, but this is subordinate to its status as a public, social and cultural good. It would seem that that it is permissible to treat water as an economic good provided that this does not interfere with the public, social and cultural benefits that accrue from water.

In respect of the question of privatisation, GC15 represents a 'careful middle road'[141] which does not establish a formal opposition between the human right to water and privatisation, but rather imposes a number of state obligations that must be met where water services are privatised. Where water services are provided privately GC15 establishes that the state has a protective role in creating an effective regulatory system (including independent monitoring, genuine public participation and the imposition of penalties for non-compliance) and ensuring that private water providers are prevented 'from comprising equal, affordable, and physical access to sufficient, safe and acceptable water'.[142] A bolder stance was taken in the original draft of GC15, which called for the deferral of privatisation until an effective regulatory system has been put in place.[143] One can only speculate as to why the CESCR retreated from this more rigorous standard[144] but perhaps adopting a precautionary standard *only* in respect to privatisation could have been regarded as indicative of a normative bias that was inconsistent with the CESCR's neutral stance on the question of which economic model States Parties ought to adopt to fulfil their obligations under the ICESCR.

In addition to effectively regulating the private sector, states must adopt a number of measures and policies to ensure equal access to affordable, sufficient, safe and acceptable water. States Parties are under an obligation to ensure that all aspects of the right to water are enjoyed without discrimination, including on the basis of social origin, property

---

[138] ibid, para.1.   [139] ibid.   [140] ibid, para.11.
[141] Malcolm Langford, 'Ambition that Overleaps Itself? A Response to Stephen Tully's Critique of the General Comment on the Right to Water' (2006) 24(3) *Netherlands Quarterly of Human Rights* 433, 453.
[142] GC15 (n 135) para.44(b)(ii).
[143] CESCR, 'General Comment 15, The Right to Water, DRAFT' (Twenty-ninth session, 2002) (29 July 2000) UN. Doc. E/C.12/2002/11, para.18.
[144] For discussion see Stephen Tully, 'A Human Right to Access Water? A Critique of General Comment No. 15' (2005) 23(1) *Netherlands Quarterly of Human Rights* 35, 53.

or other status.¹⁴⁵ Measures a State Party can take include using low-cost techniques and technologies, appropriate pricing policies for free or low-cost water and income supplements.¹⁴⁶ States also have core obligations to ensure equitable distribution of all available water facilities and services and to adopt relatively low-cost targeted water programmes to protect vulnerable and marginalised groups.¹⁴⁷

Where does GC15 stand in relation to the neo-liberal project? The comment was produced at a time when privatisation programmes were globally in retreat due to underestimated risks, overestimated profits and contractual difficulties in investment by TNCS in the Global South.¹⁴⁸ It was also a period in which a discernible discursive shift was taking place within neo-liberal water governance. Since 2000, the proponents of neo-liberal water governance have been advocating 'public-private partnerships' in which firms contribute resources and expertise, whereas governments satisfy public interest objectives. Given the difficulties associated with profit extraction in water service investments in the Global South, the requirements on State Parties to supplement the private supply of water with measures supplying basic services to the poor could minimise the prospects of civil disobedience and social strife and thereby be beneficial for enhancing the risk-return ratios of private TNC investments in the Global South. The right to water and GC15 in this sense can be understood as compatible with the imperatives of neo-liberal water governance.

On the other hand, GC15 also contains a number of provisions that can be utilised by the counter-hegemonic global water justice movement to challenge neo-liberal water policies. The Comment clearly privileges a conception of water as a public good and grants the state wide discretion with regard to the means that it can deploy to ensure equitable access to water. Of potential concern for private investors, GC15 requires that states must ensure that the private sector is monitored independently

---

[145] GC 15 (n 135) para.13.This entails obligations to take steps to remove *de facto* discrimination through, *inter alia*, the adoption of relatively low-cost targeted programmes, investments that benefit the largest parts of the population and the provision of necessary water and water facilities to those that do not have sufficient means. See paras.13–15.

[146] ibid, para.27.   [147] ibid, paras.37(e) and (h).

[148] UN Secretary General, 'Report on Freshwater Management: Progress in meeting the goals, targets and commitments of Agenda 21, the Programme for the further implementation of Agenda 21 and the Johannesburg Plan of Implementation' (11 April 2004) UN Doc. E/CN.17/2004/4, para.63.

and penalties are imposed upon it for non-compliance with the right to water.[149] The Comment also requires that water services must be affordable for all, whether provided for publicly or privately[150] and indeed makes it a violation for states to fail to effectively regulate and control water services provided by third parties.[151] These measures, coupled with the Comment's clear prioritisation of rights of access to water over the investment rights of private companies, can be used to challenge restrictions to access associated with the private provision of water services.

In summary, whilst GC15 adopts a formally neutral stance on the question of privatisation it nevertheless conceptualises water as a public good over an economic good. The Comment also requires the state to adopt a protective role in ensuring that water is equitably provided for everyone in a manner that is sufficient, safe, acceptable and affordable, whether the water provider is the state or a non-state entity. In some respects this approach is congruent with the 'social turn' in neo-liberal water policy, which requires a more proactive role for the state. However, the clear subordination of property rights to rights of access in GC15 can also be used to challenge neo-liberal water governance.

5.5.4 DEVELOPMENTS IN RELATION TO THE RIGHT TO WATER SINCE GC15

GC15 acted as a catalyst for 'a right to water norms cascade'[152] that has generated a number of national and international instruments, legal commentaries and strategies for implementing the right to water.[153] Whilst these documents at times raise concerns about aspects of privatisation,[154] they all follow GC15 in remaining formally neutral on the question, whilst

---

[149] GC15 (n 135) para.24. As Langford points out, this is the strongest language yet from the Committee on the regulation of the private sector. See Langford (n 141) 279.
[150] ibid, para.27.   [151] ibid, para.44(b)(ii).
[152] Anna FS Russell, 'Incorporating Social Rights in Development: Transnational Corporations and the Right to Water' (2011) 7(1) *International Journal of Law in Context* 1, 9.
[153] See, e.g., UNCHR, 'Final report of the Special Rapporteur, El Hadji Guissé' (14 July 2004) UN Doc. E/CN.4/Sub.2/2004/20; UNCHR, 'Report of the Special Rapporteur, El Hadji Guissé' (11 July 2005) UN Doc. E/CN.4/Sub.2/2005/25; UN Human Rights Council 'Report of the United Nations High Commissioner for Human Rights on the scope and content of the relevant human rights obligations related to equitable access to safe drinking water and sanitation under international human rights instruments' (16 August 2007) UN Doc. A/HRC/6/3.
[154] See, e.g., Guissé (2004) ibid, paras.57–60.

requiring states parties to assume protective, regulatory roles with respect to either the public or private provision of water services.[155]

On 28 July 2010 the UN General Assembly adopted a resolution (presented by Bolivia) which, for the first time, formally recognised the right to water and sanitation and acknowledged that clean water and sanitation are essential to the realisation of all human rights.[156] Whilst the resolution does not address the question of how water and sanitation services are provided, the fact that it was introduced by Bolivia, which has been at the forefront of the international campaign against water privatisation, is revealing in that it indicates that opponents of privatisation continued to regard international recognition of the right to water as an important strategy.

In September of 2010 the Human Rights Council also adopted a resolution affirming that the rights to water and sanitation are part of existing international law and also confirmed that they are legally binding upon states.[157] Consistent with previous UN jurisprudence on the right to water the resolution recognises that states 'may opt to involve non-state actors in the provision of safe drinking water and sanitation services'[158] but requires that the former to regulate the latter to ensure that they act in a manner consistent with the right to water.[159]

On 21 November 2013 the UN General Assembly's Third Committee adopted a resolution on 'The human right to safe drinking water and sanitation', confirming that the rights to water and sanitation are derived from the right to an adequate standard of living.[160] However, as a result of US pressure and opposition to 'expansive' conceptions of the right to water, the final resolution omitted any affirmation of the content of that right.[161]

---

[155] UN Human Rights Council, 'Report of the United Nations High Commissioner for Human Rights' (2007) (n 153), para.52 (noting that the 'approach of United Nations treaty bodies and special procedures has been to stress that the human rights framework does not dictate a particular form of service delivery and leaves it to States to determine the best ways to implement their human rights obligations. While remaining neutral as to the way in which water and sanitation services are provided, and therefore not prohibiting the private provision of water and sanitation services, human rights obligations nonetheless require States to regulate and monitor private water and sanitation providers.').
[156] UNGA Res 292 (28 July 2010) UN Doc. A/RES/64/292.
[157] HRC Res15/9 (6 October 2010) UN Doc. A/HRC/ES/15/9, paras.3 and 6.
[158] ibid, para.7. [159] ibid, para.9.
[160] (19 November 2013) UN Doc. A/C.3/68/L.34/Rev.1.
[161] UNGA, 'Third Committee Approves Text on Human Right to Safe Drinking Water and Sanitation, 10 Other Resolutions' (21 November 2013) GA/SHC/4092. www.un.org/News/Press/docs/2013/gashc4092.doc.htm accessed 14 December 2013.

In a press release, Amnesty International noted that the omission of these 'essential elements' of the right to water, contained in early draft resolutions, could downgrade the status of international commitments to the right to water to mere 'hollow promises'.[162]

In summary, since GC15 developments within the UN have confirmed that the right to water exists as a free-standing right under international law. The calls of the water justice movement for international recognition of the right to water have therefore been met. However, whereas the water justice movement has articulated the right to water as prohibiting the commodification of water and the privatisation of water services, the right to water within the UN framework has been articulated in a manner that is open on the question of how water services are provided. Furthermore, the hostility and opposition to the right to water by some states, notably the USA, has resulted in weak and vague formulations of the right to water at the international level, which risk rendering the right to water as an 'aspirational' norm rather than a binding obligation. This has opened up opportunities for the utilisation of the right to water not only by the water justice movement, but also by actors within the global water regime.

5.6 (RE)INCORPORATING THE RIGHT TO WATER INTO THE NEO-LIBERAL FRAMEWORK

As discussed above, the UN human rights framework effectively decoupled the right to water from opposition to the privatisation, commodification and commercialisation of water services. This provided the basis for the discursive (re)incorporation of the right to water into the neo-liberal framework. It is particularly noteworthy that TNCs with commercial interests in water services have actively promoted a particular vision of the right to water.[163] For example, AquaFed, the International Federation of Private Water Operators, recently produced a public statement urging the European Union to include the human right to safe drinking water and sanitation in the European Charter of

---

[162] Amnesty International, 'United Nations: General Assembly Makes Progress on the Human Rights to Water and Sanitation, but Only so far as the USA Permits' (*Amnesty International*, 26 November 2013) www.amnesty.org/en/library/asset/IOR40/005/2013/en/0a1e54d0-e725-4033-bbf9-6a1ecaae97dc/ior400052013en.pdf accessed 14 November 2013.

[163] See generally, Russell (n 152).

Fundamental Rights.[164] Citing various UN reports and declarations, the statement stresses the 'positive contribution' of private water operators in expanding water services, ensuring taxpayer value and making the right to water a 'reality'.[165] The World Bank and the WWC have also adopted positions in relation to the right to water which will now be examined.

5.6.1 THE WORLD BANK

In 2004 the World Bank published a detailed analysis of the legal basis of the right to water in international law, with a particular focus on GC15.[166] The World Bank publication acknowledges the existence of an 'emerging human right to water' under international law[167] and recognises GC15 as the clearest and most explicit recognition of that right.[168] After outlining the legal basis for the right to water and the obligations arising from it, the publication goes on to consider the policy dimensions arising from the Comment.[169] With respect to private sector participation in water resources management, the Bank acknowledges that it has often led to an increase in tariffs that have threatened the realisation of the human right to water.[170] It even acknowledges participation by users in the management of water facilities as a viable alternative to privatisation.[171]

Whilst the Bank acknowledges failures associated with privatisation, it also identifies problems associated with the failure to treat water as an economic resource. Of particular concern for the Bank is GC15's stipulation that State Parties may consider introducing free and low cost water policies. Against this, the World Bank commentary warns that most, if not all, water specialists would argue strongly against free water.[172] Drawing on the Dublin Principles and the World Commission for Water in the twenty-first century, the Bank stresses the need for cost recovery and full water pricing to promote the sustainable use of water resources.[173] In this respect, the World Bank reiterates the neo-liberalised

---

[164] AquaFed, 'Implement the Human Right to Safe Drinking Water and Sanitation Fully in Europe' (*AquaFed*, 10 December 2013) www.aquafed.org/pages/fr/admin/UserFiles/pdf/2013-12-10_AquaFed_HRWSinEurope_PR_EN_Pd_2013-12-09.pdf accessed 18 December 2013.
[165] ibid, 2.
[166] Salman MA Salman and Siobhan McInerney-Lankford, *The Human Right to Water: Legal and Policy Dimensions* (World Bank, 2004).
[167] ibid, 85.   [168] ibid, 86.   [169] ibid, 68.   [170] ibid, 72.   [171] ibid, 75.
[172] ibid, 70.   [173] ibid, 70–71.

discourse of sustainable development by tying environmental conservation to economic rationalism. At the same time, the Bank calls upon the state to introduce measures to ensure that basic water supplies are affordable for the poor. The Bank approvingly identifies a number of 'innovative approaches' that it regards as combining full-cost pricing with affordability for the poor. It provides three examples: the provision of six free kilolitres of water per month in South Africa; water stamps for those living below the poverty line in Chile; and the use of subsidies to needy users and tax benefits to water service providers in Armenia.[174]

Water is primarily an economic good and subsidised prices should be progressively eliminated due to their distorting impact on the end-user price of water. However, limited forms of means-tested subsidisations are regarded as sufficient to secure affordable access to water to the poor and are hence compliant with human rights norms. Such an approach is consistent with the PWC 'social turn' as well as transnational capital's reliance upon 'accumulation by subsidisation' in the water service sector, that is, the shifting of the social costs of water provision onto the state whilst keeping profits private.

The Bank also criticises the state centrism of the GC15. It argues that the 'issues surrounding the use and protection of water resources are complex, and responsibilities for such issues cannot be placed solely on the states. Individuals should bear an equal, if not a larger, portion of such responsibilities.'[175] Against the state-centric approach to duties the World Bank proposes that:

> rather than placing emphasis on the recognition of a human right to water, a more pragmatic approach would be to address the right to manage, or participate in the management of, the water resources ... [t]his participatory approach strengthens the empowerment of users, vesting them with both rights and corresponding duties with respect to water.[176]

The World Bank's approach in this respect represents a radical departure from GC15 and traditional human rights doctrine more generally. Within the traditional framework individuals have legal entitlements vis-à-vis the state, which is regarded as the ultimate guarantor of human rights. The individual's right is therefore correlated to the state's duties with respect to that right. By contrast, the World Bank's approach suggests that an individual's right to water is correlated with substantively parallel duties on the part of those individuals. Indeed, the World

---

[174] ibid, 71–72.   [175] ibid, 74.   [176] ibid, 75.

Bank argues that the individual should have 'equal, if not larger' responsibilities than the state in relation to the right to water. The concept of empowerment advanced by the World Bank thus corresponds with the idea of the 'minimization of the social role of government'[177] in neoliberal discourse and shifts attention away from a proactive state towards the economic activity of the 'active subject' free from government interference. This 'empowerment' approach has been identified as highly problematic as it assumes that 'all individuals "produce and consume" in the market equally, which obscures the classed, racialised and gendered divisions of labour and consumption practices'.[178]

The World Bank criticises GC15 for failing to include the procedural right to participate in the actual operation and management of water services.[179] This criticism could be shared by actors within the global water justice movement who also advocate active community involvement in the operationalisation of the right to water. Notwithstanding this convergence, there is a clear qualitative difference between the neoliberal conception of participation envisioned by the World Bank and the counter-hegemonic conception of 'water democracy'. Whereas the World Bank envisions a less active role for the state with regard to the right to water, the counter-hegemonic vision involves an equal partnership of local communities and government in the protection of water. The global water justice movement's insistence that water is not only a human right but also a 'public trust', that is, a resource preserved for public use that the government is required to maintain, would suggest the need to strengthen the social role of state governments whilst at the same time building alternative local models of resource management. Here the state still plays a central role, and may be understood as the ultimate guarantor of the human right to water.

There may appear to be certain dissonances in the World Bank's commentary on GC15. On the one hand it advocates more government intervention in the water service sector to ensure that services are affordable to the poorest, and on the other it calls for individuals to assume many of the responsibilities in relation to the right to water that the traditional human rights paradigm would assign to the state. In fact, both positions are consistent with the Post Washington Consensus approach adopted by the World Bank, which involves the provision of social safety

---

[177] Richard Falk, *Human Rights Horizons: The Pursuit of Justice in a Globalizing World* (Routledge, 2000) 47.
[178] Roberts (n 47) 533.   [179] Salman and McInerney-Lankford (n 166) 74.

nets for 'unfit market participants' in conjunction with traditional neo-liberal economic policy prescriptions entailing privatisation, commodification and the withdrawal of the state from social provision.

It should also be noted that the World Bank's commitment to the human right to water is far from consistent. For example, Malcolm Langford notes that whilst the research arm of the Bank has endorsed General Comment 15, it has not mainstreamed a rights-based approach into its investment arm.[180] He cites the example of the World Bank's decision to privatise Ghana's urban water supplies, linking the price of water to the exchange rate to satisfy overseas investors rather than to the income levels of Ghanaians.[181] In Gramscian terms, when the World Bank is operating 'consensually' in global civil society, it evokes the right to water for its symbolic value, but when it is using its coercive economic leverage as a member of global political society, it will often not operationalise these norms, instead prioritising the interests of transnational capital. Therefore, whatever economic and ethico-political compromises the World Bank has made, they bolster the interests of the TCC and reinforce the neo-liberal discursive framework rather than contest it.

5.6.2 THE WORLD WATER COUNCIL

The WWC is a core component of the global neo-liberal water governance regime. Its triennial World Water Forum (WWF) events create opportunities for networking between Governments, TNCs, think tanks, NGOs and International Institutions to formulate global water policy. The WWF's lack of accountability, its domination by corporate interests and its commitment to neo-liberal policies of commodification and privatisation have made it a central target for the Global Water Justice Movement.[182] A key component of the Global Water Justice Movement's critique of the WWF has been its failure to officially recognise water as a human right. This failure is interpreted as a deliberate strategy consistent with the WWC's corporate agenda.[183] At the Fourth World Water Forum held in Mexico City in 2006 the water justice movement organised

---

[180] Langford (n 141) 454.
[181] ibid. However, Langford also notes that human rights considerations were taken into account by the International Centre for Settlement of Investment Disputes – the World Bank's international disputes tribunal – in the *Suez/Vivendi vs. Argentina* case. Langford (n 141) 448.
[182] Elver (n 21) 117.   [183] Barlow (n 36) 54.

a parallel summit entitled the International Forum on the Defence of Water.[184] At the parallel event Bolivia's new water minister, Abel Mamani, pledged to support the right to water within his country, Latin America and the UN.[185] That year Bolivia had elected a new president, Evo Morales, who stood on a radical programme of social and democratic reforms presented as anti-neoliberal and anti-imperialist in nature.[186] The Morales administration would play a leading role in the struggle for the realisation of the right to water under international law in the following years.

By contrast, at the official forum, the Ministerial Declaration once again did not include the right to water. However, according to the official synthesis of the forum 'never has the right to water received as much coverage at an international meeting'.[187] At the event the President of the WWC declared that 'the right to water is an indispensable element of human dignity'.[188] In addition, three official forums addressed the right to water as a key issue and the WWC released a report entitled *The Right to Water: From Concept to Implementation*.[189] The synthesis rejected interpretations of the right to water that opposed privatisation outright:

> from the financial viewpoint, the discussion on water services should not focus on whether the supplier should be public or private. What is important is to identify who can provide this service in the most efficient manner and at the lowest cost.[190]

It is clear from such statements that the WWC wished to reclaim the mantle of the right to water from the water justice movement and its associated anti-privatisation perspective.

The WWC's report on the right to water is described by Maude Barlow as 'a bland restatement of many UN documents with almost no mention of the private sector ... and with no reference to the public-private debate raging around it'.[191] However, the report also contains a number of policy recommendations to achieve the right to water. In a section entitled

---

[184] See International Forum for the Defense of Water, 'Report' (Oxfam 2006) www.comda.org.mx/files/documentos/memoingles.pdf accessed 18 December 2013.
[185] ibid, 58.
[186] See generally, Sven Harten, *The Rise of Evo Morales and the MAS* (Zed Books, 2011).
[187] Polioptro Martinez Austria and Paul van Hofwegen (eds.), *Synthesis of the 4th World Water Forum* (Comisión Nacional de Agua, 2006), 87.
[188] ibid.
[189] Céline Dubreuil, *The Right to Water: From Concept to Implementation* (World Water Council, 2006), 21.
[190] Martinez Austria and van Hofwegen (n 187) 93. [191] Barlow (n 36) 170.

'solidarity', the report addresses the question of how to ensure that water services are affordable for the poor. The report, in line with the 'social turn' in neo-liberal water governance, contains a series of case studies and recommendations to make water services affordable to the poor through measures such as differentiated prices and subsidies for poor households.[192]

What is particularly noteworthy about the report, however, is its recognition of the right of authorities to 'cut off water supply for those who do not pay who are in a position to'.[193] This position is controversial. From England and Wales to South Africa to the USA, cutting off households' water supplies has led to deleterious – and even fatal – consequences.[194] True, the WWC states that this would only apply to those 'who could afford to pay' but it is not clear who falls into this category: does it mean people who have enough money/credit to pay the water bill or does it extend to households where the cost of water bills would take up an excessive amount of their household budget?[195] At any rate, it is arguable that cutting off a household's water supply will rarely be compatible with the human right to water when other more proportionate measures such as the imposition of fines can be applied instead.[196]

At the 2012 sixth World Water Forum held in Marseille France, the Ministerial Declaration at this conference went further than previous ones in relation to the right to water by committing its members 'to

---

[192] Dubreuil (n 189) 32–38. [193] ibid, 12.
[194] In England and Wales the disconnection of water services was banned in 1999 on the basis of public health concerns. See House of Commons Research Paper 98/117 10 December 1998 Water Industry Bill. Water Industry Act 1999. According to South Africa's Human Sciences Research Council, water cut offs forced people to obtain water from polluted streams, which led to an outbreak of cholera, infecting 250,000 people and killing nearly 300. See Jaques Pauw, 'Metered to Death: How Water Caused Riots and a Cholera Epidemic' (*Centre for Public Integrity*, 5 February 2003) www.publicintegrity.org/2003/02/05/5713/metered-death accessed 21 June 2013.
[195] Water has been estimated to be 'unaffordable' if the cost exceeds 2 percent of household expenditure or 1.25 percent for poorer households. See US Environmental Protection Agency, *Information for States on Developing Affordability Criteria for Drinking Water* (Washington DC, 1997) 45.
[196] The policy of imposing cut offs in England and Wales for non-payment was criticised by medical and nursing professions, who argued that a clean water supply was essential for human life, hygiene and health: 'there was no reason why the companies should have access to a remedy for non-payment of debt that was not open to other creditors seeking to recover debts'. Cited in Emanuele Lobina, *UK Water Privatization: A Briefing* (Public Services International Research Unite, 25 June 2001), 19. www.archives.gov.on.ca/en/e_records/walkerton/part2info/partieswithstanding/pdf/CUPE18UKwater.pdf accessed 21 June 2013.

accelerate the full implementation of the human rights obligations relating to access to safe and clean drinking water and sanitation'.[197] For the first time the WWF produced a Declaration that formally committed the conference to human rights principles in relation to water. However, the language of the Declaration was criticised by a number of human rights activists for falling short of recognising the right to water, instead only recognising human rights obligations *related* to access to water.[198] The Declaration was criticised as being 'a step backwards for water justice and the UN process that has begun to enforce the human right to water'.[199]

Indeed other parts of the Ministerial Declaration show a continued commitment to the principles of private financing, public-private partnerships and sustainable cost recovery, suggesting that the conception of the human right to water is wedded to the Post Washington Consensus 'social turn' in neo-liberal water governance.[200]

However, the 2015 Ministerial Declaration for the seventh World Water Forum held in Gyeongju, Korea, went a step further and, mirroring the language of the 2012 Rio +20 Summit, 'reaffirmed' its 'commitment to the human right to safe drinking water and sanitation and ensuring progressive access to water and sanitation for all'.[201] This latest development shows just how far the right to water has been embedded in the language of global governance. In the early days of the world water forum, the right to water was conspicuously absent from the Ministerial Declarations. In 2015 it featured prominently in both the preamble and in the second paragraph of the text. Should the mainstreaming of the

---

[197] Ministerial Declaration of the Sixth World Water Forum, Marseille 13 March 2009, para.3. www.worldwaterforum6.org/en/news/single/article/the-ministerial-declaration-of-the-6th-world-water-forum/ accessed 21 June 2013.

[198] Cited in Brent Patterson, 'NEWS: Opposition to the "Davos of water" Ministerial Statement' (*Blue Planet Project*, 13 March 2013) www.blueplanetproject.net/index.php/news-opposition-to-the-davos-of-water-ministerial-statement/ accessed 21 June 2013.

[199] Brent Patterson, 'NEWS: Governments Back Track on Right to Water at "Davos of water" Forum' (*Blue Planet Project*, 14 March 2013) www.blueplanetproject.net/index.php/news-governments-back-track-on-right-to-water-at-davos-of-water-forum/ accessed 21 June 2013; Brent Patterson, 'UPDATE: Marseille Was the Staging Ground for Rio+20' (*Blue Planet Project*, 18 March 2013) www.blueplanetproject.net/index.php/update-marseille-was-the-staging-ground-for-rio20/ accessed 21 June 2013. See also the critical remarks of Bolivia's minister for water cited in Claire Provost, 'World Water Forum Falls Short on Human Rights, Claim Experts' *Guardian* (London, 14 March 2013).

[200] See Ministerial Declaration of the Sixth World Water Forum (n 197) paras.22 and 26.

[201] Ministerial Declaration of the Seventh World Water Forum, Gyeongju 13 April 2015, para.2. www.worldwatercouncil.org/fileadmin/world_water_council/documents/press_releases/Ministerial_Declaration_7th_World_Water_Forum_1304_Final.pdf accessed 11 February 2017.

right to water be understood as a victory for the water justice movement and their campaigning efforts or the co-option of a potentially radical ideal, or perhaps both of these things?

5.7 CASE STUDY: THE RIGHT TO WATER IN SOUTH AFRICA

Given that leading actors in the Global Water Regime recognise the right to water, acknowledge the mistakes of the past and are incorporating more pro-poor social policies into their models of water governance, the question must be asked: what, if anything, stands between the global water regime and the water justice movement? Whilst anti-privatisation activism has undoubtedly led to shifts away from some of the simplistic assumptions of neo-liberal approaches to water, a gulf still exists between the reformed neo-liberal approach to water governance and the vision of the water justice movement. Whilst the principles of community participation, subsidisation and price differentiation sound compatible with the demands of the water justice movement, their marriage to the principles of privatisation and full cost recovery have negative consequences on the ways in which they are enforced.

To provide an illustration of the limits of the 'social turn' in neo-liberal water governance, this section will draw on a case study from South Africa, which has constitutionally enshrined the right to water and has introduced a Free Basic Water (FBW) policy praised by the World Bank and the WWC.[202] Despite this praise, the FBW has come under considerable criticism from an array of social movements, human rights organisations and academics.[203] This section will provide a brief evaluation of the implementation of the FBW policy to illustrate the critical differences between the counter-hegemonic conception of the right to water and the conception of the right to water associated with the social turn in neo-liberal water governance.

5.7.1 BACKGROUND TO THE FREE BASIC WATER POLICY

In 1994 approximately 12 million South Africans lacked access to any form of safe water supply.[204] The new African National Congress (ANC)-

---

[202] See Salman and McInerney-Lackford (n 166) 71 and 79–80; Dubreuil (n 189) 17–19.
[203] Discussed *infra* this section.
[204] Mike Muller, 'Parish Pump Politics: The Politics of Water Supply in South Africa' (2004) 7(1) *Progress in Development Studies* 33, 34.

led government's initial approach to this was outlined in its welfare-orientated Reconstruction and Development Program (RDP).[205] The RDP required that in poor and rural areas a minimum amount of water necessary for health and hygiene would be provided free of charge, followed by 'a progressive block tariff to ensure that long-term costs of supplying large-volume users are met and that there is cross-subsidy to promote affordability for the poor'.[206] However, following pressure exerted by the World Bank, Western governments and TNCs such as Suez and Biwater, the government drastically decreased grants and subsidies to local municipalities, and water supply was re-organised on the principles of full cost recovery, the corporatisation of water services and credit control measures.[207]

The principle of full cost recovery entailed a conception of water as primarily an economic good and a fiscal pressure on municipalities to maximise profits from water services.[208] The corporatisation of public services involved the transfer of services to a unit or department which is managerially or financially ring-fenced.[209] For example, in 2001 in the City of Johannesburg water services were corporatised under the auspices of a single ring-fenced corporation whose only shareholder is the City of Johannesburg.[210] Whilst still formally publicly owned, these services were fragmented and organised along similar lines to the private sector.[211] Credit control measures were aimed at curtailing loss in water revenues and reducing fiscal deficits in water services through the imposition of punitive measures for non-payment of water bills. Such measures have involved the disconnection and restriction of water through water cut-offs, water management devices, collateral and collective service deprivations

---

[205] Lyla Mehta with Oriol Mirosa, *Financing Water for All: Behind the Border Policy. Convergence in Water Management* (IDS, 2004) 19.
[206] African National Congress, *Reconstruction and Development Programme: A Policy Framework* (Umanyano, 1994) 30.
[207] Dale T McKinley, 'The Struggle against Water Privatization in South Africa' in Balanyá et al. (eds.) (n 68) 182.
[208] Jackie Dugard, 'Civic Action and the Legal Mobilisation: The Phiri Water Meters Case' in Jeff Handmaker and Remko Berkhout (eds.) *Mobilising Social Justice in South Africa: Perspectives from Researchers and Practitioners* (ISS and Hivos, 2010) 78.
[209] David McDonald and Laïla Smith, 'Privatizing Cape Town: From Apartheid to Neo-Liberalism in the Mother City' (2004) 41(8) *Urban Studies* 1461, 1470.
[210] Dugard (n 208) 79.
[211] Laila Smith, 'The Murky Waters of Second Wave Neoliberalism: Corporatization as a Service Delivery Model in Cape Town' in David McDonald and Greg Ruiters (eds.), *The Age of Commodity: Water Privatization in Southern Africa* (Earthscan, 2005) 169.

and pre-paid metres (PPMs).[212] All of these measures conform to a neo-liberal organisation of water services with water treated as a commodity, water provision as a business and the consumers of water as customers rather than citizens. The principle of financial sustainability becomes the predominant concern over other social aspects of water governance.

5.7.2 THE FREE BASIC WATER POLICY

However, by the end of the 1990s these measures had resulted in high levels of non-payment, vast amounts of municipal arrears and large-scale water cut-offs in poor communities, in one instance causing a cholera epidemic infecting more than 250,000 people and killing nearly 300.[213] In response to the catastrophic consequences of the imposition of full cost recovery policies, the South African government formalised the FBW policy in May 2001.[214] This policy called for every household to be provided with six kilolitres (6000 litres) of free water per month.[215] This model reflects a shift from 'full cost recovery' to 'sustainable cost recovery' in line with the 'social turn' in neo-liberal water governance. It was argued in the previous section that sustainable cost recovery was related to an attempt to (re)produce stable markets in highly sensitive areas such as the provision of water. It seems that such economic rationality was a driving force behind the implementation of FBW policy, as this policy was calculated to be administratively cheaper to apply than it would have been to continue to service the accounts of those who defaulted on their water bills.[216]

While the FBW hints at some of the demands raised by the water justice movement in providing many households in South Africa with a lifeline supply of water, a number of shortcomings have been identified with the policy. First, the amount of free water provided under the scheme was insufficient to meet basic needs. The amount was based on the assumption that in an average household of eight people this will allow for the provision of an average of 25 litres per day per person.[217] According to the WHO, between 50 and 100 litres of water per person per day are needed to ensure that most basic needs are met and few health concerns arise, with 25 litres being the bare minimum for survival

---

[212] Sean Flynn and Danwood Mzikenge Chirwa, 'The Constitutional Implication of Commercializaing Water in South Africa' (2005) in McDonald and Ruiters (eds.) ibid, 67–71.
[213] See Jaques Pauw (n 194).   [214] Bond and Dugard (n 63) 8.   [215] ibid.
[216] ibid.   [217] ibid, 9.

in the short term.²¹⁸ Furthermore, the policy discriminated against large households with multi-unit dwellings that are common in poor areas of South Africa, meaning that the persons in such households receive even less than 25 litres. For some poor households the FBW supply often only lasted until the middle of the month, leaving them without a water supply for two weeks in some instances.²¹⁹

Second, driven by the imperatives of cost recovery, water service providers have often adopted non-progressive tariff structures that have made the cost of water above the free tariff prohibitively costly for many households. As Patrick Bond and Jackie Dugard point out: '[t]hese tariffs, typically, provide the six kilolitre FBW, followed by a sharp convex curve, such that the next consumption block is unaffordable to many households, leading to even higher rates of water disconnections in many settings'.²²⁰ Whilst cost recovery has been justified under the rubric of water conservation, in reality it has been the poor who have been punitively disconnected because they cannot afford to pay their water bills whilst 'hedonistic water consumption in … richer, swimming-pooled (and predominantly white) suburbs' has been largely unaffected.²²¹ In many instances the FBW policy has exacerbated water inequalities and re-entrenched many of the social and economic divisions of the old apartheid regime.

Third, the FBW policy has often been accompanied by harsh enforcement measures for credit control on water charges in excess of the FBW tariff. PPMs have been the most controversial of these measures. These mechanisms were installed in poor areas of Johannesburg and provided each household with 6 kilolitres of free water per month, with the requirement that any more than that be paid in advance to access water services. In one instance, two children in the impoverished Phiri township in Soweto died in a shack fire after their pre-paid meter's supply automatically disconnected.²²² In less dramatic instances, PPMs have forced impoverished residents in large households to make undignified and unhealthy choices such as not flushing their toilets or washing their clothes.²²³ PPMs represent the ultimate neo-liberal approach to

---

²¹⁸ Guy Howard And Jamie Bartram, *Domestic Water Quantity, Service Level and Health* (World Health Organization, 2003) 22.
²¹⁹ Dugard (n 209) 84.   ²²⁰ Bond and Dugard (n 63) 9.   ²²¹ Dugard (n 209) 74.
²²² Patrick Bond and Jackie Dugard, 'The Case of Johannesburg Water: What Really Happened at the Pre-Paid "Parish Pump"' (2008) 12 *Law, Democracy and Development* 1, 1.
²²³ Dugard (n 208) 84.

access to water: responsibility is shifted to the individual household to calculate and economise their water consumption whilst little attention is paid to the realities of poverty and social, racial and gender inequalities.[224]

5.7.3 THE FREE BASIC WATER POLICY CONTESTED: THE MAZIBUKO RULING

Whilst the enforcement measures that often accompanied the FBW policy were supposed to foster individual responses to water supply, they often generated collective forms of resistance such as the Johannesburg-based Anti-Privatization Forum (APF), which formed the Coalition against Water Privatization (CAWP).[225] The organisational efforts of CAWP also led to the famous litigation case of *Mazibuko*, in which five residents of Phiri challenged the FBW policy on the basis that it violated the right to water under section 27(1) of the Constitution.[226] The applicants also challenged the City of Johannesburg's installation of PPMs on the basis that it was unlawful, administratively unfair and unfairly discriminatory under section 9 of the Constitution. The applicants were also supported by the COHRE and the Centre for Applied Legal Studies (CALS).

Initially, the applicants were successful. The South Gauteng High Court found in favour of the applicants.[227] Judge Tsoka declared the installation of PPMs in Phiri unconstitutional on the basis that, *inter alia*, it violated the right to equality as PPMs were only installed in poor black areas of Johannesburg.[228] Furthermore, Tsoka declared that the FBW amount was insufficient to meet the basic needs of Phiri residents.[229] It was ruled that the basic minimum should be increased to 50 litres per person a day to comply with section 27.[230] The respondents appealed to the Supreme Court of Appeal on 25 March 2009, which partially upheld

---

[224] See generally, Antina von Schnitzler, 'Citizenship Prepaid: Water, Calculability, and Techno-Politics in South Africa' (2008) 34(4) *Journal of Southern African Studies*; Prishani Naidoo, 'Eroding the Commons: Prepaid Water Meters in Phiri, Soweto' (Public Citizen, 2008) www.citizen.org/cmep/article_redirect.cfm?ID=11991 accessed 24 June 2013.
[225] Dugard (n 209) 87–89.
[226] Constitution of the Republic of South Africa 1996 Act No. 108 of 1996, Government Gazette No. 176778, vol.378, 18-12-1996. Section 27(1)(b) states that '[e]veryone has the right to have access to ... sufficient ... water'.
[227] *S. v. Mazibuko*, [2008] ZAGPHC 106 (Wit. Loc Div.) [2008] 4 All S.A. 471.
[228] ibid, 155.   [229] ibid, 179.   [230] ibid, 183.

the High Court ruling but held that the minimum of amount of water required for the FBW policy should be 42 litres per person a day, and that while the installation of PPMs was unlawful, the declaration of unlawfulness should be suspended for two years to give the City time to bring its water policy in line with the reasonableness requirement of the Constitution.[231]

The respondents counter-appealed against the Supreme Court of Appeal's judgment to the Constitutional Court (South Africa's highest court) requesting an order to reinstate the high court order.[232] The Constitutional Court dealt a heavy blow to the applicants by delivering a unanimous judgment dismissing the applicants' two grounds of appeal and overturning the previous two court judgments.[233] Contrary to the rulings of the lower courts, the Constitutional Court found that the introduction of the PPMs was not unfair discrimination because it had been done for the legitimate purpose of increasing water revenue.[234] The Court argued that the measures were not discriminatory because PPMs had not been applied in other black townships where the failure to collect water payments was less acute.[235] The fact that PPMs had been installed exclusively in poor black townships and not applied in the context of other poor debtors – such as businesses and government departments – does not seem to be evidence to the court of the discriminatory nature of the PPMs. By focusing on the *purpose* for the introduction of PPMs rather than the *consequences* of their introduction for poor black communities, the court adopts a narrow formal conception of equality that ignores power relations and structural inequalities: 'while people in Johannesburg's richer suburbs with conventional meters continue to enjoy substantive protections prior to water disconnection, poverty-stricken people in Phiri with pre-paid meters have been forced to forgo such procedural protections'.[236]

The Court also found that the FBW policy was reasonable under section 27(1).[237] On this point, the Court rejected the applicants' argument that the reasonableness of state measures should be determined by reference to minimum core standards (as identified by the CESCR

---

[231] *City of Johannesburg and others v. Mazibuko and others* [2009] ZASCA 20 (S. Afr. S.C.), [2009] 8 B. Const. L. R. 791 [62].
[232] *Mazibuko and others v. City of Johannesburg and others* [2009] ZACC 28 (C. Afr. Const. Ct.), [2010] 3 B. Const. L. R. 239.
[233] ibid, 171.   [234] ibid, 150.   [235] ibid, 149.   [236] Bond and Dugard (n 63) 2.
[237] Mazibuko and others (n 232) [83].

in GC15) and instead adopted a highly deferential approach, arguing that there was no constitutional obligation to provide any particular amount of free water and that it was institutionally inappropriate for a court to pronounce on the precise steps required to realise socioeconomic rights.[238] This is a procedural, rather than a substantive conception of socioeconomic rights associated with the Constitutional Court's more conservative reasonableness jurisprudence.[239] Rather than assessing the concrete impact of state (in)action on citizens' dignity, equality or wellbeing, the court merely assesses whether the state has adopted policies that conform to certain procedural standards and ostensibly aim to address socioeconomic rights obligations. In the absence of any grounding in minimum core obligations or benchmarks and indicators, such procedural interpretations forestall the use of socioeconomic rights as vehicles for substantive socioeconomic empowerment and transform them into apologia for the status quo. The FBW policy and the constitutionally enshrined right to health in South Africa have therefore in fact served as symbolic legitimisers of the neo-liberal regime of water governance.

In summary, the introduction of FBW in South Africa is consistent with broader shifts in neo-liberal water governance to incorporate 'pro-poor' policies within a commercialised and privatised framework of water services delivery. However, the provision of free water has been insufficient to address the needs of many of South Africa's poorest and most marginalised communities, particularly those in large households and informal settlements in urban townships. The policy has also often been accompanied by harsh forms of credit control such as PPMs and non-progressive tariff blocks that have in some instances exacerbated inequalities. The FBW policy has been implemented in the context of a neo-liberal model of water governance in which water is treated primarily as a commodity and is fixated on cost recovery and profit maximisation at the expense of the social dimensions of water provision. By ruling that these policies are compatible with section 27 of the South African Constitution, the Constitutional Court judgment provides an illustration of a hegemonic conception of the right to water congruent with the imperatives of neo-liberalism. The FBW policy also illustrates the shortcomings of

---

[238] ibid, 61, 68.
[239] For a criticism of the reasonableness standard in the Constitutional Court's jurisprudence see David Bilchitz, *Poverty and Fundamental Rights: The Justification and Enforcement of Socio-Economic Rights* (Oxford University Press, 2007).

5.8 THE RIGHT TO WATER AS A COUNTER-HEGEMONIC STRATEGY

this conception, and how it differs from the egalitarian vision of the right to water that the water justice movement advocates.[240]

Whilst the privatisation, commodification and commercialisation of water services has been regarded as inimical to the right to water by the water justice movement, the right to water has also been utilised by key actors within the water governance regime in a manner that is congruent with neo-liberal policy prescriptions. At the international level, the human right to water is formally neutral with regard to privatisation. The case study from South Africa illustrates that even in the absence of formal privatisation, the right to water can be interpreted in a manner consistent with wider neo-liberal policy prescriptions – such as the commodification of water and the commercialisation of water services.

Would this suggest that Bakker's critique of the limitations of human rights discourse as a means to challenge privatisation is correct? Whilst it is the case that the right to water discourse developed by the UN is neutral on privatisation and open with regard to the best method of water services delivery, its focus on universal access to water regardless of ability to pay nevertheless arguably still 'potentially serves to challenge some of the simplistic bases of rapid neoliberalisation shifts, and also highlight several critical issues that are sidelined in the push forward with market approaches'.[241]

This is not merely a theoretical point. In surveying an array of legal strategies to challenge the privatisation of water services in Uruguay, Colombia, Germany, Italy, France and Indonesia, Jackie Dugard and Katherine Drage conclude that 'Whether or not "rights" frameworks are invoked, pro-public activists derive authority, legitimacy and solidarity in their legal campaigns from the recent international recognition of a right to water'.[242] This was witnessed in the recent victory of a lawsuit

---

[240] However, it should be noted that the process of contestation in the courts led to Johannesburg Water voluntarily expanding the free basic allocation of water from 6000 to 10,000 litres per month for indigent households. The resort to the courts should not therefore be regarded as a complete failure for right to water advocates. See Mirosa and Harris (n 80) 940.

[241] Mirosa and Harris (n 80) 944.

[242] Jackie Dugard and Katherine Drage, *Shields and Swords: Legal Tools for Public Water* (Municipal Services Project, 2012) 3.

filed by the Coalition of People Rejecting the Privatisation of Water in Jakarta, which resulted in the Central Jakarta District Court ruling that the privatisation of Jakarta's water system was illegal and ordering that it be taken back into public control.[243]

And even in instances of judicial defeat such as *Mazibuko*, litigation around the right to water actually produced a number of positive outcomes from the water justice movement's perspective.[244] First, as the APF founder put it, the litigation strategy 'provided something to organize around ... it became a center of mobilization and reinvigorated the struggle'. Second, it attracted high profile media coverage and provided a platform for water justice activists to air their grievances to a wider audience and promote awareness. Third, as a direct result of the politicisation of these issues during the legal mobilisation process the City raised the amount of the FBW to 50 litres per day per person (the amount the applicants asked for). And finally, the City agreed not to prosecute anybody for bypassing the PPMs.[245] Thus, whilst the right to water continues to serve as a legal and moral anchor in anti-privatisation strategies, it would seem foolhardy to retreat from the discourse and thus allow those who advance a pro-privatisation agenda to progressively colonise its meaning.

What of Bakker's deeper critique of human rights discourse as suffering from individualistic, anthropocentric and state-centric biases that make it ill-suited for the counter-hegemonic vision of water as commons? I would argue that whilst these limitations can be found in certain readings of human rights, and arguably the dominant readings, it is too simplistic to reduce human rights discourse to such interpretations. My claim here is two-fold. First, whilst the UN framework for the right to water and the approach adopted by various national courts is not co-extensive with the demands of the water justice movement, neither is it inherently incompatible with these demands either. Second, at any rate, the right to water's discursive meaning is not limited to these aforementioned frameworks. In this section I will seek to demonstrate that there is nothing inherently individualistic, anthropocentric or state-centric about evoking the right to water.

---

[243] Corry Elyada, 'Court Decision Ends Privatization in Jakarta' (*Jakarta Post* 24 March 2015).
[244] Jackie Dugard, Urban Basic Services: Rights, Reality, Resistance' in Malcolm Langford et al., *Socioeconomic Rights in South Africa: Symbols or Substance?* (Cambridge University Press, 2014) 303.
[245] ibid, 301–302.

A purely individualistic conception of the right to water would be one framed in negative and procedural terms, requiring the state to create an enabling environment in which citizens can access water services of their choosing without arbitrary interference.[246] However, the international legal right to water is much broader than this. It requires that the state use its maximum available resources to ensure that everyone has sufficient, safe, acceptable, physically accessible and affordable water.[247] To put this into practice requires the imposition of collective cost through taxation/cross-subsidised user-fees and arguably imposes obligations on citizens to be collectively responsible for each other. As former Special Rapporteur on the right to water El Hadji Guissé notes:

> As regards the participation of taxpayers, contributions may be adjusted to ensure that every person contributes to ensuring access to water and sanitation in accordance with his or her financial means ... Progressive tariffs or cross-subsidies should be established in accordance with the economic and financial capacities of users.[248]

Furthermore, throughout GC15 it is repeatedly stipulated that States Parties owe obligations to both individuals *and* groups/communities.[249] Finally, outside of the official domain of international law, the water justice movement have conceptualised the right to water in a communitarian and collectivist manner.[250] Therefore, the charge that human rights are necessarily individualistic is at the very least an oversimplification.

The charge of anthropocentrism with regard to human rights might seem undeniable at first: the welfare and dignity of human beings is at the centre of human rights concerns and therefore other considerations such as the environment become at best secondary. This does not mean that environmental concerns are absent from a human rights framework, however. GC15 recognises that water is a 'limited natural resource'[251] and that 'the realization of the right to water must be ... sustainable, ensuring that the right can be realized for present and future

---

[246] For such an account see Fredrik Segerfeldt, 'Private Water Saves Lives' *Financial Times* (London 25 August 2005).
[247] ICESCR (n 76) article 2(1); GC15 (n 135) para.2.
[248] Final report of the Special Rapporteur, El Hadji Guissé (2004) UN Doc. E/CN.4/Sub.2/2004/20, para.53.
[249] GC15 (n 135) paras.12(a), 14, 16, 25, 33, 48.
[250] For a selection of such declarations see Sierra Club, 'Water Is a Human Right, Not a Commodity' www.sierraclub.org/committees/cac/water/human_right/ accessed 28 June 2013.
[251] GC15 (n 135) para.2.

generations'.²⁵² The Comment further stipulates that States Parties should adopt 'comprehensive and integrated strategies and programmes' to ensure the sustainable use of water such as reducing and eliminating practices of unsustainable extraction, pollution, desertification and loss of biodiversity.²⁵³

It might nevertheless still be objected that the environmental measures in GC15 are framed on the basis of the rights of future *human* generations to water and do not consider the rights of other sentient beings, let alone the inherent value of the environment. However, at the very least it can still be argued that the environmental dimensions of GC15 demonstrate that human rights are not necessarily incompatible with non-anthropocentric concerns.

Moreover, more radically non-anthropocentric visions of the right have been articulated in the water justice movement. This is most strikingly witnessed in the emerging discourse of earth rights, which declares that nature itself is a subject holder of rights.²⁵⁴ A decade after the Cochabamba Water War, activists gathered again in that city for the World People's Conference on Climate Change and the Rights of Mother Earth held in April 2010.²⁵⁵ The Conference produced the People's Agreement of Cochabamba, which states that 'To guarantee human rights and to restore harmony with nature, it is necessary to effectively recognize and apply the rights of Mother Earth'.²⁵⁶ With respect to water it states that 'We demand recognition of the right of all peoples, living beings, and Mother Earth to have access to water, and we support the proposal of the Government of Bolivia to recognize water as a Fundamental Human Right'.²⁵⁷ The emergence of such discourses reveals that human rights are not necessarily anthropocentric and can be rearticulated in ways that can fully embrace environmental concerns.

Finally, on the question of whether human rights are problematically state-centric, this requires some unpacking. Bakker's criticism here is that '"rights talk" resuscitates a public/private binary that recognises only two unequally satisfactory options – state or market control: twinned

---

²⁵² ibid, para.11.   ²⁵³ ibid, para.28.
²⁵⁴ See, e.g., Thomas Berry, 'Rights of the Earth: Recognising the Rights of All Living Things' (2002) *Resurgence* 214.
²⁵⁵ Climate and Capitalism, 'Documents of the World People's Conference on Climate Change and the Rights of Mother Earth Bolivia, April 2010 (South Branch Publications, 2010) available at http://readingfromtheleft.com/PDF/CochabambaDocuments.pdf accessed 29 June 2013.
²⁵⁶ ibid, 4.   ²⁵⁷ ibid, 8.

corporatist models from which communities are equally excluded'.[258] Bakker's argument that the public/private binary simply maps on to a state/market binary in human rights discourse is somewhat of an oversimplification. A binary that is apparent in international human rights law is that between state and non-state actors.[259] It is the former – which comprises all branches of government (executive, legislative, judicial) and other national, regional or local public or governmental authorities[260] – that are directly bound by the ICESCR and other international human rights treaties.[261] Non-state actors (or 'third parties' as they are referred to in GC15)[262] can include 'individuals, groups, corporations and other entities'.[263] When water services are operated or controlled by third parties the state is under an obligation to ensure that those third parties do not violate the right to water. Third parties therefore only have indirect responsibility in relation to the human right to water. Thus, while states are viewed as the ultimate guarantors and enforcers of the human right to water, there is nothing in international human rights law that suggests that the *only* alternative to direct state provision of water services is the market.

Nevertheless, it might be legitimate to ask if having the state as the ultimate guarantor of the right to water is desirable given the problems of corruption, inefficiency, lack of investment, lack of public participation and failure to reach poor communities often associated with centralised, bureaucratic state provision.[264] Whilst privatisation has generally failed to address these problems, it is important to bear in mind that 90 percent of the world's water and sanitation services are publicly owned and multiple problems exist in that sector too.[265] Moreover, as the example of South Africa demonstrates, neo-liberal

---

[258] Bakker (n 39) 440.
[259] See generally, Manisuli Ssemyonjo, 'The Applicability of International Human Rights Law to Non-State Actors: What Relevance to Economic, Social and Cultural Rights?' (2008) 12(5) *The International Journal of Human Rights* 725–760.
[260] Human Rights Committee, General Comment 31, UN Doc. CCPR/C/21/Rev.1/Add.13 (2004) para.4.
[261] ICESCR (n 76) article 2(1) refers to the obligations of 'Each State Party to the present Covenant'.
[262] GC15 (n 135) para.23.   [263] ibid.
[264] For a critique of the public sector provision of water see generally, Fredrik Segerfeldt, *Water For Sale: How Business and the Market Can Resolve the World's Water Crisis* (CATO Institute, 2005).
[265] David Hall, Emanuele Lobina and Victoria Correl, *Replacing Failed Private Water Contracts* (Public Services International Research Unit, 2010) 2.

policies such as full cost recovery and corporatisation can also be pursued under the guise of formal public ownership. Against both the state and market, Bakker proposes the 'community' as a third model of water services delivery.[266] On this point she is supported by the Transnational Institute (TNI), who advocate the role of citizens' participation and civil society movements in water service delivery as 'viable alternatives to both privatized water delivery and inadequate, state-run water utilities'.[267]

Whilst the TNI provide compelling arguments that diverse forms of participation and community management of water services have often improved access to water and the affordability of services, it is nevertheless apparent that the 'community' should no more be idealised or romanticised than the state as a provider of water services. Bakker herself acknowledges this and points out that inequitable power relations and resource allocation exist within communities.[268] These inequalities can lead to a range of different outcomes, some of which may in fact bolster neo-liberalism. Indeed, as Ben Page documents, communities can serve as agents of commodification.[269] Community participation is often encouraged as a method of co-option within the context of neo-liberal water governance, often in the limited form of using consultants to assess the willingness to pay in order to assist private investment decisions[270] or on the basis of disseminating information on largely predetermined policies, often with the goal of (re)producing citizens as responsible water consumers by making them aware of their obligations in relation to cost recovery mechanisms.[271]

As already documented, the solutions posed by most of the global justice movement lie with reforming state governance whilst fostering alternative local models of resource management.[272] Likewise, whilst the TNI promotes increased community management, it also acknowledges the 'crucial role of national governments in providing finance and support for water utilities'.[273] As such, counter-hegemonic visions of water governance do not seek to displace state provision, but rather to democratise and supplement it. There is nothing inherent in human rights discourse that is incompatible with this.

---

[266] Bakker (n 39) 444. [267] Balanyá et al. (eds) (n 68) 247. [268] Bakker (n 39) 444.
[269] Ben Page, 'Communities as Agents of Commodification: The Kumbo Water Authority in North-West Cameroon' 34 (4) Geoforum 483–498.
[270] Balanyá et al. (n 68) 254. [271] Roberts (n 47) 120. [272] Bakker (n 39) 446.
[273] Balanyá et al. (eds.) (n 68) 263.

Having established that the right to water does not inherently suffer from any of the limitations identified by Bakker, it might still be argued that it is precisely the open-endedness of human rights discourse that opens it up to the potential of appropriation, and therefore the water justice movement should express their demands in terms of discourses less open to co-option, such as 'water as commons'.[274] Leaving aside the question of whether commons discourse could also be co-opted, I would argue that the error in this line of reasoning is viewing 'water as commons' and 'the right to water' as competing either/or categories. As Oriol Mirosa and Leila Harris argue, whilst the human right to water is relatively agnostic in terms of prescriptive methods of water governance, it is absolutely clear in its end-goal commitment to the universal access of every individual to clean and safe water regardless of ability to pay.[275] This approach challenges the neo-liberal bias towards the end goals of efficiency, cost recovery and profitability, and also provides a framework to measure and critique neo-liberal policies against.[276] Conversely, whilst 'water as commons' is clearer in terms of its prescriptive methods of water governance, it is 'somewhat fuzzy' with respect to its end goals.[277] Water as commons and the right to water therefore serve two different, and potentially complementary, discursive elements in a 'portfolio' of counter-hegemonic articulations (which might also include indigenous, spiritual, environmentalist, socialist and other discourses).

5.9 CONCLUSION

Recent developments at the UN and WWC reveal that whilst the use of the 'right to water' is now more ubiquitous than ever, the same divisions around the issues of commodification, commercialisation and privatisation continue to persist. These events also demonstrate the continued resilience to co-option by the water justice movement. Perhaps what is most striking about the difference between the proponents of 'chastened neo-liberal' reforms to water governance and the water justice movement is that the latter believe that radical social transformation is required to realise the right to water. The water scarcity crisis is not viewed as primarily rooted in the economic undervaluation of water as neo-liberals insist, but rather in the commodification of water and the environment more generally in the context of massive economic, social and cultural

---

[274] Bakker (n 39) 447.   [275] Mirosa and Harris (n 80) 936.   [276] ibid.   [277] ibid.

inequalities. As the TNI put it 'It is hard to see how water for all can be achieved without far more ambitious policies to fight poverty and redistribute wealth'.[278]

The neo-liberal approach to water governance and the right to water ignores these broader issues of socioeconomic inequality which mean that it tolerates the hedonistic and profit-driven consumption, extraction, diversion and pollution of water by the rich on the one hand, and forces impoverished populations to make choices between washing themselves and flushing their toilets or buying clothes and food for themselves and their families, on the other. Without addressing these inequalities, neo-liberal approaches become fixated on cost recovery, profitability and efficiency combined with limited ameliorative policies of subsidisation and differential pricing aimed at the poor. Such strategies can only go so far in the context of poverty and inequality and can, as the case study of South Africa illustrates, end up reproducing patterns of exclusion, repression and hierarchy.

In some ways, right to water discourse has been important in providing legitimacy for the Global Water Regime, which had come under much criticism following the failures of past privatisation policies. It also has the potential to serve the interests of capital investment by requiring states to adopt social measures that could offset some of the risks associated with water services in the Global South. However, whilst rhetorically evoking the right to water, there has been limited operationalisation of it in the practice of neo-liberal governance. For example, the World Bank has not integrated the right to water into its investment arm and northern states such as the Canada and the USA continue to oppose an international right to water.[279] Such precautions reveal that the right to water has not been wholeheartedly embraced within global governance.

The right to water continues to be an important discursive strategy for the water justice movement, who have articulated it in a radically different way to the actors in the water governance regime. It is used to challenge the legitimacy of the WWC, has been used in Uruguay and Bolivia to prevent the privatisation of water services, and has been

---

[278] Balanyá et al. (n 68) 258.
[279] See, e.g., Gregson (Canada), Summary Record of the 56th Meeting, 22 April 2003, UN Doc. E/CN.4/2003/SR.56, para.49; U.S. Mission to the United Nations Press Release, Explanation of Vote by John F Sammis, U.S. Deputy Representative to the Economic and Social Council, on Resolution A/64/L.63/Rev.1, The Human Right to Water (July 28, 2010) http://usun.state.gov/briefing/statements/2010/145279.htm. accessed 30 June 2013.

invoked to lend legal and moral legitimacy to movements fighting inequities associated with neo-liberal approaches to water governance. In Ireland a mass movement against the introduction of water charges has been mobilised under the banner of the 'Right2Water'. Paul O'Connell, a human rights activist involved in the Irish water protests, has noted that while the right to water, as a legal right, can be rendered in a 'market friendly' way, the assertion of water as a right by the Irish protesters is much more than a formal claim; it is 'the rejection of the idea that there is no alternative to the commodification of essential services and resources'.[280] Similarly, in Detroit protestors against water cut offs have marched under the slogan 'Water is a human right. Turn on the Water! Tax Wall Street'.[281] Such protestors, emboldened by a UN statement declaring that water shut offs are in breach of the human right to water,[282] are, like the Irish protesters, making more than a formal legal claim: they are drawing attention to the basic entitlements of citizens and the inequitable public spending decisions of government.

Whilst the right to water has been successful in challenging privatisation in a number of contexts, the South African example demonstrates that even where the right to water is formally recognised it can be interpreted in a manner congruent with neo-liberalism. Its radical potential therefore lies with it being wedded to alternative models of water governance, such as 'water as commons' and 'public-public partnerships' as well as to broader programmes of wealth distribution and social transformation.

---

[280] Paul O'Connell, 'Demand the Future: The Right to Water and Another Ireland' (*Critical Legal Thinking*, 29 September 2014) www.criticallegalthinking.com/2014/09/29/demanding-future-right2water-another.ireland/.

[281] Drew Gibson, 'Let Them Drink Pop: Detroit's Water Crisis and the Fight for Basic Human Rights in the Motor City' (*Truthout*, 26 July 2014). www.truth-out/org/opinion/item/25183-let-them-drink-pop-detroits-water-crisis-and-the-fight-for-basic-human-rights-in-the-motor-city/.

[282] UN News, 'In Detroit, city-backed water shut-offs "contrary to human rights,"' (n 6).

CONCLUSION

'How to weld the present to the future, satisfying the urgent necessities of the present and working usefully to create and "anticipate" the future?'[1]

This book has set out to consider the counter-hegemonic potential of socioeconomic rights discourse within the domain of global civil society. It has sought to do so through adopting a neo-Gramscian analytic framework to critically examine three global justice movements that have mobilised in part around socioeconomic rights recognised under international law. Mindful of the pitfalls of reaching overgeneralised conclusions drawn from a fairly limited range of source material, it is nevertheless possible to offer some tentative general observations about the possibilities and limitations of evoking socioeconomic rights discourse within the context of broader counter-hegemonic praxis.

The three case studies identified a number of advantages that adopting socioeconomic rights discourse afforded to global justice movements. First, socioeconomic rights can be used to redefine the boundaries of what is considered just and unjust. Whilst international legal standards associated with socioeconomic rights are fairly open ended in terms of their prescriptive methods of political and economic governance, they are nevertheless absolutely clear in their end-goal commitment to the universal access of every individual to basic goods and services required for their human dignity, regardless of ability to pay. This end-goal commitment provides a framework against which neo-liberal policies can be critiqued and the neo-liberal bias towards the goals of efficiency, cost recovery and profitability can be challenged.

Second, appeals to international human rights standards allowed social movements to make the necessary counter-hegemonic step from promoting particular and sectorial needs to advancing projects framed in

---

[1] Antonio Gramsci, *Selections from Political Writings (1910–1920)* (International Publishers, 1987) 67.

universal terms. In response to the hegemony of neo-liberalism, a viable counter-hegemony, spanning south and north, needs to draw together 'subaltern social forces around an alternative ethico-political conception of the world, constructing a common interest that transcends narrower interests situated in the defensive routines of various groups'.[2] Reliance on socioeconomic rights discourse facilitated the integration of geographically dispersed movements with divergent ideological, political and cultural references into unified global campaigns. Evoking socioeconomic rights discourse also enabled the global justice movements to enter into alliances with actors within the human rights movement, and in doing so opened up the possibilities for informing alternative understandings of human rights.

Third, socioeconomic rights were used as a form of immanent critique of the extant order. Social movements and transnational campaigns were able to highlight discrepancies between the widespread ratification and rhetorical acceptance of socioeconomic rights norms on the one hand, and the negative impact of policies and tendencies associated with neo-liberal globalisation on the other. Gramsci's insistence that counter-hegemonic critique should base itself in part on concrete, existing and accepted politico-legal standards is premised upon an understanding of the contradictory nature of hegemony. As the hegemonic bloc requires the support of at least some sections of the subaltern classes, hegemonic discourse is couched in appeals to universal norms such as justice, fairness and human rights. Counter-hegemony involves, in part, appealing to these same values whilst demonstrating their unrealisable nature under extant relations and structures of power. Subaltern movements have been able to demonstrate that neo-liberal policies such as dismantling social programmes, imposing regressive tax structures and withdrawing regulation of corporate activity are incompatible with obligations contained within the Universal Declaration of Human Rights and other international treaties that require states to use their maximum available resources to progressively realise socioeconomic rights.

Despite these aforementioned strengths, socioeconomic rights discourse is not immune to the risks of neo-liberal appropriation. The case study chapters documented that this was particularly the case in relation to the outcome of inter-governmental negotiations such as those within the Food and Agriculture Organization, the World Health Organization,

---

[2] William K Carroll, 'Hegemony, Counter-Hegemony, Anti-Hegemony' (2006) 2(2) *Socialist Studies* 9, 21.

the World Water Forum and the United Nations General Assembly. Socioeconomic rights standards were watered down, marginalised or recast in ways that suppressed their subversive oppositional tendencies. On a number of occasions neo-liberal actors argued that the adoption of pro-market and pro-corporate positions provided the optimal basis for the enjoyment of socioeconomic rights. In other instances socioeconomic rights were reduced to vague or aspirational norms so that they could not operate as legal restraints on the market. It is noteworthy that in all three case studies the various global justice movements were unsuccessful in achieving globally binding socioeconomic rights standards to match the hard-edged enforcement regimes that currently exist for international trade and investment.[3]

Ultimately, the taming of socioeconomic rights discourse in these settings was made possible by the very nature of the unequal power relations within the domain of global civil society. The influence of, *inter alia*, corporate lobbying, European and American military and economic dominance, NGO conservatism and the disconnection between global institutional structures and popular political bases significantly constrains the capacity of socioeconomic rights discourse to serve a transformative function in these settings. Even UN human rights bodies that appear to exercise relative autonomy from the neo-liberal order are hampered in what they can achieve within global civil society. It is noteworthy, for example, that prior to drafting General Comment 15 on the right to water the CESCR did not consult with any of the leaders or representatives of grassroots movements resisting water privatisation, instead limiting consultation to inter-governmental organisations and professional human rights NGOs. It is this general disconnection between subaltern movements and the structures of global governance that can allow the transformative messages of grassroots rights struggles to get 'lost in translation' and be converted into the language of apolitical narrow legalism.

Being divorced from cites of popular struggle – and concomitantly being dependent upon winning and maintaining legitimacy in the eyes of States Parties – also increases the likelihood that human rights bodies will gravitate towards incremental reformism. The remarks of former CESCR member (1997–2012) Eibe Riedel in a recent interview are telling in this respect:

---

[3] Here we are reminded of Marx's observation that 'Right can never be higher than the economic structure of society and the cultural development thereby determined'. Karl Marx, *Critique of the Gotha Program* (International Publishers, 1973) 10.

the Committee should take great care not to overstep its role once the Optional Protocol (to the ICESCR) is in force ... It would be wise to choose micro-level issues first and keep away from macro-issues like extraterritorial application of ICESCR rights, or poverty generally, or environmental protection issues on a large scale. This would definitely frighten off many states from ratifying.[4]

This cautious and incremental approach is clearly pragmatic from the vantage point of the CESCR. Nevertheless, it is not difficult to see how this approach excludes wider structural critique of neo-liberalism that is necessary to build a counter-hegemonic movement. When socioeconomic rights campaigns become divorced from broader challenges to systemic injustice there are dangers that they can easily become co-opted into the neo-liberal hegemonic framework through the processes of passive revolution and *trasformismo*. Given the nature of these risks, Eoin Rooney and Colin Harvey have expressed doubts about the 'mainstreaming' of socioeconomic rights as a counter-hegemonic strategy. As they argue:

> The progressive ideal of rights envisages (socioeconomic rights) transforming the mainstream but needs to be mindful of how the mainstream can transform and co-opt rights-based discourse. Given current structural realities and power relations, the latter is often more likely. Such co-option undermines the capacity of (socioeconomic rights) to provide a vehicle through which orthodox approaches to governing can be scrutinised and challenged.[5]

Rooney and Harvey end their article by posing the rhetorical question: 'might the counter-hegemonic ambitions of human rights be better preserved on the margins?'[6] Whilst entirely sympathetic to the concerns motivating Rooney and Harvey's position, I feel that their notion of 'counter-hegemonic ambitions preserved on the margins' is contradictory, at least in the Gramscian understanding of the term used in this thesis. Counter-hegemonic praxis – the war of position – involves creating alternative institutions and intellectual resources within existing society in order to slowly build up the 'strength of the social foundations

---

[4] Quoted in Iilias Bantekas and Lutz Oette, *International Human Rights Law and Practice* (Cambridge University Press, 2013) 217.
[5] Eoin Rooney and Colin Harvey, 'Better on the Margins? A Critique of Mainstreaming Economic and Social Rights' in Colin Harvey, Aoife Nolan and Rory O'Connell (eds.), *Human Rights and Public Finance* (Hart, 2013) 134.
[6] ibid, 135.

of a new state'.⁷ In other words, it means transforming the counter-hegemony on the margins into a new form of hegemony. As Robert Cox points out, this gives rise to complex strategic implications: 'It means actively building a counter-hegemony within an established hegemony while resisting the pressures and temptations to relapse into pursuit of incremental gains for subaltern groups within the framework of bourgeois hegemony'.⁸ The pertinent question in relation to socioeconomic rights discourse is how to harness its counter-hegemonic potential within global civil society whilst minimising, or at any rate mitigating, the risks of *trasformismo*. It is this question that will now be addressed.

C.1 A TRIPARTITE MODEL OF COUNTER-HEGEMONIC SOCIOECONOMIC RIGHTS PRAXIS

Drawing upon the praxis of the global justice movements documented in the three case study chapters, I argue that engagement with socioeconomic rights discourse within the domain of global civil society can bolster counter-hegemonic projects through a combination of different tactical and strategic orientations in relation to three different contexts: (1) participation in inter-governmental and other 'official' settings primarily to gain visibility, co-ordinate movement activity and advance incremental discursive shifts in global governance; (2) strategic alliances with UN agencies, human rights bodies and special rapporteurs that are marginalised or peripheral to the neo-liberal global order so as to gain legitimacy, expertise and resources; and (3) connecting socioeconomic rights standards to counter-hegemonic models of governance within 'subaltern counterpublics' (i.e. transnational spaces outside of the domain of global officialdom) in order to guard against the co-option and dilution of oppositional ideologies.

C.1.1 TACTICAL PARTICIPATION IN INTER-GOVERNMENTAL SETTINGS

Engagement with inter-governmental forums as sites to advance socioeconomic rights requires an awareness of the limitations of these spaces given current structural realities and power relations. Nevertheless, it is

---

⁷ Robert W Cox, 'Gramsci, Hegemony and International Relations' in Stephen Gill (ed.), *Gramsci, Historical Materialism and International Relations* (Cambridge University Press, 1993) 53.
⁸ ibid.

possible to discern three opportunities presented by engagement in such forums. The first advantage is an 'in-process' one. The participation in inter-governmental forums by progressive states, actors within the UN human rights system and civil society organisations (informally or through civil society mechanisms) provides a transnational space where these forces can co-ordinate strategy, exchange ideas and foster solidarity whilst promoting oppositional interpretations of rights and governance.

The second advantage can be understood negatively in the sense that the disappointing outcomes of inter-governmental forums can be used to raise consciousness and oppositional tendencies through revealing the ways in which key actors within neo-liberal governance have obstructed efforts to advance socioeconomic rights standards. This in turn can be used to highlight the contradictions between the realisation of socioeconomic rights and the imperatives of neo-liberal globalisation. Recall, for example, the remarks of former UN Special Rapporteur on the Right to Food, Jean Ziegler, after successful US efforts to block a binding convention on the right to food within the FAO. Ziegler argued that the US position highlighted the 'profound ... contradictions' between 'social justice and human rights' on the one hand and 'the Washington Consensus, which emphasises liberalisation, deregulation, privatization and the compression of State domestic budgets' on the other.[9]

Third, movements may be able to invoke some of the incremental advances in socioeconomic rights standards won through inter-governmental forums to advance important victories in the context of national and international struggles against neo-liberal governance. For example, despite the limitations of the Doha public health declaration in relation to TRIPS, it nevertheless did provide a moral authority that health activists could use to contest intellectual property rights (IPRs) protection, both in domestic courts and in international forums. Furthermore, whilst efforts to advance the right to water through the UN General Assembly were undermined by successful US efforts to omit any reference to the content of that right, the international recognition of the right to water has nevertheless been a legal and moral anchor that pro-public activists have been able to derive authority, legitimacy and solidarity

---

[9] Quoted in Peter Russet, 'The US Gets Its Way' (*APRN*, 30 June 2002) www.aprnet.org/concerns/50-issues-a-concerns/146-us-opposes-right-to-food-at-world-summit accessed 3 November 2013.

from in their campaigns against water privatisation.[10] Declarations and other statements are not inert or reified matter but rather are living instruments which can be skilfully appealed to by subaltern forces. It is precisely through exploiting the chinks in the armour of hegemony that counter-hegemonic strategy can be advanced.

In short, whilst attempts to advance socioeconomic rights norms through inter-governmental forums will usually result in the significant watering down of these norms, participation in these forums may nevertheless create points of visibility and pressure on governmental and inter-governmental institutions in ways that can indirectly benefit counter-hegemonic activity.

C.1.2 INVOKING THE JURISPRUDENCE OF INTERNATIONAL HUMAN RIGHTS BODIES

International socioeconomic rights standards provide a basis for contesting neo-liberal globalisation. However, they are also subject to neo-liberal appropriation. Whilst the International Covenant on Economic, Social and Cultural Rights (ICESCR) is fairly vague with regard to the content of socioeconomic rights and the concomitant obligations of States Parties, the CESCR has been able to clarify these issues to some extent through its general comments and state reporting procedures (and in the future through the complaints and enquiry mechanisms of the Optional Protocol). This is also true of other human rights treaty bodies such as the Committee on the Rights of the Child as well as Special Rapporteurs in relation to the rights to food, health, water and other socioeconomic rights. The possibilities for this jurisprudence to help generate counter-hegemonic praxis are limited by a number of structural features of international legal discourse, notably its state centrism and formal neutrality on substantive political and economic questions.

Nevertheless, the socioeconomic rights jurisprudence of the CESCR contains a number of frames that can be incorporated into counter-hegemonic discourse. The approach generally adopted by the CESCR emphasises the state's proactive and protective role in ensuring the material wellbeing of its citizens in ways that challenge neo-liberal calls for withdrawal of the state from economic activity. A number of other frames within the CESCR's jurisprudence are worth considering: the

---

[10] Jackie Dugard and Katherine Drage, *Shields and Swords: Legal Tools for Public Water* (Municipal Services Project, 2012) 3.

presumption against retrogressive measures can be used to challenge the logic of austerity; the prohibition against discrimination can challenge privatisation and the introduction of user-fees that disproportionately impact on poor and marginalised groups; and the goal of progressive realisation of universal access to certain material entitlements condemns widespread poverty and material deprivation and opens up legal and other institutional channels to challenge them.

Of course, the neutrality of the human rights framework is a potential stumbling block to the counter-hegemonic capacity of socioeconomic rights discourse under international law. Where socioeconomic rights discourse is emphatic in emphasising particular outcomes to be achieved, namely that all individuals have access to the adequate level of resources required for their dignity, it has very little to say about the governance *processes* required to realise those rights. It is precisely this ambivalence that opens socioeconomic rights discourse to neo-liberal appropriation. For example, it can be argued that market liberalisation and privatisation create the optimal conditions for the enjoyment of socioeconomic rights. However, after over three decades of neo-liberal governance there is now a wealth of evidence concerning the negative impact that these policies can have, particularly on the most disadvantaged members of society. This is increasingly being noted, either explicitly or implicitly, in the reporting processes of human rights bodies and special rapporteurs.[11]

The strengths of the UN human rights framework can be utilised, whilst its limitations minimised, through the complementary use of such discourse in tandem with counter-hegemonic models for governance, such as food sovereignty, open-knowledge and water as commons. Whereas the latter models are clearer in terms of their prescriptive methods of governance, they are less clear with respect to their end goals. Socioeconomic rights can play a complementary role in tandem with these alternative models of governance within a 'portfolio' of counter-hegemonic articulations. The end-goal requirements of socioeconomic rights can be used negatively to critique existing neo-liberal models and positively to argue for the superiority of alternative models of governance. However, this 'chain of equivalence' between socioeconomic rights and counter-hegemonic models of governance cannot be constructed within the UN human rights framework due to the obstacles of

---

[11] Paul O' Connell, 'On Reconciling Irreconcilables: Neo-Liberal Globalisation and Human Rights' (2007) 7(3) *Human Rights Law Review* 483, 501–507.

state-centrism and apoliticism. This task must take place within the context of 'subaltern counterpublics'.

C.1.3 BUILDING SUBALTERN COUNTERPUBLICS

Whilst the aforementioned socioeconomic rights engagements can assist counter-hegemonic activity, they also run the risk of channelling activism into narrow legalism and incrementalism. In order to mitigate these threats, a third strategy is required that involves synthesising socioeconomic rights norms with counter-hegemonic articulations of alternative forms of governance. The appropriate institutional setting for this activity is the 'subaltern counter-public', which, to recall, is a parallel discursive area 'where members of the subordinated social groups invent and circulate counter discourses, which in turn permit them to formulate oppositional interpretations of their identities, interests and needs'.[12] It is within such settings that oppositional understandings of socioeconomic rights can be articulated and connected to struggles for new modes of social relations that transcend neo-liberalism.

The case study chapters all contained examples of subaltern counterpublics within the domain of global civil society operating in the manner just outlined. Many of these forums emerged in parallel to official global governance gatherings. Recall, for example, the parallel NGO/CSO Forums for food sovereignty that took place at the same time as the FAO World Food Summits. At these events the dominant neo-liberal approach to food security framework was criticised and the alternative food sovereignty model was presented as the viable means through which to realise the right to food. Similarly, the International Forum on the Defence of Water parallel events that took place during the official tri-annual World Water Forum (WWF) conferences provided a platform for criticising the policies of privatising, commercialising and commodifying water services whilst arguing that the right to water could be realised through alternative models of water governance based around the ideas of 'water as commons' and 'public-public partnerships'.[13] These parallel forums are an illustration of what Richard Falk

---

[12] Nancy Fraser, 'Rethinking the Public Sphere: A Contribution to the Critique of Actually Existing Democracy' in Francis Barker, Peter Hulme and Margaret Iversen (eds.), *Post Modernism and the Re-Reading of Modernity* (Manchester University Press, 1992) 84.

[13] Oriol Mirosa and Leila M Harris, 'Human Right to Water: Contemporary Challenges and Contours of a Global Debate' (2012) 44(3) *Antipode* 932, 942–943.

calls 'globalization-from-below': the subaltern alternative to neo-liberal 'globalization-from-above'.[14] Falk notes 'the creative tactics used by transnational participating groups' denied formal access to global conferences. These tactics enabled these movements to impact upon global agendas whilst strengthening transnational links through new forms 'of participatory politics that ... could be regarded as fledgling attempts to constitute "global democracy"'.[15]

The case study chapters also documented the creation of alternative declarations containing socioeconomic rights standards advanced by subaltern movements. Recall, for example, FoodFirst Information and Action Network's Draft International Code of Conduct on the Human Right to Adequate Food, La Via Campesina's Universal Declaration on the Rights of Peasants, the public health NGO coalition's Medical Research and Development Treaty or the International Forum on Globalization's Cochabamba Declaration. All of these treaties and declarations could be credited with catalysing further rights-based processes within the domain of 'official' global governance, but more profoundly they also helped to unite activists from diverse contexts, and provided a shared basis for discussions and articulations of alternative frameworks. This also allowed for the process of reticulation between socioeconomic rights discourses and alternative models of governance such as food sovereignty, access to knowledge and water as commons.

***

In analysing the transnational power of human rights norms, Thomas Risse and Kathryn Sikkink identify a 'boomerang' pattern of influence which involves domestic groups in repressive states bypassing their states directly through searching out international allies to try to bring pressure on their states from outside: 'National opposition groups, NGOs, and social movements link up with transnational networks and INGOs who then convince international human rights organizations, donor institutions, and/or great powers to pressure norm-violating states'.[16] Risse and

---

[14] Richard Falk, 'Resisting "Globalization-from-Above" through "Globalization-from-Below"' in Barry K Gills (ed.), *Globalization and the Politics of Resistance* (Macmillan, 2000) 46.
[15] ibid, 53–55.
[16] Thomas Risse and Kathryn Sikkink 'The Socialization of International Human Rights Norms into Domestic Practices' in Thomas Risse, Stephen C Ropp and Kathryn Sikkink (eds.), *The Power of Human Rights: International Norms and Domestic Change* (Cambridge University Press, 1999) 18.

Sikkink are concerned here with civil and political rights rather than socioeconomic rights. In relation to counter-hegemonic socioeconomic rights praxis, I would observe a process that is in some ways similar and in other ways distinct from Risse and Sikkink's 'boomerang' model. The similarity lies in the dialectic between the national and transnational, whereby national struggles for rights give rise to global forms of activism, which in turn help to create new frameworks of solidarity, leverage and legitimacy to assist human rights struggles of national movements. The difference is that the struggles of national movements were in many instances not directed primarily against national governments but rather against transnational actors such as transnational corporations (TNCs), northern states and institutions such as the World Bank, International Monitory Fund and the World Trade Organization. In some instances, national governments were even tactical allies in these struggles. For example, South African and Brazilian health activists rallied in support of efforts by their national governments to circumvent strict IPRs rules in the face of pressures from corporations and northern states.

In addition to establishing networks of international solidarity to support national struggles, the convergence of progressive states and civil society organisations at the international level created spaces where common problems could be identified and alternative solutions proposed. This in turn fed back in to the national context, where normative frameworks formulated at the level of global civil society provided the basis for national development strategies. Consider, for example, the enshrining of the concept of food sovereignty within a number of national constitutions in the Global South, increasing forms of south-south co-operation in the field of medical research and health rights or the constitutional entrenchment of the right to water in a manner that proscribed the privatisation of water services in Bolivia and Uruguay. Of course, none of the above-mentioned developments should be idealised. The path towards building alternatives to neo-liberalism will be a long and arduous one riddled with setbacks and contradictions. The point, however, is that these processes contain transformative potential that point beyond mere modifications to the existing neo-liberal paradigm.

This book has attempted to illustrate the ways that socioeconomic rights in global civil society may be able to play a positive role in struggles involving transformative hegemony, as well as highlight some of the potential limitations of evoking this discourse. Ultimately, the counter-hegemonic potential of socioeconomic rights discourse lies in its interaction with modes of analysis and forms of governance that seek to challenge the underlying structural logic of neo-liberal globalisation. The final word will be given to Robin Blackburn, who captured this truth so eloquently:

> 'Human rights' can serve as a valuable watchword and measure. But because inequality and injustice are structural, constituted by multiple intersecting planes of capitalist accumulation and realization, more needs to be said— especially in relation to financial and corporate power and how these might be curbed and socialized. The plight of billions can be represented as a lack of effective rights, but it is the 'property question'— the fact that the world is owned by a tiny elite of expropriators—that is constitutive of that plight. The slogan of rights takes us some way along the path; but it alone cannot pose the property question relevant to the 21st century.[17]

---

[17] Robin Blackburn, 'Reclaiming Human Rights' (2011) 69 *New Left Review* 126, 137–138.

BIBLIOGRAPHY

Aalbers MB, 'Neoliberalism is Dead ... Long Live Neoliberalism!' (2013) 37(3) *International Journal of Urban and Regional Research* 1083

Abbott KW, 'The Concept of Legalization' (2000) 54(3) *International Organization* 401

Adouharb MR and Cingranelli D, *Human Rights and Structural Adjustment* (Cambridge University Press, 2007)

Albo G, Gindin S and Panitch L, *In and Out of the Crisis: The Global Financial Meltdown and Left Alternatives* (PM Press, 2010)

Albrecht K, Germann J and Ratjen S, *How to Use the Voluntary Guidelines on the Right to Food: A Manual for Social Movements, Community Based Organisations and Non-Governmental Organisations* (FIAN, 2007)

Albuquerque C, 'Chronicle of an Announced Birth: The Coming into Life of the Optional Protocol to the International Covenant on Economic, Social and Cultural Rights – The Missing Piece of the International Bill of Human Rights' (2010) 32(1) *Human Rights Quarterly* 144

Alston P, 'Human Rights and Basic Needs: A Critical Assessment' (1979) 12 *Human Rights Journal* 19

— 'International Law and the Human Right to Food' in Alston P and Tomasevki K (eds.), *The Right to Food* (Martinus Nijhoff, 1984)

— 'Resisting the Merger and Acquisition of Human Rights by Trade Law: A Reply to Petersmann' (2002) 13(4) *European Journal of International Law* 815

— 'Foreword' in Langford M (ed.), *Social Rights Jurisprudence: Emerging Trends in International and Comparative Law* (Cambridge University Press, 2008)

Alter KJ, 'The Multiple Roles of International Courts and Tribunals: Enforcement, Dispute Settlement, Constitutional and Administrative Review' (2012) Faculty Working Papers, Paper 212, 3 http://scholarlycommons.law.northwestern.edu/cgi/viewcontent.cgi?article=1211&context=facultyworkingpapers accessed 16 August 2013

Althusser L, 'Ideology and Ideological State Apparatuses' in *Louis Althusser, Lenin and Philosophy and other Essays* (Monthly Review Press, 1972)

Amoore L and Langley P, 'Ambiguities of Global Civil Society' (2004) 30 *Review of International Studies* 89

Anderson P, 'Renewals' (2000) 1 *New Left Review*
Anderson RD and Wager H, 'Human Rights, Development and the WTO: The Cases of Intellectual Property and Competition Policy' (2006) 9(3) *Journal of International Economic Law* 707
Araghi F, 'The Invisible Hand and the Visible Foot: Peasants, Globalization and Dispossession', in Haroon Akram-Lodhi A and Kay C (eds.), *Peasants and Globalisation: Political Economic, Rural Transformation and the Agrarian Question* (Routledge, 2009)
Arendt H, *The Origins of Totalitarianism* (Schocken, 1958)
Armaline T, Glasberg DS and Purkayastha B, *The Human Rights Enterprise* (Polity, 2015)
Austria PM and Hofwegen P (eds.), *Synthesis of the 4th World Water Forum* (Comisión Nacional de Agua, 2006)
Baker G, 'Problems in the Theorisation of Global Civil Society' (2002) 50 *Political Studies* 928
Bakker I and Gill S, 'Global Political Economy and Social Reproduction' in Bakker I and Gill S (eds.), *Power, Production and Social Reproduction* (Palgrave, 2003)
Bakker K, "Neoliberalizing Nature? Market Environmentalism in Water Supply in England and Wales' (2005) 95(3) *Annals of the Association of American Geographers* 542
'The "Commons" versus the "Commodity": Alter-Globalization, Anti-Privatization and the Human Right to Water in the Global South' (2007) 39(3) *Antipode* 430
Balanyá B et al. (eds.), *Reclaiming Public Water: Achievements, Struggles and Visions for around the World* (Transnational Institute, 2005)
Balkin JM and Seigel RB, 'Principles, Practices and Social Movements' (2006) 154 *University of Pennsylvania Law Review* 926
Bantekas I and Oette L, *International Human Rights Law and Practice* (Cambridge University Press, 2013)
Barlow M and Clarke T, *Blue Gold: The Battle against the Corporate Theft of the World's Water* (Earthscan, 2002)
Barlow M, *Blue Covenant: The Global Water Crisis and the Coming Battle for the Right to Water* (The New Press, 2007)
Bartholomew A and Hunt A, "What's Wrong with Rights?" (1991) 9 *Law and Inequality* 1
Baxi U, *The Future of Human Rights* (Oxford University Press, 2002)
The Future of Human Rights (Oxford University Press, 2006)
'Adjudicatory Leadership in a Hyper-Globalizing World' in Gill S, *Global Crisis and the Crises of Global Leadership* (Cambridge University Press, 2012) 61
Bayles J and Smith S, *The Globalization of World Politics* (Oxford University Press, 2001)
Bello W, 'How to Manufacture a Global Food Crisis' (2008) 51(4) *Development* 450

The Food Wars (Verso, 2009)

Benford RD and Snow D, 'Ideology, Frame Resonance, and Participant Mobilization' (1988) 26 *Annual Review of Sociology* 197

Benhabib S, *Dignity in Adversity: Human Rights in Troubled Times* (Polity, 2011)

Bentham J, 'Anarchical Fallacies' in Waldron J (ed.), *Nonsense upon Stilts: Bentham, Burke and Marx on the Rights of Man* (Methuen, 1987)

Berry T, 'Rights of the Earth: Recognising the Rights of All Living Things' (2002) *Resurgence* 214

Beuchelt T and Virchow D, 'Food Sovereignty or the Human Right to Adequate Food: Which Concept Serves Better as International Development Policy for Global Hunger and Poverty Reduction?' (2012) 29 *Agriculture and Human Values* 259

Bieler A and Morton AD, 'The Gordian Knot of Agency–Structure in International Relations: A Neo-Gramscian Perspective' (2001) 7(1) *European Journal of International Relations* 5, 19

Bilchitz D, *Poverty and Fundamental Rights: The Justification and Enforcement of Socio-Economic Rights* (Oxford University Press, 2007)

Blackburn R, 'Crisis 2.0' (2011) 72 *New Left Review* 33

Block DR et al., 'Food Sovereignty, Urban Food Access, and Food Activism: Contemplating the Connections through Examples from Chicago' (2011) *Agriculture and Human Values*. Available at http://cis.uchicago.edu/outreach/summerinstitute/2012/documents/sti2012-block-sovereignty-access-activism-chicago.pdf accessed 1 December 2012

Boden M, 'Primitive Accumulation, Neo-Liberalism and Counter-Hegemony in the Global South: Re-Imagining the State' (n/d) www.nottingham.ac.uk/shared/shared_cssgj/Documents/smp_papers/boden.pdf accessed 3 July 2013

Boerma AH, *A Right to Food: A Selection from Speeches* (FOA, 1976)

Bond P and Dugard J, 'Water, Human Rights and Social Conflict: South African Experiences' (2008) *Law, Social Justice & Global Development (An Electronic Law Journal)* 1

'The Case of Johannesburg Water: What Really Happened at the Pre-Paid "Parish Pump"'(2008) 12 *Law, Democracy and Development* 1

Boyer R, *The Future of Economic Growth: As Old Becomes New* (Edward Elgar, 2004)

Boyle, 'The Second Enclosure Movement and the Construction of the Public Domain' (2003) 66 *Law and Contemporary Problems* 33

Brand U, 'Order and Regulation: Global Governance as a Hegemonic Discourse of International Politics?' (2005) 12(1) *Review of International Political Economy* 155–176

Briscoe J and Ferranti D, *Water for Rural Communities: Helping People Help Themselves* (World Bank, 1988)

Brown R, 'Unequal Burden: Water Privatization and Women's Human Rights in Tanzania' (2010) 18(1) *Gender and Development* 59

Brown W, *States of Injury: Power and Freedom in Late Modernity* (Princeton University Press, 1995)

——'"The Most We Can Hope For . . ." Human Rights and the Poverty of Fatalism' (2004) 103(2/3) *South Atlantic Quarterly* 451, 461–462

Buchanan JM, *Property as a Guarantor of Liberty* (Edward Elgar, 1993)

Cameron A and Palan R, 'The Imagined Economy: Mapping Transformations in the Contemporary State' (1999) 28(2) *Millennium* 267

Carroll T, 'Auctioning off Manila's Water Services: Market Extension, the World Bank and Socio-Institutional Neoliberalism' (2006) Asia Research Centre Working Paper No. 138 19 wwwarc.murdoch.edu.au/publications/wp/wp138.pdf accessed 9 June 2013

Carroll WK, 'Hegemony, Counter-Hegemony, Anti-Hegemony' (2006) 2(2) *Socialist Studies* 9

——'Hegemony and Counter-Hegemony in a Global Field' (2007) 1(1) *Studies in Social Justice* 36

——'Crisis, Movements, Counter-hegemony: In Search of the New' (2010) 2(2) *Interface* 168

Castro JE, 'Neoliberal Water and Sanitation Policies as a Failed Development Strategy: Lessons from Developing Countries' (2008) 8 *Progress in Development Studies* 63

Cerny P, 'Embedding Neoliberalism: The Evolution of a Hegemonic Paradigm' (2008) 2(1) *The Journal of International Trade and Diplomacy* 1

Chamas C, Prickril B and Sarnoff JD, 'Intellectual Property and Medicine: Towards Global Health Equity' in Wong T and Dutfield G (eds.), *Intellectual Property and Human Development: Current Trends and Future Scenarios* (Cambridge University Press, 2011)

Chandra R, *Knowledge as Property: Issues in the Moral Grounding of Intellectual Property Rights* (Oxford University Press, 2010)

Chang H, 'Company Profits Depend on the 'Welfare Payments' They Get from Society' *Guardian*, 5 April 2013

Charvet J and Kaczynska-Nay E, *The Liberal Project and Human Rights: The Theory and Practice of a New World Order* (Cambridge University Press, 2008)

Chaturvedi S and Thorsteinsdóttir H, 'BRICS and South-South Cooperation in Medicine: Emerging Trendsin Research and Entrepreneurial Collaborations' (RIS Discussion Paper no.177, March 2012) www.ris.org.in/images/RIS_images/pdf/dp177_pap.pdf accessed 1 December 2013

Cheru F, 'Debt, Adjustment and the Politics of Effective Response to HIV/AIDS in Africa' (2002) 23(2) *Third World Quarterly* 299

Chimni BS, 'International Institutions Today: An Imperial State in the Making' (2004) 15 *European Journal of International Law* 1

Chon M, 'A Review of Intellectual Property, Human Rights and Development: The Role of NGOs and Social Movements by Duncan Matthews' (2012) 2(2) *The IP Law Book Review* 63

Chong DPL, *Freedom from Poverty: NGOs and Human Rights Praxis* (University of Pennsylvania Press, 2010)

Chow H, 'Food Sovereignty in Venezuela' (*World Development Movement*, 1 August 2009) www.wdm.org.uk/food-sovereignty/food-sovereignty-venezuela accessed 2 November 2013

Claeys P, 'From Food Sovereignty to Peasants' Rights: An Overview of Via Campesina's Struggle for New Human Rights' (*La Via Campesina*, 15 May 2013) 2 http://viacampesina.org/downloads/pdf/openbooks/EN-02.pdf accessed 3 January 2014

— 'The Creation of New Rights by the Food Sovereignty Movement: The Challenge of Institutionalizing Subversion' (2014) 46(5) *Sociology* 844

Clarke S, 'The Neoliberal Theory of Society' in Saad-Fiho A and Johnson D (eds.), *Neoliberalism: A Critical Reader* (Pluto Press, 2005) 50

Colas A, 'Neoliberalism, Globalisation and International Relations' in Saad-Fiho A and Johnson D (eds.), *Neoliberalism: A Critical Reader* (Pluto Press, 2005)

Cornwall A and Brock K, 'Beyond Buzzwords: "Poverty Reduction", "Participation" and "Empowerment" in Development Policy' (2005) United Nations Research Institute for Social Development, Program Paper No. 10, 8

Correa C M, *Intellectual Property Rights, the WTO and Developing Countries: The TRIPS Agreement and Policy Options* (Zed, 2000)

— *Trade Related Aspects of Intellectual Property Rights: A Commentary on the TRIPS Agreement* (Oxford University Press, 2007)

Correa CM and Matthews D, *The Doha Declaration Ten Years on and Its Impact on Access to Medicines and the Right to Health: Discussion Paper* (UNDP, 2011)

Couch C, *The Strange Non-Death of Neoliberalism* (Polity, 2011)

Cox RW, 'Gramsci, Hegemony and International Relations' in Gill S (ed.), *Gramsci, Historical Materialism and International Relations* (Cambridge University Press, 1993)

— 'The Way Ahead: Toward a New Ontology of World Order' in Eschle C and Maiguashea B (eds.) *Critical Theory and World Politics* (Boulder, 2001)

Cranston M, 'Human Rights Real and Supposed' in David Daiches Raphael (ed.), *Political Theory and the Rights of Man* (Indiana University Press, 1967)

— *What Are Human Rights?* (Bodley Head, 1973)

Craven M, *The International Covenant on Economic, Social and Cultural Rights: A Perspective on Its Development* (Clarendon Press, 1995)

— 'The Justiciability of Economic, Social and Cultural Rights' in Burchill R Harris D and Owers A (eds.), *Economic, Social and Cultural Rights: Their Implementation in United Kingdom Law* (1999)

'The Violence of Dispossession: Extra-Territoriality and Economic, Social and Cultural Rights' in Baderin Mashood and McCorquodale Robert (eds.) *Economic, Social and Cultural Rights in Action* (Oxford University Press, 2007)

Cutler AC, 'Gramsci, Law, and the Culture of Global Capitalism' (2005) 8(4) *Critical Review of International Social and Political Philosophy* 527

"Towards a Radical Political Economy Critique of Transnational Economic Law" in Marks S (ed.) *International Law on the Left: Re-Examining Marx Legacies* (Cambridge University Press, 2008) 216

Davidson-Harden A, *Local Control and Management of Our Water Commons: Stories of Rising to the Challenge* (Council of Canadians 2007)

Davidson-Harden A, Naidoo A and Harden A, 'The Geopolitics of the Water Justice Movement' (2007) 11 *Peace, Conflict and Development* 1

De Schutter O, '*Food Commodity Speculation and Food Price Crises: Regulation to Reduce the Risks of Price Volatility*' (United Nations Office for the High Commissioner for Human Rights, 2010)

'Food Crisis: Five Priorities for the G20', Guardian (Manchester, 16 June 2011)

'The Right to Food in Times of Crisis' in Just Fair Report, *Freedom from Hunger: Realising the Right to Food in the UK* (Just Fair 2013) 8–9 www.edf.org.uk/blog/wp-content/uploads/2013/03/Freedom-from-Hunger.Just-Fair-Report.FINAL_.pdf accessed 7 October 2013

Desmaraisi A, 'The Vía Campesina: Consolidating an International Peasant and Farm Movement' (2002) 29(2) *Journal of Peasant Studies* 91

De Souza R, 'Liberal Theory, Human Rights and Water-Justice: Back to Square One?' (2008) *Law, Social Justice & Global Development Journal* 1

Dean J, *Democracy and Other Neoliberal Fantasies: Communicative Capitalism and Left Politics* (Duke University Press, 2009)

Dean M, *Governmentality: Power and Rule in Modern Society* (2nd edn., Sage, 2010)

Dijkstra G, 'The PRSP Approach and the Illusion of Improved Aid Effectiveness: Lessons from Bolivia, Honduras and Nicaragua' (2011) 29(1) *Development Policy Review* 111

Donnelly J, *Universal Human Rights in Theory and Practice* (2nd edn., Cornell University Press 2003)

Doty TP, 'Healthcare as a Commodity: The Consequences of Letting Business Run Healthcare' (March 2008) www.ucalgary.ca/familymedicine/system/files/Resident+Research+Review+Report.pdf accessed 2 December 2013

Douzinas C, *The End of Human Rights* (Hart, 2000)

Human Rights and Empire: The Political Philosophy of Cosmopolitanism (Routledge, 2007)

Downes C, 'Must the Losers of Free Trade Go Hungry? Reconciling WTO Obligations and the Right to Food' (2007) 47 *Virginia Journal of International Law* 619

Drahos P, 'Thinking Strategically about Intellectual Property Rights' (1997) 21(3) *Telegraph Communications Policy* 201

Dubreuil C, *The Right to Water: From Concept to Implementation* (World Water Council, 2006)

Dugard J, 'Civic Action and the Legal Mobilisation: The Phiri Water Meters Case' in Handmaker J and Berkhout R (eds.) *Mobilising Social Justice in South Africa: Perspectives from Researchers and Practitioners* (ISS and Hivos, 2010)

Urban Basic Services: Rights, Reality, Resistance' in Langford M et al. (eds), *Socioeconomic Rights in South Africa: Symbols or Substance?* (Cambridge University Press, 2014)

Dugard J and Drage K, *Shields and Swords: Legal Tools for Public Water* (Municipal Services Project, 2012)

Duggan L, *The Twilight on Equality* (Beacon Press, 2003)

Dumenil G and Levy D, 'The Neoliberal (Counter-)Revolution' in Saad-Fiho A and Johnson D, *Neoliberalism: A Critical Reader* (Pluto Press, 2005) 9

Dyck A, *Rethinking Rights and Responsibilities: The Moral Bonds of Community* (Georgetown University Press, 1994)

Dyer-Witheford N and Peuter G, 'Commons and Cooperatives' (2010) 4(1) *Affinities: A Journal of Radical Theory, Culture and Action* 30

Edelman M and James C, 'Peasants' Rights and the UN System: Quixotic Struggle? Or Emancipatory Idea Whose Time Has Come?' (2011) 38(1) *The Journal of Peasant Studies* 81, 96

Eide A, 'Realization of Social and Economic Rights and the Minimum Threshold Approach' (1989) 10 *Human Rights Law Journal* 35

'The International Human Rights System' in Asbjørn E, Krause C and Rosas A (eds.) *Economic, Social and Cultural Rights: A Textbook* (Martinus Nijhoff Publishers, 1995)

'Globalization, Universalization and the Human Right to Adequate Food' in Ogunrinade A, May JD and Oniang RO (eds.), *Not by Bread Alone: Food Security and Governance in Africa* (University of Witwatersrand Press, 1999)

'The Importance of Economic and Social Rights in an Age of Economic Globalization' in Eide WB and Kracht U (eds.), *Food and Human Rights in Development Volume 1: Legal and Institutional Dimensions in Selected Topics* (Intersentia, 2005)

Elver H, 'The Emerging Global Freshwater Crisis and the Privatization of Global Leadership in Gill S (ed.), *Global Crises and the Crisis of Global Leadership* (Cambridge University Press, 2012)

Elyada C, 'Court Decision Ends Privatization in Jakarta' (*Jakarta Post* 24 March 2015).

Evans T, 'The Human Right to Health?' (2002) 201 23(2) *Third World Quarterly* 197
The Politics of Human Rights: A Global Perspective (Pluto Press, 2005)

Human Rights in the Global Political Economy: Critical Perspectives (Lynne Reinner, 2011)

Evans T and Ayers AJ, 'In the Service of Power: The Global Political Economy of Citizenship and Human Rights' (2006) 10(3) *Citizenship Studies* 289

Evenson D, 'Cuba's Biotechnology Revolution' (2007) 9 (1) *MEDICC Review* 8

Fabre C, *Social Rights under the Constitution: Government and the Decent Life* (Oxford University Press, 1999)

Faini R and Grill E, 'Who Runs the IFIs?' (2004) Centre for Economic and Policy Research, Discussion Paper no. 4666, 21 www.dagliano.unimi.it/media/WP2004_191.pdf accessed 16 August 2013

Falk R, 'Resisting "Globalization-from-Above" through "Globalization-from-Below"' in Gills BK (ed.), *Globalization and the Politics of Resistance* (Macmillan, 2000) 46

Human Rights Horizons: The Pursuit of Justice in a Globalizing World (Routledge, 2000)

Farmer P, *Pathologies of Power: Health, Human Rights and the New War on the Poor* (University of California Press, 2005)

Femia JV, *Gramsci's Political Thought: Hegemony, Consciousness, and the Revolutionary Process* (Clarendon Press, 1981)

Ferraz OLM, 'Poverty and Human Rights' (2008) 28(3) *Oxford Journal of Legal Studies* 585

'The Right to Health in the Courts of Brazil: Worsening Health Inequities?' (2009) 11(2) *Health and Human Rights* 33

Fine B and Milonakis D, '"Useless but True": Economic Crisis and the Peculiarities of Economic Science' (2011) 19(2) *Historical Materialism* 3

Finnegan W, 'Letter from Bolivia: Leasing the Rain' *New Yorker* (New York, 8 April 2002) 43

Fisher WF and Ponniah T (eds.) *Another World Is Possible: Popular Alternatives to Globalization at the World Social Forum* (Zed Books, 2003)

Flynn S and Chirwa DM 'The Constitutional Implication of Commercializing Water in South Africa' in McDonald D and Ruiters G (eds.), *The Age of Commodity: Water Privatization in Southern Africa* (Earthscan, 2005)

Ford L, 'Challenging Global Environmental Governance: Social Movement Agency and Global Civil Society' (2003) 3(2) *Global Environmental Politics* 120

Forman L, 'Trade Rules, Intellectual Property and the Human Right to Health' (2007) 21(3) *Ethics & International Affairs* 337

'"Rights" and "Wrongs": What Utility for the Right to Health in Reforming Trade Rules on Medicines?' (2008) 10(2) *Health and Human Rights* 37

'From TRIPS-PLUS to Rights Plus? Exploring the Right to Health Impact Assessment of Trade-Related Intellectual Property Rights through the Thai Experience' (2012) 7(2) *Asian Journal of WTO and International Health Law and Policy* 347

Fraser N, 'Rethinking the Public Sphere: A Contribution to the Critique of Actually Existing Democracy' in Barker F, Hulme P and Iversen M (eds.), *Post Modernism and the Re-Reading of Modernity* (Manchester University Press, 1992)

Freeman A, 'Racism, Rights and the Quest for Equality of Opportunity' (1988) 23 *Harvard Civil Rights and Civil Liberties Law Review* 295

Fried C, *Right and Wrong* (Harvard University Press, 1978)

Friedman M, *Capitalism and Freedom* (University of Chicago Press, 1962)

Friedman M and Friedman R, *Free to Choose: A Personal Statement* (Secker & Warburg, 1980)

Friedmann H, 'International Regimes of Food and Agriculture since 1870' in Shanin T (ed.) *Peasants and Peasant Societies* (Basil Blackwell, 1987)

— 'The Political Economy of Food: A Global Crisis' (1993) 197(1) *New Left Review* 29

— 'Discussion: Moving Food Regimes Forward: Reflections on Symposium Essays'. (2009) 26(4) *Agriculture and Human Values* 335–344

Friedmann H and McMichael P, 'Agriculture and the State System: The Rise and Decline of National Agricultures, 1870 to the Present' (1989) 29(2) *Sociologia Ruralis* 93

Fukuyama F, *The End of History and the Last Man* (Penguin, 1992)

Furlong K, 'Neoliberal Water Management: Trends, Limitations, Reformulations' (2010) 1 *Environment and Society: Advances in Research* 46

Gabel P, 'The Phenomenology of Rights – Consciousness and the Pact of the Withdrawn Selves' (1984) 62 *Texas Law Review* 1563

Gargarella R, Domingo P and Roux T, 'Courts, Rights and Social Transformation: Concluding Reflections' in Gargarella R, Domingo P and Roux T (eds.), *Courts and Social Transformation in New Democracies: An Institutional Voice for the Poor?* (Ashgate, 2006)

Gathii JT, The Neoliberal Turn in Regional Trade Agreements (2011) 86 *Washington Law Review* 421

Gearty C, *Can Human Rights Survive?* (Cambridge University Press, 2006)

George E, 'The Human Right to Health and HIV/AIDS: South Africa and South-South Cooperation to Reframe Global Intellectual Property Principles and Promote Access to Essential Medicines' (2011) 18(1) *Indiana Journal of Global Legal Studies* 176

George S, 'How to Win the War of Ideas: Lessons from the Gramscian Right' (1997) *Dissent* 47

Germain R and Kenny M, 'Engaging Gramsci: International Relations Theory and the New Gramscians' (1998) 24(3) *Review of International Studies* 3

Germann J, 'The Human Right to Food: "Voluntary Guidelines" Negotiations' in Atasoy Y (ed.), *Hegemonic Transitions, the State and Crisis in Neoliberal Capitalism* (Routledge, 2009)

Gervais D, 'International Intellectual Property and Development: A Roadmap to Balance?' (2005) 2(4) *Journal of Generic Medicines* 327

Gewirth A, *Reason and Morality* (University of Chicago Press, 1978)

Gianviti F, 'Economic, Social and Cultural Rights and the International Monetary Fund' (2002) www.imf.org/external/np/leg/sem/2002/cdmfl/eng/gianv3.pdf accessed 12 December 2013

Gibson D, 'Let Them Drink Pop: Detroit's Water Crisis and the Fight for Basic Human Rights in the Motor City' (*Truthout*, 26 July 2014). www.truth-out.org/opinion/item/25183-let-them-drink-pop-detroits-water-crisis-and-the-fight-for-basic-human-rights-in-the-motor-city/

Gill R, 'Discourse Analysis' in Bauer MW and Gaskell GD (eds.), *Qualitative Researching with Text, Image and Sound: A Practical Handbook for Social Research* (Sage, 2000) 172

Gill S, *American Hegemony and the Trilateral Commission* (Cambridge University Press, 1990)

'Globalisation, Market Civilisation, and Disciplinary Neoliberalism' (1995) 24 *Millennium: Journal of International Studies* 399

'New Constitutionalism, Democracy and Global Political Economy' (1998) 10(1) *Pacifica Review* 23

'Toward a Postmodern Prince? The Battle in Seattle as a Moment in the New Politics of Globalisation' (2000) 29(1) *Millennium: Journal of International Studies* 131

'Constitutionalizing Inequality and the Clash of Globalizations' (2002) 4(2) *International Studies Review* 47

'Introduction: Global Crisis and the Crises of Global Leadership' in Gill S (ed.) *Global Crisis and the Crises of Global Leadership* (Cambridge University Press, 2012)

Gill S and Claire Cutler A. (eds.), *New Constitutionalism and World Order* (Cambridge University Press, 2014)

Gill S and Law D, 'Global Hegemony and the Structural Power of Capital' in Gill S (ed.) *Gramsci, Historical Materialism and International Relations* (Cambridge University Press, 1993) 93–124

Gills BK, 'Introduction: Globalization and the Politics of Resistance 'in Gills BK (ed.), *Globalization and the Politics of Resistance* (Macmillan, 2000) 3

Gilpin R, *Understanding the Global Political Economy* (Princeton University Press, 2001)

Gispin R, *War and Change in World Politics* (Cambridge University Press, 1981)

Gleick P, 'The Right to Water' (2007) 2 www.pacinst.org/reports/human_right_may_07.pdf accessed 28 May 2013

Glendon MA, *Rights Talk: The Impoverishment of Political Discourse* (Free Press, 1991)

Godoy AG, *Of Medicines and Markets: Intellectual Rights and Human Rights in the Free Trade Era* (Stanford University Press, 2013)

Goldman M, 'How "Water for All!" Policy Became Hegemonic: The Power of the World Bank and Its Transnational Policy Networks' (2007) 38 *Geoforum* 786

Goldstein J, 'Introduction: Legalization and World Politics' (2000) 54(3) *International Organization* 385

Gonzalez CG, 'Institutionalizing Inequality: The WTO Agreement on Agriculture, Food Security and Developing Countries' (2002) 27 *Colombia Journal of Environmental Law* 433

'An Environmental Justice Critique of Comparative Advantage: Indigenous Peoples, Trade Policy and the Mexican Neoliberal Reforms' (2011) 32(3) *University of Pennsylvania Journal of International Law* 723

Gonzalez M and Yanes M, *The Last Drop: The Politics of Water* (Pluto Press, 2015)

Goodale M and Merry SE (eds.), *The Practice of Human Rights: Tracking Law between the Global and the Local* (Cambridge University Press, 2007)

Gramsci A, in Hoare Q and Smith G (eds. and trans.), *Selections from the Prison Notebooks of Antonio Gramsci* (Lawrence & Wishart, 1971)

Gramsci A, in Hoare Q (ed. and trans.), *Selections from Political Writings (1921–1926)* (Lawrence & Wishart, 1978)

Gramsci A, *Selections from Political Writings (1910–1920)* (Lawrence & Wishart, 1987)

Gramsci A, in Bootham D (ed. and trans.), *Further Selections from the Prison Notebooks* (Lawrence & Wishart, 1995)

Grass AK, 'How to Crash a Nestlé Waters Press Conference' (*foodandwaterwatch.org*, April 5 2013) www.foodandwaterwatch.org/blogs/how-to-crash-a-nestle-waters-press-conference/ accessed 27 May 2013

Gross A, 'Is There a Human Right to Private Health Care?' (2013) 41(1) *The Journal of Law, Medicine and Ethics* 138

Grusky S, 'The IMF, the World Bank and the Global Water Companies: A Shared Agenda' (2001) available at www.internetfreespeech.org/documents/sharedagenda.pdf accessed 6 June 2013

Hale SL, 'Water Privatization In the Philippines: The Need to Implement the Human Right to Water' (2006) 15 *Pacific Rim & Policy Journal* 765

Hall D and Lobina E, *Pipe Dreams: The Failure of the Private Sector to Invest in Developing Countries* (World Development Movement 2006) http://gala.gre.ac.uk/3601/1/PSIRU_9618_%2D_2006%2D03%2DW%2Dinvestment.pdf accessed 11 December 2013

Hall D et al., 'Controlling the Agenda at WWF – the Multinationals' Network' (2009) www.waterjustice.org/uploads/attachments/wwf5-controlling-the-agenda-at-wwf.pdf accessed 30 March 2013

Public-public partnerships (PUPs) in water (PSI-TNI-PSIRU, 2009) 2 www.tni.org/sites/www.tni.org/files/download/pupinwater.pdf accessed 10 December 2013

Hall D, Lobina E and Correl V, *'Replacing Failed Private Water Contracts* (Public Services International Research Unit, 2010)

Hall S, 'The Work of Representation' in Hall S (ed.), *Representations: Cultural Representations and Signifying Practices* (Sage, 1997)

Hall S, Massey D and Rustin M, 'After Neoliberalism: Analysing the Present' in Hall S, Massey D and Rustin M (eds.) *After Neoliberalism? The Kilburn Manifesto* (Surroundings, 2013) 4 available at http://lwbooks.co.uk/journals/soundings/pdfs/manifestoframingstatement.pdf accessed 8 August 2013

Hardin G, 'Tragedy of the Commons' (1968) 162 *Science* 1243

Haroon Akram-Lodhi A, 'Land Grabs, the Agrarian Question and the Corporate Food Regime' (2015) 2(2) *Canadian Food Studies* 233.

Harten GV, 'Private Authority and Transnational Governance: The Contours of the International System of Investor Protection' (2005) 21(4) *Review of International Political Economy* 600

Harten S, *The Rise of Evo Morales and the MAS* (Zed Books, 2011)

Harvey D, *A Brief History of Neoliberalism* (Oxford University Press, 2005)

 'Neoliberalism as Creative Destruction' (2007) 610 *The Annals of the American Academy of Political and Social Science* 21

Harvey F, 'Global Majority Faces Water Shortages "Within Two Generations"' *Guardian* (24 May 2013)

Harvey P, 'Aspirational Law' (2004) 52 *Buffalo Law Review* 701

Hathaway O, 'Do Human Rights Treaties Make a Difference?' (2002) 111(8) *Yale Law Journal* 1935, 1944

Hayek FA, *The Road to Serfdom* (first published 1944, ARK, 1986)

 The Constitution of Liberty (Routledge & Kegan Paul, 1960)

 Law, Legislation and Liberty Volume 2: The Mirage of Social Justice (Routledge and Kegan Paul, 1976)

Held D and McGrew A, *Governing Globalization: Power, Authority and Global Governance* (Wiley, 2002)

Helfer LR, 'Regime Shifting: The TRIPs Agreement and New Dynamics of International Intellectual Property Lawmaking' (2004) 29(1) *Yale Journal of International Law* 1

Helfer LR and Austin GW, *Human Rights and Intellectual Property: Mapping the Global Interface* (Cambridge University Press 2011)

 'Parmaceutical Patents and the Human Right to Health: The Contested Evolution of the Transnational Legal Order on Access to Medicines' in Haliday TC and Shaffer G (eds.) *Transnational Legal Orders* (Cambridge University Press, 2014)

Heller KJ, 'Bechtel Abandons Its ICSID Claim against Bolivia' (*Opinio Juris*, 10 February 2006) http://lawofnations.blogspot.co.uk/2006/02/bechtel-abandons-its-icsid-claim.html accessed 16 December 2013

Hermann RM, 'World Health Assembly: Members Debate US Proposed Advisory Meeting On Health R&D' (IP Watch, 24 May 2014) www.ip-watch.org/2013/05/24/world-health-assembly-members-debate-us-proposed-advisory-meeting-on-health-rd/ accessed 1 December 2013

Hertel S and Minkler L, 'Economic Rights: The Terrain' in Shareen Hertel and Lanse Minkler (eds.), *Economic Rights: Conceptual, Measurement, and Policy Issues* (Cambridge University Press, 2007)

Hertzlinger R, *Market-Driven Health Care: Who Wins, Who Loses, in the Transformation of America's Largest Service Industry* (Addison-Wesley, 1997)

Hettinger EC, 'Justifying Intellectual Property' in Moore AD (ed.), *Intellectual Property: Moral Legal and International Dimensions* (Rowman & Littlefield, 1997)

Heywood M, 'Debunking "Conglomo-talk": A Case Study of the Amicus Curie as an Instrument for Advocacy, Investigation and Mobilisation' (2002) 6 *Law Democracy and Development* 12

Hiskes RP, 'Missing the Green: Golf Course Ecology, Environmental Justice, and Local "Fulfillment" of the Human Right to Water' (2010) 32(2) *Human Rights Quarterly* 326

Hohfield WN, 'Some Fundamental Legal Concepts as Applied to Judicial Reasoning' (1913) 23 *Yale Law Journal* 16

'Some Fundamental Legal Concepts as Applied to Judicial Reasoning' (1917) 26 *Yale Law Journal* 710

Holt-Giménez E and Patel R, *Food Rebellions! Crisis and the Hunger for Justice* (Pambazuka Press, 2009)

Holt-Giménez E and Shattuck A, 'Food Crises, Food Regimes and Food Movements: Rumblings of Reform or Tides of Transformation?' (2011) 38 *Journal of Peasant Studies* 109

Holt-Giménez E and Altieri MA, 'Agroecology, Food Sovereignty, and the New Green Revolution' (2013) 37(1) *Agroecology and Sustainable Food Systems* 90

Horwitz MJ, 'Rights' (1988) 23 *Harvard Civil Rights-Civil Liberties Law Review* 393

Houtart F, 'Knowledge, Copyright and Patents (ii): Conference Synthesis' in Fisher WF and Ponniah T (eds.) *Another World Is Possible: Popular Alternatives to Globalization at the World Social Forum* (Zed, 2003)

Howard G and Bartram J, *Domestic Water Quantity, Service Level and Health* (World Health Organization, 2003)

Howard-Hassmann RE, 'The Right to Food under Hugo Chavez' (2015) 37(4) *Human Rights Quarterly* 1024

Hughes J, 'The Philosophy of Intellectual Property' (1988) 77 *Georgetown Law Journal* 287, 288

Hunt A, 'Rights and Social Movements: Counter-Hegemonic Strategies' (1990) 17(3) *Journal of Law and Society* 311

Ignatieff M, *Human Rights as Politics and Idolatry* (Princeton University Press, 2001)

Ishay MR, *The History of Human Rights: From Ancient Times to the Globalization Era* (University of California Press, 2004)

Jansen K and Gupta A, 'Anticipating the Future: "Biotechnology for the Poor" as Unrealized Promise?' (2009) 41 *Futures* 436

Jarosz L, 'Comparing Food Security and Food Sovereignty Discourses' (2014) 4(2) *Dialogues in Human Geography* 168.

Johnston J and Laxer G, 'Solidarity in the Age of Globalization: Lessons from the anti-MAI and Zapatista Struggles' (2003) 32 *Theory and Society* 39

Joseph S, 'Trade to Live or Live to Trade: The World Trade Organization, Development, and Poverty' in Baderin M and McCorquodale R (eds.) *Economic, Social and Cultural Rights in Action* (Oxford University Press, 2007)

Juhasz A, 'Cochabamba Water War Presents Globalization Alternative to the World' (IFG Newsletter, February 2001) www.tyrannyofoil.org/article.php?id=90 accessed 5 June 2013

Kapczynski A, 'The Access to Knowledge Mobilization and the New Politics of Intellectual Property' (2008) 117 *Yale Law Journal* 804

Kapoor T, 'Is Successful Water Privatization a Pipe Dream?: An Analysis of Three Global Case Studies' (2015) (40) *Yale Journal of International Law* 157.

Keck M and Sikkink K, 'Transnational Advocacy Networks in the Movement Society in Meyer DS and Tarrow S (eds.) *The Social Movement Society* (Rowman and Littlefield, 1998)

Activists beyond Borders (Cornell University Press, 1998)

Kelley D, *A Life of One's Own: Individual Rights and the Welfare State* (Cato Institute, 1998)

Kennedy D (David), 'A New Stream of International Law Scholarship' in Beck RJ, Arend AC and Lugt RDV (eds.) *International Rules: Approaches from International Law and International Relations* (Oxford University Press, 1996)

(Duncan), 'The Critique of Rights in Critical Legal Studies' in Brown W and Halley J (eds.), *Left Legalism/Left Critique* (Duke University Press, 2002)

The Dark Side of Virtue: Reassessing International Humanitarianism (Princeton University Press, 2004)

Keohane R, *After Hegemony: Cooperation and Discord in the World Political Economy* (Princeton University Press, 1984)

Kerr TV, 'Supplying Water Infrastructure to Developing Countries via Private Sector Project Financing' (1995) 8 *Georgetown international Environmental Law Review* 91

Kerry VB and Lee K, 'TRIPS, the Doha Declaration and paragraph 6 decision: what are the remaining steps for protecting access to medicines?' (2007) *Globalization and Health* http://researchonline.lshtm.ac.uk/9769/1/1744-8603-3-3.pdf accessed 13 April 2013

Khor M, 'South American Ministers Vow to Avoid TRIPS-plus Measures' (*Third World Network*, 1 June 2006) www.twnside.org.sg/title2/twninfo414.htm accessed 2 December 2013

Kiddell-Monroe R, Iversen JH and Gopinathan U, 'Medical R&D Convention Derailed: Implications for the Global Health System' (2013) *Journal of Health Diplomacy* 1. www.hsph.harvard.edu/global-health-and-population/files/2013/06/Medical-R-and-D-Convention-Derailed-Implications-For-The-Global-Health-System-3.pdf accessed 1 December 2013

Kindleberger C, *The World in Depression, 1929–39* (University of California Press, 1973)

Kipnis A, 'Neoliberalism Reified: Suzhi Discourse and Tropes of Neoliberalism in the People's Republic of China' (2007) 13 *Journal of the Royal Anthropological Institute* 383

Kirkpatrick C and Parker D, 'Domestic Regulation and the WTO: The Case of Water Services in Developing Countries' (2005) 28(10) *World Economy* 1491

Klein N, 'Bush's AIDS Test', *The Nation* (27 October 2003) www.thenation.com/article/bushs-aids-test# accessed 28 November 2013

The Shock Doctrine: The Rise of Disaster Capitalism (Penguin, 2007)

Klug H, 'Campaigning for Life: Building a New Transnational Solidarity in the Face of HIV/AIDS and TRIPS' in De Sousa Santos B and Rodriguez-Garavito CA (eds.) *Law and Globalization from Below: Towards a Cosmopolitan Legality* (Cambridge University Press, 2005)

Klugman J et al. (eds.), *A Sourcebook for Poverty Reduction Strategies, Volume II: Macroeconomic and Sectoral Approaches* (World Bank, 2002)

Kohl B, 'Challenges to Neoliberal Hegemony in Bolivia' (2006) 38(2) *Antipode* 305

Koskenniemi M, *From Apology to Utopia: The Structure of International Legal Argument* (Cambridge University Press, 2005)

'A Response' (2006) 7(12) *German Law Journal* 1103, 1103

Kothari M, 'Privatizing Human Rights – The Impact of Globalization on Access to Adequate Housing, Water and Sanitation' (*Social Watch*, 2003) http://unpan1.un.org/intradoc/groups/public/documents/apcity/unpan010131.pdf accessed 30 June 2013

Kotz DM, *The Rise and Fall of Neoliberal Capitalism* (Harvard University Press, 2015)

Krasner S, 'Structural Causes and Regime Consequences: Regimes as Intervening Variables' (1982) 36(2) *International Organizations* 185.

Kunnermann R and Epal-Ratjen S, *The Right to Food: A Resource Manual for NGOs* (AAAS Science and Human Rights Program, 1999) 49

Laclau E and Mouffe C, *Hegemony and Socialist Strategy: Towards a Radical Democratic Politics* (Verso, 1985)

Lamy P, 'Towards Shared Responsibility and Greater Coherence: Human Rights, Trade and Macroeconomic Policy' (2010) www.wto.org/english/news_e/sppl_e/sppl146_e.htm accessed 13 December 2013

Lang A, *World Trade Law after Neoliberalism: Re-Imagining the Global Economic Order* (Oxford University Press, 2011)

Langford M, 'Ambition that Overleaps Itself? A Response to Stephen Tully's Critique of the General Comment on the Right to Water' (2006) 24(3) *Netherlands Quarterly of Human Rights* 433

—— et al. (eds.), *Global Justice, State Duties: The Extraterritorial Scope of Economic, Social, and Cultural Rights in International Law* (Cambridge University Press, 2013)

Larner W, 'Neo-Liberalism: Policy, Ideology, Governmentality' (2000) 63 *Studies in Political Economy* 5

Lee T, 'Water Management since the Mar Del Plata Action Plan: Lessons for the 1990s' (1992) 16(3) *Natural Resources Forum* 202

Lenin VI, *The State and Revolution* (Penguin, 1992)

Leys C, *Market-Driven Politics* (Verso, 2001)

Litowitz D, 'Gramsci, Hegemony and the Law' (2000) *Brigham Young University Law Review* 515, 527

Little D, 'The Nature and Basis of Human Rights', in Outka G and Reeder JP, *Prospects for a Common Morality* (Princeton University Press, 1993)

Lobina E, UK Water Privatization: A Briefing (Public Services International Research Unite, 25 June 2001) www.archives.gov.on.ca/en/e_records/walkerton/part2info/partieswithstanding/pdf/CUPE18UKwater.pdf accessed 21 June 2013

Locke J, *Two Treatises of Government* (first published 1690, Cambridge University Press, 1994)

Love J, 'WHO negotiators propose putting off R&D treaty discussions until 2016' (*Knowledge Ecology International*, 28 November 2012) http://keionline.org/node/1612 accessed 1 December 2013

MacIntyre A, *After Virtue: A Study of Moral Theory* (Duckworth, 1981)

MacLean J, 'Cuba Creates Four Anti-Cancer Vaccines, Media Ignores It' (*Green Left Review*, 25 February 2013) www.greenleft.org.au/node/53426 accessed 10 May 2013

Macpherson CB, *The Political Theory of Possessive Individualism: Hobbes to Locke* (Oxford University Press, 1964)

Malpas J, 'The Forms of Water: In the Land and in the Soul' (2006) 1(2) *Transforming Cultures eJournal*. http://epress.lib.uts.edu.au/journals/index.php/TfC/article/view/257

Manahan MA, Zanzanaini G and Campero C, '*Future of Water Movement Session: A Summary*' (2012) www.fame2012.org/en/2012/04/10/future-of-water-movement/ accessed 30 September 2013

Marcus D, 'The Normative Development of Socioeconomic Rights through Supranational Adjudication' (2006) 42 *Stanford Journal of International Law* 53

Marin P, *Public-Private Partnerships for Urban Water Utilities: A Review of Experiences in Developing Countries* (World Bank, 2009)

Marks S, 'Exploitation as an International Legal Concept' in Marks S (ed.) *International Law on the Left: Re-Examining Marxist Legacies* (Cambridge University Press, 2008) 281

'Human Rights and Root Causes' (2011) 41(1) *Modern Law Review* 57

Marks SP, 'From the "Single Confused Page" to the "Decalogue for Six Billion Persons": The Roots of the Universal Declaration of Human Rights in the French Revolution' (1998) 20(3) *Human Rights Quarterly* 459

Marks SP and Benedict AL, 'Access to Products, Vaccines and Medical Technologies' in Zuniga JM, Marks SP and Gostin LO (eds.), *Advancing the Human Right to Health* (Oxford University Press, 2013)

Marshall TH, 'Citizenship and Social Class' in Marshall TH and Bottomore T (eds.) *Citizenship and Social Class* (Pluto Press, 1992)

Marx K, *Capital: A Critical Analysis of Capitalist Production Volume 1* (Lawrence & Wishart, 1961)

A Contribution to the Critique of Political Economy (International Publishers Company, 1970)

Critique of the Gotha Program (International Publishers, 1973)

'On the Jewish Question' reproduced in O'Malley J (ed.), *Marx: Early Political Writings* (Cambridge University Press, 1993)

Marx K and Engels F, *The German Ideology: Part One* (Lawrence & Wishart, 1970)

Matthews D, *Globalising Intellectual Property Rights: The TRIPS Agreement* (Routledge, 2002)

'Is History Repeating Itself? The Outcome of the Negotiations on Access to Medicines, The HIV/AIDS Pandemic and Intellectual Property Rights in the World Trade Organisation' (2004) *Law, Social Justice & Global Development* 1

Intellectual Property, Human Rights and Development (EE, 2011)

Mayne R, 'The Global Campaign on Patents and Access to Medicines: An Oxfam Perspective' in Drahos P and Mayne R (eds.), *Global Intellectual Property Rights: Knowledge, Access, Development* (Palgrave Macmillan, 2002)

McDonald D and Smith L, 'Privatizing Cape Town: From Apartheid to Neoliberalism in the Mother City' (2004) 41(8) *Urban Studies* 1461

McKinley DT, 'The Struggle against Water Privatization in South Africa' in Balanyá B et al. (eds.), *Reclaiming Public Water: Achievements, Struggles and Visions for Around the World* (Transnational Institute, 2005)

McMichael P, 'Food Security and Social Reproduction: Issues and Contradictions' in Bakker I and Gill S (eds.) *Power, Production and Social Reproduction* (Palgrave Macmillan, 2003)

'Banking on Agriculture: A Review of the World Development Report 2008' (2009) 9(2) *Journal of Agrarian Change* 235

'A Food Regime Genealogy' (2009) 36(1) *Journal of Peasant Studies* 139

'The Land Grab and Corporate Food Regime Restructuring' (2012), 39(3–4), *The Journal of Peasant Studies* 681

Mechem K, 'Food Security and the Right to Food Discourse in the United Nations' (2004) 10 *European Journal of International Law* 631

Mehta L and Mirosa O, *Financing Water for All: Behind the Border Policy. Convergence in Water Management* (IDS, 2004)

Meier BM, 'Employing Health Rights for Global Justice: The Promise of Public Health in Response to the Insalubrious Ramifications of Globalization' (2006) 39 *Cornell International Law Journal* 711

Mendonca ML, 'Human Rights: Conference Synthesis on Economic, Social and Cultural Rights' in Fisher WF and Ponniah T (eds.) *Another World Is Possible: Popular Alternatives to Globalization at the World Social Forum* (Fernwood, 2003) 309

Merges RP, *Justifying Intellectual Property* (Harvard University Press, 2011)

Merry SE, *Human Rights and Gender Violence: Translating International Law into Local Justice* (Chicago University Press, 2005)

Mertus J, 'Doing Democracy "Differently": The Transformative Potential of Human Rights NGOs in Transnational Civil Society' (1999) 15 *Third World Legal studies* 205

Michelman FI, 'The Constitution, Social Rights, and Liberal Political Justification' (2003) 1(1) *International Journal of Constitutional Law* 13

Miliband R, 'Counter-Hegemonic Struggles' (1990) 26 *Socialist Register* 346

Miller D, 'How Neoliberalism Got Where It Is: Elite Planning, Corporate Lobbying and the Release of the Free Market' in Birch K and Mykhnenko V (eds.) *The Rise and Fall of Neoliberalism* (Zed Books, 2010) 24

Milne S, *The Enemy Within: The Secret War against the Miners* (Verso, 2004)

Minogue K, 'The History of the Idea of Human Rights' in Laquer W and Rubin B (eds.), *The Human Rights Reader* (New Amsterdam Library, 1979)

Mirosa O and Harris LE, 'Human Right to Water: Contemporary Challenges and Contours of a Global Debate' (2012) 44(3) *Antipode* 932

Mittelman JH, *Wither Globalization?* (Routledge, 2004)

Monbiot G, 'Property, Theft and How We Must Breach This Sacred line' *Guardian*, 25 March 2013

Moon G, 'Trading in Good Faith? Importing States' Economic Human Rights Obligations into the WTO's Doha Round Negotiations' (2013) 13(2) *Human Rights Law Review* 245

Moore AD, 'Introduction' in Moore AD (ed.), *Intellectual Property: Moral Legal and International Dimensions* (Rowman & Littlefield, 1997)

'Towards a Lockean Theory of Intellectual Property' Moore AD (ed.), *Intellectual Property: Moral Legal and International Dimensions* (Rowman & Littlefield, 1997)

Morgan B, 'The Regulatory Face of the Human Right to Water' (2004) 15 *Journal of Water Law* 179

Morgenthau H and Thompson K, *Politics among Nations: The Struggle for Power and Peace* (McGraw-Hill, 2005)

Morsink J, *The Universal Declaration of Human Rights: Origins, Drafting and Intent* (University of Pennsylvania Press) 157–238

Morton AD, 'Mexico, Neoliberal Restructuring and the EZLN: A Neo-Gramscian Analysis' in Gills BK (ed.), *Globalization and the Politics of Resistance* (Macmillan, 2000)

Mosher L, 'Welfare Reform and the Re-Making of the Model Citizen' in Young M et al. (eds.), *Poverty: Rights, Social Citizenship and Legal Activism* (UBC Press, 2007) 119

Moshman R, 'The Constitutional Right to Water in Uruguay' (2005) 5(1) *Sustainable Development Law & Policy*, Winter 65

Mowbray J, 'The Right to Food and the International Economic System: An Assessment of the Rights-Based Approach to the Problem of World Hunger'. (2007) 20(3) *Leiden Journal of International Law* 545

Moyn S, *The Last Utopia: Human Rights in History* (Harvard University Press, 2010)
'A Powerless Companion: Human Rights in the Age of Neoliberalism' (2014) 77(4) *Law and Contemporary Problems* 147.

Mueller J, 'The Tiger Awakens: The Tumultuous Transformation of India's Patent System and the Rise of Indian Pharmaceutical Innovation' (2007) 68 *University of Pittsburgh Law Review* 491

Muhr T, 'Counter-Hegemonic Regionalism and Higher Education for all: Venezuela and the ALBA' (2010) 8(1) *Globalisation, Societies and Education* 39

Muller M, 'Parish Pump Politics: The Politics of Water Supply in South Africa' (2004) 7(1) *Progress in Development Studies* 33

Munck R, 'Neoliberalism and Politics, and the Politics of Neoliberalism' in Saad-Fiho A and Johnson D, *Neoliberalism: A Critical Reader* (Pluto Press, 2005) 60

Murray K, 'Whose Right to Water?' (2003) *Dollars & Sense* (November/December edition) 23

Mutua M, 'Standard Setting in Human Rights: Critique and Prognosis' (2007) 29(3) *Human Rights Quarterly* 547

Naegele J, 'What Is Wrong with Fully Fledged Privatization?' (2004) 6 *Journal of Law and Social Challenges* 99

Naidoo P, 'Eroding the Commons: Prepaid Water Meters in Phiri, Soweto' (Public Citizen, 2008) www.citizen.org/cmep/article_redirect.cfm?ID=11991 accessed 24 June 2013

Narula S, 'International Financial Institutions, Transnational Corporations and Duties of States' in Langford et al. (eds.) *Socioeconomic Rights in South Africa: Symbols or Substance?* (Cambridge University Press, 2014)

Nash F, 'Participation and Passive Revolution: The Reproduction of Neoliberal Water Governance Mechanisms in Durban, South Africa' (2012) 45(1) *Antipode* 101

Nash K, *The Political Sociology of Human Rights* (Cambridge University Press, 2015)

Neier A, 'Social and Economic Rights: A Critique' (2006) 13(2) *Human Rights Brief* 1

Nelson PJ and Dorsey E, *New Rights Advocacy: Changing Strategies of Development and Human Rights NGOs* (Georgetown University Press, 2008)

Nemeth T, *Gramsci's Philosophy: A Critical Study* (Harvester Press, 1981)

Nozick R, *Anarchy, State and Utopia* (first published 1974, Blackwell, 2001)

O'Connell P, 'On Reconciling Irreconcilables: Neo-Liberal Globalisation and Human Rights' (2007) 7(3) *Human Rights Law Review* 483

'The Human Rights to Health in an Age of Market Hegemony' in Harrington J and Stuttaford M (eds.) *Global Health and Human Rights: Legal and Philosophical Perspectives* (Routledge, 2010)

Vindicating Socioeconomic Rights: International Standards and Comparative Experiences (Routledge, 2011)

'The Death of Socio-Economic Rights' (2011) 74(4) *Modern Law Review* 532

'Let Them Eat Cake: Socio-Economic in an Age of Austerity' in Harvey C, Nolan A and O'Connell R (eds.), *Human Rights and Public Finance* (Hart, 2013)

'Demand the Future: The Right to Water and Another Ireland' (*Critical Legal Thinking*, 29 September 2014) www.criticallegalthinking.com/2014/09/29/demanding-future-right2water-another.ireland/

Odell JS and Sell SK, 'Reframing the Issue: The WTO Coalition on Intellectual Property and Public Health, 2001' in Odell JS (ed.), *Negotiating Trade: Developing Countries in the WTO and NAFTA* (Cambridge University Press, 2006)

Okogbule NO, 'Modest Harvests: Appraising the Impact of Human Rights Norms on International Economic Institutions in Relation to Africa' (2011) 15(5) *The International Journal of Human Rights* 728

Oprea L et al., 'Ethical Issues in Funding Research and Development of Drugs for Neglected Tropical Diseases' (2009) 35 *Journal of Medical Ethics* 310

Ordinkalu CA, 'Analysis of Paralysis or Paralysis by Analysis? Implementing Economic, Social and Cultural Rights under the African Charter on Human and Peoples' Rights' (2001) 23 *Human Rights Quarterly* 327

Oshaug A and Eide WB, 'The Long Process of Giving Contention to an Economic, Social and Cultural Right: Twenty Five Years with the Case of the Right to

Adequate Food' in Bergsmo M (ed.), *Human Rights and Criminal Justice for the Downtrodden: Essays in Honour of Asbjorn Eide* (Martinus Nijhoff, 2003)

Oshaug A, 'Developing Voluntary Guidelines for Implementing the Right to Adequate Food: Anatomy of an Intergovernmental Process' in Eide WB and Krache U (eds.) *Food and Human Rights in Development: Volume 1: Legal and Institutional Dimensions and Selected Topics* (Intersentia, 2005)

Oshuag A, 'The Netherlands and Making of the Voluntary Guidelines on the Right to Food' in Hospes O and Van Der Meulen B (eds.), *Fed up with the Right to Food? The Netherlands Policies and Practices Regarding the Human Right to Adequate Food* (Wageningen, 2009)

Owen T, 'From "Pirates" to "Heroes": New Discourse Change, and the Contested Legitimacy of Generic HIV/AIDS Medicines' (2013) 8(3) *The International Journal of Press/Politics* 259

Pace N, Seal A and Costello A, 'Food Commodity Derivatives: A New Cause of Malnutrition?' (2008) 371 *The Lancet* 1648

Page B, 'Communities as Agents of Commodification: The Kumbo Water Authority in North-West Cameroon' 34(4) *Geoforum* 483

Paine E, 'The Road to the Global Compact: Corporate Power and the Battle over Global Public Policy at The United Nations' (*Global Policy Forum*, October 2000) http://dspace.cigilibrary.org/jspui/bitstream/123456789/17581/1/The%20Road%20to%20the%20Global%20Compact.pdf?1) accessed 17 August 2013

Palley T, 'From Keynesianism to Neoliberalism: Shifting Paradigms in Economics' in Saad-Fiho A and Johnson D, *Neoliberalism: A Critical Reader* (Pluto Press, 2005) 20

Patel R, *Stuffed and Starved: Markets, Power and the Hidden Battle for the World Food System* (Black, 2007)

'Commentary: The Hungry of the Earth' (2008) *Radical Philosophy* 151 www.radicalphilosophy.com/commentary/the-hungry-of-the-earth accessed 31 October 2013

Patterson B, 'NEWS: Opposition to the "Davos of water" Ministerial Statement' (*Blue Planet Project*, 13 March 2013) www.blueplanetproject.net/index.php/news-opposition-to-the-davos-of-water-ministerial-statement/ accessed 21 June 2013

'NEWS: Governments Back Track on Right to Water at "Davos of water" Forum'(*Blue Planet Project*, 14 March 2013) www.blueplanetproject.net/index.php/news-governments-back-track-on-right-to-water-at-davos-of-water-forum/ accessed 21 June 2013.

'UPDATE: Marseille Was the Staging Ground for Rio+20'(*Blue Planet Project*, 18 March 2013) www.blueplanetproject.net/index.php/update-marseille-was-the-staging-ground-for-rio20/ accessed 21 June 2013.

Pauw J, 'Metered to Death: How Water Caused Riots and a Cholera Epidemic' (*Centre for Public Integrity*, 5 February 2003) www.publicintegrity.org/2003/02/05/5713/metered-death accessed 21 June 2013

Pearson T, 'Venezuela Initiates Production of Generic Medicines and Cell Phones' (*Venezuela Analysis*, 5 May 2009) http://venezuelanalysis.com/news/4422 accessed 2 December 2013

Peck J, *Constructions of Neoliberal Reason* (Oxford University Press, 2010)

Peck J and Tickell A, 'Neoliberalizing Space' (2002) 34(3) *Antipode* 380

Peet R, 'Ideology, Discourse and the Geography of Hegemony: From Socialist to Neoliberal Development in Postapartheid South Africa' (2002) 34(1) *Antipode* 54

Petersmann E, 'The WTO Constitution and Human Rights' (2000) 3(1) *Journal of International Economic Law* 19

—— 'Time for a United Nations "Global Compact" for Integrating Human Rights into the Law of Worldwide Organizations: Lessons from European Integration' (2002) 13 *European Journal of International Law* 621

Petrova V, 'At the Frontiers of the Rush for the Blue Gold: Water Privatization and the Human Right to Water' (2005–2006) 31 *Brooklyn Journal of International Law* 577

Philips L, 'Taxing the Market Citizen: Fiscal Policy and Inequality in an Age of Privatization' (2000) 63 *Law and Contemporary Problems* 111

Pieterse M, 'Beyond the Welfare State: Globalization of the Neo-Liberal Culture and the Constitutional Protection of Social and Economic Rights in South Africa' (2003) 14 *Stellenbosch Law Review* 3

—— 'Eating Socioeconomic Rights: The Usefulness of Rights Talk in Alleviating Social Hardship Revisited' (2007) 29 *Human Rights Quarterly* 796

Plant R, 'Social and Economic Rights Revisited' (2003) 14 *Kings College Law Journal* 1

—— *The Neo-liberal State* (Oxford University Press, 2010)

Plomer A, 'The Human Rights Paradox: Intellectual Property Rights and Rights of Access to Science' (2013) 35 *Human Rights Quarterly* 143

Pogge T, *World Poverty and Human Rights* (Polity, 2002)

—— 'Human Rights and Global Health: A Research Program' (2005) 36 *Metaphilosophy* 182

—— 'Recognized and Violated by International Law: The Human Rights of the Global Poor' (2005) 18(4) *Leiden Journal of International Law* 717

—— 'The First UN Millennium Development Goal: A Cause for Celebration?' in Follesdal A and Pogge T (eds.), *Real World Justice* (Springer, 2005)

—— 'Severe Poverty as a Human Rights Violation' in Pogge T (ed.), *Freedom from Poverty as a Human Rights* (Oxford University Press, 2007)

—— 'The Health Impact Fund: Enhancing Justice and Efficiency in Global Health' (2012) 13(4) *Journal of Human Development and Capacities* 537

Pogge T and Polanyi K, *The Great Transformation: The Political and Economic Origins of Our Time* (Beacon Press, 1944)

Pollin R, *Contours of Descent* (Verso, 2005) 173

Provost C, 'World Water Forum Falls Short on Human Rights, Claim Experts' *Guardian* (London, 14 March 2013)

Rajagopal B, *International Law from Below: Development, Social Movements and Third World Resistance* (Cambridge University Press, 2004)

  'Counter-Hegemonic International Law: Rethinking Human Rights and Development as a Third World Strategy' (2006) *Third World Quarterly* 767

Read J, 'A Genealogy of Homo-Economicus: Neoliberalism and the Production of Subjectivity' (2009) 6 *Foucault Studies* 25

Renique G, 'Latin America: The New Neoliberalism and Popular Mobilization' (2009) 23(1) *Socialism and Democracy* 1

Risse T and Sikkink K, 'The Socialization of International Human Rights Norms into Domestic Practices' in Risse T, Ropp SC and Sikkink K (eds.), *The Power of Human Rights: International Norms and Domestic Change* (Cambridge University Press, 1999)

Robbins P, *Stolen Fruit: The Tropical Commodities Disaster* (Zed Books, 2003)

Robbins PR, 'Transnational Corporations and the Discourse of Water Privatization' (2003) 15 *Journal of International Development* 1073

Roberts A, 'Privatizing Social Reproduction: The Primitive Accumulation of Water in an Era on Neoliberalism' (2008) 40(4) *Antipode* 535

  The Logic of Discipline: Global Capitalism and the Architecture of Government (Oxford University Press, 2010)

Robinson F, 'Human Rights Discourse and Global Civil Society: Contesting Globalisation' Prepared for the 2002 Annual Meeting of the International Studies Association, New Orleans, Louisiana, March, 2002 http://isanet.ccit.arizona.edu/noarchive/robinson.html accessed 30 September 2013

Robinson RE, 'Measuring Compliance with the Obligation to Devote the "Maximum Available Resources" to Realising Economic, Social and Culture Rights' (1994) 16 *Human Rights Quarterly* 693

Robinson WI and Harris J, 'Towards a Global Ruling Class? Globalization and the Transnational Capitalist Class' (2000) 64(1) *Science & Society* 11

Robinson WI, 'Social Theory and Globalisation: The Rise of the Transnational State' (2001) 30 *Theory and Society* 157

  'Gramsci and Globalisation: From Nation-State to Tansnational Hegemony' in Bieler A and Morton AD (eds.) *Images of Gramsci: Connections and Contentions in Political Theory and International Relations* (Routledge, 2006)

  'The Global Capital Leviathan' (2011) 165 *Radical Philosophy* 2

  Global Capitalism and the Crisis of Humanity (Cambridge University Press, 2014)

Rooney E and Harvey C, 'Better on the Margins? A Critique of Mainstreaming Economic and Social Rights' in Harvey C, Nolan A and O'Connell R (eds.), *Human Rights and Public Finance* (Hart, 2013)

Rose NJ, 'Optimism of the Will: Food Sovereignty as Transformative Counter-Hegemony of the 21st Century' (PhD Thesis, RMIT University 2013)

Rothbard M, *The Ethics of Liberty* (New York University Press, 1998)

Rowden R, *The Deadly Ideas of Neoliberalism: How the IMF Has Undermined Public Health and the Fight against AIDS* (Zed Books, 2009)

Roy A, *Power Politics* (South End Press, 2001)

Ruckert A, 'Towards an Inclusive-Neoliberal Regime of Development: From the Washington to the Post-Washington Consensus' (2006) 39(1) *Labour, Capital and Society* 35

— 'Producing Neoliberal Hegemony? A Neo-Gramscian Analysis of the Poverty Reduction Strategy Paper (PRSP) in Nicaragua' (2007) 79 *Studies in Political Economy* 91

Ruggie JG, 'International Regimes, Transactions, and Change: Embedded Liberalism in the Postwar Economic Order' (1982) 36(2) *International Regimes* 379

Rupert M, 'The New World Order: Passive Revolution or Transformative Process?' in Amoore L (ed.), *The Global Resistance Reader* (Routledge, 2005) 194

— 'Reading Gramsci in an Era of Globalising Capitalism' in Bieler A and Morton AD (eds.) *Images of Gramsci: Connections and Contentions in Political Theory and International Relations* (Routledge, 2006)

Russell AFS, 'Incorporating Social Rights in Development: Transnational Corporations and the Right to Water' (2011) 7(1) *International Journal of Law in Context* 1, 9

Russet P, 'The US Gets Its Way' (*APRN*, 30 June 2002) www.aprnet.org/concerns/50-issues-a-concerns/146-us-opposes-right-to-food-at-world-summit accessed 3 November 2013

Saad-Fiho A and Johnson D, 'Introduction' in Saad-Fiho A and Johnson D, *Neoliberalism: A Critical Reader* (Pluto Press, 2005) 1

Salman SMA and McInerney-Lankford S, *The Human Right to Water: Legal and Policy Dimensions* (World Bank, 2004)

Salomon ME, *Global Responsibility for Human Rights: World Poverty and the Development of International Law* (Oxford University Press, 2007)

— 'Poverty, Privilege and International Law: The Millennium Development Goals and the Guise of Humanitarianism' (2008) 51 *German Yearbook of International Law* 39

— 'Why Should It Matter that Some Have More? Poverty, Inequality and the Potential of International Human Rights Law' (2011) 37 *Journal of International Studies* 2137

Sampat B and Lichtenberg F, 'What Are the Respective Roles of the Public and Private Sectors in Pharmaceutical Innovation?' (2011) 30(2) *Health Affairs* 332

Sassoon AS, 'Hegemony' in Bottomore T (ed.), *A Dictionary of Marxist Thought* (2nd edn., Blackwell, 2006)

Schanbacher WD, *The Politics of Food: The Global Conflict between Food Security and Food Sovereignty* (ABC-CLIO, 2010)

Scheingold S, *The Politics of Rights* (Yale University Press, 1974)

Scherr S, *Halving Global Hunger: Background Paper of Task Force on Hunger* (UNDP, 2003)

Schnitzler A, 'Citizenship Prepaid: Water, Calculability, and Techno-Politics in South Africa' (2008) 34(4) *Journal of Southern African Studies*

Schoofs M, 'AIDS: The Agony of Africa: South Africa Acts Up' (1999) 44 *The Village Voice* 2

Segerfeldt F, 'Private Water Saves Lives' Financial Times, (London 25 August 2005)
  Water For Sale: How Business and the Market Can Resolve the World's Water Crisis (CATO Institute, 2005)

Sell SK, 'Access to Medicines in the Developing World: International Facilitation or Hindrance?' (2002) 20 *Wisconsin International Law Journal* 481
  Private Power, Public Law: The Globalization of Intellectual Property Rights (Cambridge University Press, 2003)

Selznick P, 'Law in Context Revisited' (2003) 30 *Journal of Law and Society* 177

Sen A, 'The Right Not To Be Hungry' in Alston P and Tomasevki K (eds.), *The Right to Food* (Martinus Nijhoff, 1984)

Sepulveda M, *The Nature of Obligations under the International Covenant on Economic, Social and Cultural Rights* (Intersentia, 2003)

Shamir R, 'Corporate Social Responsibility: A Case of Hegemony and Counter-hegemony' in Sousa Santos B and Rodriguez-Garivito CA (eds.) *Law and Globalization from Below: Towards a Cosmopolitan Legality* (Cambridge University Press, 2005)

Shaw M, *Theory of the Global State: Globality as an Unfinished Business* (Cambridge University Press, 2000)

Shestack J J, 'The Philosophic Foundations of Human Rights' (1998) 20(2) *Journal of Human Rights* 201

Shihata IFI (ed.), *The World Bank Inspector Panel in Practice* (2nd edn., Oxford University Press, 2000)

Shills E and Finch H (trans. and ed.) *The Methodology of the Social Sciences (1903–17)* (Free Press, 1997)

Shiva V, *Water Wars: Privatization, Pollution and Profit* (South End Press, 2002)
  'Water Democracy' (Foreword) in Olivera O, *Cochabamba! Water War in Bolivia* (South End Press, 2004)

Shrybman S, 'Thirst for Control: New Rules in the Global Water Grab' (2002) Prepared for the Council of Canadians. www.canadians.org/water/documents/campaigns-tfc.pdf accessed 30 May 2013

Shue H, *Basic Rights: Subsistence, Affluence and U.S. Foreign Policy* (2nd edn., Princeton University Press, 1996)
   'Rights in Light of Duties' in Brown PG and MacLean D (eds.) *Human Rights and U.S. Foreign Policy* (Lexington, 1979)
Simma B and Paulus AL, 'The Responsibility of Individuals for Human Rights Abuses in Internal Conflicts: A Positivist View' (1999) 93(2) *American Journal of International Law* 302
Singh K, 'Patents vs Patients: AIDS, TNCs and Drug Price Wars' (Public Interest Research Centre 2001) www.madhyam.org.in/admin/tender/Patents%20vs.%20Patients,%20%20AIDS,%20TNCs%20and%20Drug%20Price%20Wars.pdf accessed 11 November 2013
Sklair L, *The Transnational Capitalist Class* (Blackwell, 2001)
   Globalization: Capitalism & Its Alternatives (3rd edn., Oxford University Press, 2002)
Skogly S, 'The Obligation of International Assistance and Co-Operation in the International Covenant on Economic, Social and Cultural Rights' in Bergsmo M (ed.), *Human Rights and Criminal Justice for Downtrodden: Essays in Honour of Asbjørn Eide* (Kluwer Law International, 2003)
Smith L, 'The Murky Waters of Second Wave Neoliberalism: Corporatization as a Service Delivery Model in Cape Town' in McDonald D and Ruiters G (eds.), *The Age of Commodity: Water Privatization in Southern Africa* (Earthscan, 2005)
Smithers R, 'Almost half of the world's food thrown away, report finds' *Guardian* (Manchester, 10 January 2013)
Sodano V, 'Food Policy Beyond Neo-Liberalism' in Erasga D (ed.) *Sociological Landscape: Theories, Realities and Trends* (InTech, 2012) 375
Soederberg S, 'The Transnational Debt Architecture and Emerging Markets: The Politics and Paradoxes of Punishment' (2003) 26(6) *Third World Quarterly* 927
   'Recasting Neoliberal Dominance in the Global South? A Critique of the Monterrey Consensus' (2005) 30 *Alternatives* 325
   'Taming Corporations or Buttressing Market-Led Development? A Critical Assessment of the Global Compact' (2007) 4(4) *Globalizations* 500
Sousa Santos B, *Towards a New Legal Common Sense: Law, Globalization and Emancipation* (Cambridge University Press, 2002)
   'Beyond Neoliberal Governance: The World Social Forum as Subaltern Politics and Legality' in Sousa Santos B and Rodriguez-Garivito CA (eds.) *Law and Globalization from Below: Towards a Cosmopolitan Legality* (Cambridge University Press, 2005)
Spronk S, 'International Solidarity for the Struggle for Water Justice in El Alto, Bolivia' (*Znet*, 10 May 2005) www.zcommunications.org/international-solidarity-for-the-struggle-for-water-justice-in-el-alto-bolivia-by-susan-spronk.html accessed 10 December 2013

Ssenyonjo M, 'The Applicability of International Human Rights Law to Non-State Actors: What Relevance to Economic, Social and Cultural Rights?' (2008) 12 (5) *The International Journal of Human Rights* 725

Stammers N, 'Social Movements and the Social Construction of Human Rights' (1999) 21(4) *Human Rights Quarterly* 980

Human Rights and Social Movements (Pluto, 2009)

Steiner H and Alston P, *International Human Rights in Context: Law, Politics, Morals* (2nd edn., Oxford University Press, 2000)

Steiner H, Alston P and Goodman R, *International Human Rights Law in Context: Law, Politics, Morals* (3rd edn., Oxford University Press, 2007)

Stewart F and Wang M, 'Poverty Reduction Strategy Papers within the Human Rights Perspective' in Alston P and Robinson M (eds.), *Human Rights and Development: Towards a Mutual Reinforcement* (Oxford University Press, 2005)

Stiglitz J, *Globalisation and Its Discontents* (Penguin, 2002)

Strath B 'The Liberal Dilemma: The Economic and the Social, and the Need for a European Contextualization of a Concept with Universal Pretensions' in Jackson B and Stears M (eds.) *Liberalism as Ideology: Essays in Honour of Michael Freeden* (Oxford University Press, 2012)

Stuckler D and Basu S, *The Body Economic: Why Austerity Kills* (Basic Books, 2013)

Sunstein CR, 'Against Positive Rights' (1993) 2 *East European Constitutional Review* 35

Sykes AO, 'Comparative Advantage and the Normative Economics of International Trade Policy' (1998) 1 *Journal of International Economic Law* 49

t'Hoen E, 'TRIPS, Pharmaceutical Patents, and Access to Medicines: A Long Way from Seattle to Doha' (2002) 3(1) *Chicago Journal of International Law* 39

The Global Politics of Pharmaceutical Monopoly Power: Drug patents, Access, Innovation and the Application of the WTO Doha Declaration on TRIPS and Public Health (AMB, 2009) 3

Thielbörger P, 'Re-Conceptualizing the Human Right to Water: A Pledge for a Hybrid Approach' (2015) 15(2) *Human Rights Law Review* 225

Thomas P, *The Gramscian Moment: Philosophy, Hegemony and Marxism* (Brill, 2009) 105–108

Tully S, 'A Human Right to Access Water? A Critique of General Comment No. 15' (2005) 23(1) *Netherlands Quarterly of Human Rights* 35

Turley J, 'The Hitchhikers Guide to CLS: Unger, and Deep Thought' (1987) 81 *Northwestern University Law Review* 593

Tushnet M, 'An Essay on Rights' (1984) 62 *Texas Law Review* 1363

'Civil Rights and Social Rights: The Future of the Reconstruction Amendments' (1992) 25 *Loyola of Los Angeles Law Review* 1207

Unger RM, 'The Critical Legal Studies Movement' (1983) 96(3) *Harvard Law Review* 561

Valente FLS and Franko AMS, 'Human Rights and the Struggle against Hunger: Laws, Institutions, and Instruments in the Fight to Realize the Rights to Adequate Food' (2010) 13 *Yale Human Rights and Development Law Journal* 435

Varghese S, 'Transnational Led Privatization and the New Regime for the Global Governance of Water' (2003) 9/10 (1/2) *Water Nepal: Journal of Water Resources Development* 77

Vasak K, 'Pour une troisième génération des droits de l'homme' in Swinarski C (ed.) *Studies and Essays on International Humanitarian Law and Red Cross Principles in Honour of Jean Pictet* (Martinus Nijhoff, 1984)

Vidar M, 'The Right to Food in International Law' (2003) 5 www.actuar-acd.org/uploads/5/6/8/7/5687387/fao_the_right_to_food_in_international_law.pdf accessed 15 October 2013

Vierdag EW, 'The Legal Nature of the Rights Granted by the International Covenant on Economic, Social and Cultural Rights' (1978) 9 *Netherlands Year Book of International Law* 69

Wahl P, 'The Role of Speculation in the 2008 Food Price Bubble' in FIAN (eds.) *The Global Food Challenge: Towards a Human Rights Approach to Trade and Investment Policies* (2009) www.fian.org/resources/documents/others/the-global-food-challenge/pdf accessed 2 November 2013

Walzer M, 'The Civil Society Argument' in Mouffe C (ed.) *Dimensions of Radical Democracy: Pluralism, Citizenship, Community* (Phronesis, 1992)

Weede E, 'Human Rights, Limited Government, and Capitalism' (2008) 28(1) *Cato Journal* 35

Whelan DJ and Donnelly J, 'The West, Economic and Social Rights, and the Global Human Rights Regime: Setting the Record Straight' (2007) 29(4) *Human Rights Quarterly* 908

White LE and Perelman J (eds.), *Stones of Hope: How African Activists Reclaim Human Rights to Challenge Global Poverty* (Stanford University Press, 2010)

Williams M, 'Privatization and the Human Right to Water: Challenges for the New Century' (2007) 28 *Michigan Journal of International Law* 469

Williams P, 'Alchemical Notes: Reconstructing Ideals from Deconstructed Rights' (1987) 22 *Harvard Civil Rights-Civil Liberties Law Review* 401

Williams R, 'Base and Superstructure in Marxist Cultural Theory' (1973) 87 *New Left Review* 3

Wilson R, 'Afterword to "Anthropology and Human Rights in a New Key": The Social Life of Human Rights' (2006) 108(1) *American Anthropologist* 77

Windfuhr M and Jonsen J, *Food Sovereignty: Towards Democracy in Localized Food Systems* (ITDG, 2005)

Winpenny J, World Panel on Financing and Infrastructure, *Financing Water for All* (2003) www.worldwatercouncil.org/fileadmin/wwc/Library/Publications_and_reports/CamdessusSummary.pdf accessed 26 January 2014

Winstanley G, 'A Declaration from the Poor Oppressed People of England' (first printed 1649) www.bilderberg.org/land/poor.htm accessed 20 September 2013

Wittman H, 'Interview: Paul Nicholson, La Via Campesina' (2009) 36(3) *The Journal of Peasant Studies* 676

—— 'Food Sovereignty: A New Rights Framework for Food and Nature?' (2011) 2 *Environment and Society: Advances in Research* 87

—— Desmarais A and Wiebe N, 'The Origins and Potential of Food Sovereignty' 2–3 in Wittman H, Desmarais A and Wiebe N (eds), *Food Sovereignty: Reconnecting Food, Nature and Community* (Fernwood, 2013)

Wolf M, 'Seeds of its own destruction' *Financial Times* (London, 8 March 2009)

Woodiwiss A, *Human Rights* (Routledge, 2005)

Yamin AE, 'The Future in the Mirror: Incorporating Strategies for the Defense and Promotion of Economic, Social and Cultural Rights into the Mainstream Human Rights Agenda' (2005) 27 *Human Rights Quarterly* 1200

Young KG, 'The Minimum Core of Economic and Social Rights: A Concept in Search of Content' (2008) 33 *Yale Journal of International Law* 113

—— 'Securing Health through Rights' in Pogge T, Rimmer M and Rubenstein K (eds.), *Incentives for Global Health: Patent Law and Access to Essential Medicines* (Cambridge University Press, 2010)

—— *Constituting Economic and Social Rights* (Oxford University Press, 2012)

Ziegler J, Golay C, Mahon C and Way S, *The Fight for the Right to Food* (Palgrave, 2010)

INDEX

A2K movement. *See* access to knowledge movement
ABRANDH. *See* Brazilian Action for Nutrition and Human Rights
access to knowledge (A2K) movement, 163–164
access to medicine. *See also* right to health
  CESCR and, 172–173
  commodification of medicine and, 161
  ethico-political framework for, 163–166
    in A2K movement, 163–164
  as global movement, 9, 169–172
    NGOs and, 170–171
    in South Africa, 169–170
  global regime for, through IPRs, 154–158
    criticisms of, 157–158
    minimum standard settings under, 155–156
    R&D sector, expansion of, 154–155, 157–158
    TCC and, 156–157
  in Global South, 155–156
  ICESCR and, 172–173
  in India, 155–156, 161
  IPR paradigm compared to, 166
  lack of, 151–152
  ownership of IPRs and, 158–166
    economic rationalism for, 161
    ethico-political framework for, 159–162
    individualism as factor in, 161
    Lockean influences on, 159–160
  TRIPS and, 172–179
  under UN human rights system, 172–179

access to water. *See also* right to water
  in Global South, 203
  private sector involvement in, 200, 201, 200–201
  monopolies and, 205
  through public-private partnerships, 207
  water as commodity, 201–204, 206–207, 211
  right to water and, 196–198
  through social movements, 205
  water as 'commons', 204–207
  through public-public partnerships, 207
accumulation by dispossession, 5–6, 33
  IPRs and, 5
  TRIPS and, 5, 34
agency, legal positivism and, 52
Agreement on Agriculture (AoA), 97–98, 99–100
Agreement on Trade-Related Intellectual Property Rights (TRIPS), 154–158
  access to medicine and, 172–179
  accumulation by dispossession and, 5, 34
  flexibility of, 162
  in Global South, 183–188
  IPR protections under, 152, 154–155
  minimum standards setting, 155–156
  patent protection under, 155–156
  right to health and, 171–172, 193–195
  Doha Declaration, 179–183
  in Global South, 183–188
  in post-Doha landscape, 183–188
  TRIPS reform and, 179–183
  WTO response to, 174–176

Agreement on Trade-Related Investment Measures (TRIMs), 34
agricultural trade, liberalisation of, 98–101
agroecology, 104, 107
ALBA. *See* Bolivarian Alliance for the People of Our Americas
All Farmers Network, 127
Alston, Philip, 112
Althusser, Louis, 22
Altieri, Miguel A., 105
Amnesty International, 7
Amoore, Louise, 46–47
Anderson, Robert, 87
Annan, Kofi, 37
anthropocentrism, right to water and, 241–242
AoA. *See* Agreement on Agriculture
Association of Southeast Asian States (ASEAN), 34
austerity measures, 3

Bakker, Karen, 202, 242–244
Barlow, Maude, 216, 229
Baxi, Upendra, 37, 86–87, 149
Bello, Walden, 108
Bija Satyagraha Movement, 149
bilateral investment treaties (BITs), 35
bilateral trade agreements (BTAs), 35
BITs. *See* bilateral investment treaties
Blackburn, Robin, 259
Boerma, Addeke H., 109
Bolivarian Alliance for the People of Our Americas (ALBA), 186
Bond, Patrick, 235
Brazilian Action for Nutrition and Human Rights (ABRANDH), 127
BTAs. *See* bilateral trade agreements
Buchanan, James, 28

CAFTA. *See* Central American Free Trade Agreement
capitalism
  Gramsci and, 19–20
  TCC, 30
CAWP. *See* Coalition Against Water Privatization

CEDAW. *See* Convention on the Elimination of All Forms of Discrimination Against Women
Central American Free Trade Agreement (CAFTA), 185
Centre on Housing Rights and Evictions (COHRE), 7, 205
CESCR. *See* Committee on Economic, Social and Cultural Rights
Chandra, Rajshree, 164–165
CIPIH. *See* Commission on Intellectual Property Rights and Public Health
civil rights. *See* first generation civil rights
civil society, global
  common sense, 24–25
  counter-hegemony in, 46–47
  global governance and, 38–39
  hegemony and, 20–22, 35–40
  IFIs and, 36–37
  NGOs as part of, 35–36
  NTHB in, 36
  PRSPs in, 39–40
  PWC, 38–39, 47–48
  TNLB in, 47–48
  transformismo process, 40
  UN role in, 37
Claeys, Priscilla, 142
Coalition Against Water Privatization (CAWP), 236
COHRE. *See* Centre on Housing Rights and Evictions
Commission on Intellectual Property Rights and Public Health (CIPIH), 189–190
Committee on Economic, Social and Cultural Rights (CESCR), 6–7, 50, 67
  access to medicine and, 172–173
  critique of neoliberal globalisation, 75–79
  formation of, 71
  IMF and, 76–77
  obligations of, 71–73
  right to food and, 111
  right to water and, 77–78
  SAPs and, 75–76

World Bank and, 76–77
  on World Food Crisis of 2008,
    139–140
  WTO and, 76–77
common sense, 24–25
commons. *See* global commons
communication systems, globalisation
  of, 5–6
comparative advantage theory, 102–103
Consumer Project on Technology
  (CPTech), 164
Convention on the Elimination of All
  Forms of Discrimination
  Against Women (CEDAW),
  210
Convention on the Rights of the Child
  (CRC), 210
corporate power
  accumulation by dispossession, 5–6
  international monetary policy
    influenced by, 4
  TNCs and, 29
counter-hegemony, 24–25
  in global civil society, 46–47
  global justice movements and, 42–47
    construction of, obstacles to,
      46–47
    sites of resistance and, 45
    TANs and, 45
    TCCs and, 42–43, 47
    WSF and, 45
  globalisation and, 42–47
  right to food through, 134, 144–149
  right to water as, 239–245
    through litigation, 240
  socioeconomic rights and, 79–93
    as Immanent critique, 83–85
    normative obligations of, 81–83
    universalism and, 80–81
  war of position and, 24–25, 44–45
  water as Commons, 204–207
counter-movements, 5–6
Cox, Robert, 30, 43, 56
CPTech. *See* Consumer Project on
  Technology
Craven, Matthew, 113–114
CRC. *See* Convention on the Rights of
  the Child

de Peuter, Greig, 207
decadent hegemony, 23
Declaration of Cochabamba, 214–216,
  242
Declaration of Marrakech, 212
development rights, 61
disciplinary neoliberalism, 29–30
discourse analysis, 11–12
Donnelly, Jack, 69
Douzinas, Costas, 83
Draft Declaration on the Rights of
  Peasants, 142–143
Drage, Katherine, 239
Dublin Principles, 210–211, 225
Dugard, Jackie, 235, 239
Duggan, Lisa, 182–183
Dyer-Witherford, Nick, 207

ECJ. *See* European Court of Justice
economic, social and cultural (ESC)
  rights, 61–62. *See also*
  socioeconomic rights
economic inequality. *See* inequality,
  economic
economic liberalism, socioeconomic
  rights and, 85–88
economic prosperity, global declines in,
  2–5
Eide, Asbjørn, 118–119, 124
Elver, Hilal, 148, 199
embedded liberalism, 27–28
Engels, Friedrich, 20
England and Wales, right to water in,
  230
ESC rights. *See* socioeconomic rights
EU-ASEAN FTA, 186
European Court of Justice (ECJ), 34
European Free Trade Area Court, 34

Falk, Richard, 43–44
FAO. *See* Food and Agriculture
  Organisation
Farmer, Paul, 3
FBW policy. *See* Free Basic Water
  policy
Femia, Joseph, 23
FIAN. *See* Food First Information and
  Action Network

first generation civil rights, 61
first generation political rights, 61
food. *See* right to food
Food and Agriculture Organisation (FAO), 3, 6–7
   WFS and, 115–118
      Rome Declaration on World Food Security, 115, 117–118
Food and Water Watch, 205
Food First Information and Action Network (FIAN), 7, 118–119
food price crisis, 100
food security, 101–105
   comparative advantage theory for, 102–103
   defined, 102
   food sovereignty compared to, 109
   in Global South, 102–103
   through R&D, 103–104
   through reform measures, 104–105
food sovereignty
   counter-hegemonic foundations of, 107–108
   defined, 106–107
   food security compared to, 109
   historical origins of, 105–106
   at international level, 108
   as movement, 9
   political framework for, 101, 105–108
   right to food and, 144–149
Forman, Lisa, 186
Foucault, Michel, 12
foundationalism, human rights and, 50–51
Fraser, Nancy, 44
Free Basic Water (FBW) policy, 232–238
   CAWP and, 236
   historical development of, 232–234
   Mazibuko ruling and, 236–238
   problems with, 234–236, 239
Free Trade Agreement of the Americas (FTAA), 185
Free Trade Agreements (FTAs). *See also* Agreement on Trade-Related Intellectual Property Rights; Central American Free Trade Agreement; North American Free Trade Agreement; Transatlantic Trade and Investment Partnership; Trans-Pacific Partnership Agreement
   IPRs in, 152
French Declaration of the Rights of the Man and the Citizen 1789, 1–2
Friedman, Milton, 28
Friedmann, Harriet, 96–97
FTAA. *See* Free Trade Agreement of the Americas
FTAs. *See* Free Trade Agreements

G8 countries. *See* Group of 8 countries
GATS. *See* General Agreement on Trade in Services
GATT. *See* General Agreement on Trade and Tariffs
Gearty, Conor, 5
General Agreement on Trade and Tariffs (GATT), 34, 97–98
General Agreement on Trade in Services (GATS), 34
George, Erika, 192–193
George, Susan, 100
Germann, Julian, 130
Gill, Stephen, 29–30, 32
global Commons
   enclosure of, 5–6
   tragedy of the commons theory, 160
   water as, 204–207
      through public-public partnerships, 207
global governance, 38–39
global hunger. *See also* food security
   food price crisis and, 100
   food sovereignty, 101, 105–108
      counter-hegemonic foundations of, 107–108
      defined, 106–107
      food security compared to, 109
      historical origins of, 105–106
      at international level, 108
   malnutrition and, 95–96
   reduction strategies for, 94
   through agroecology, 104, 107

right to food and, 94–96
  reduction strategies, 94
global justice movements. *See also*
    access to medicine; food
    sovereignty; water justice
    movement
  counter-hegemony and, 42–47
    construction of, obstacles to,
      46–47
    sites of resistance and, 45
    TANs and, 45
    TCCs and, 42–43, 47
    WSF and, 45
  new rights and, 7, 9–10
  NGOs and, 46
  socioeconomic rights and, 9–10
  TCCs and, 42–43, 47
  transformative politics through, 10
global political society. *See* political
    society
global poor, human rights of, 1–2
Global South. *See also* India
  access to water in, 203
  food security in, 102–103
  IPRs in, 155–156
  neoliberal food regime in, 96–101
  patent protection in, 155–156
  right to health in, 183–188
  right to water in, 203
  TRIPS in, 183–188
Global Water Regime, 246
globalisation
  accumulation by dispossession from,
    5–6
  of communication systems, 5–6
  counter-hegemony and, 42–47
  counter-movements as result of, 5–6
  hegemonic, 5–6
  of neoliberalism, 25–42
    historical development of, 29–30
    TCC and, 30
    through TNCs, 29
  of socioeconomic rights, 74–79
Godoy, Angelina Snodgrass, 158, 187
Gonzalez, Carmen G., 102–103
Gramsci, Antonio, 17, 20, 53, 56.
    See also hegemony; Neo-
    Gramscianism

  capitalism and, 19–20
  on civil society, 20–22
  on common sense, 24–25
  on counter-hegemony, 24–25
    war of position and, 24–25, 44–45
  on historical materialism, 18–20
  passive revolution for, 23, 105
  on political society, 20–22
  on state power, 20–22
  transformismo process, 23, 40
Group of 8 (G8) countries, 34
Guissé, El-Hadji, 217–218, 241

HAI. *See* Health Action International
Hall, Stewart, 12
Harris, Leila, 245
Harvey, Colin, 251–252
Harvey, David, 5, 26, 28, 33. *See also*
    accumulation by dispossession
Health Action International (HAI),
    164
hegemonic globalisation, 5–6
hegemonic stability theory (HST), 31
hegemony, 17–25. *See also* counter-
    hegemony
  civil society and, 20–22, 35–40
  common sense and, 24–25
  decadent, 23
  defined, 17–18, 20
  internal, 23
  Marxism and, 17–18
  minimal, 23, 42
  neoliberalism and, 31–40
    global civil society and, 35–40
    global political society and, 32–35
    HST, 31
    Neo-Gramscianism and, 32
  passive revolution and, 23
  paths to, 22–23
  political society and, 20–22, 32–35
  by ruling class, 22–23
  transformismo process, 23
  water as commodity, 201–204,
    206–207, 211
  world, 56
Heinrich Boil Foundation, 205
Hertel, Shareen, 82
Heywood, Mark, 170

historical materialism, 18–20
  structure-superstructure metaphor and, 18–20
Hobbes, Thomas, 59–60
Holt-Giménez, Eric, 105
HST. *See* hegemonic stability theory
Human Rights Council, 223
Human Rights First, 7
human rights regimes
  access to medicine and, 172–179
  foundationalist approach to, 50–51
  under French Declaration of the Rights of the Man and the Citizen 1789 1–2
  of global poor, 1–2
  legal positivist approach to, 51–53
    disadvantages of, 52–53
    human agency in, 52
    as moral rights, 52
  Neo-Gramscian framework for, 50–60
    constructivist perspectives in, 56–58
    realist perspectives in, 56–58
    in social constructivist approach, 53–55
  neoliberalism and, 8–9
  Rights of Man, 59–60
  scope of, 50–55
  social change and, 55–60
    constructivist perspective on, 55–59
    realist perspective on, 55–59
    TNHB and, 57
  social constructivist approach to, 53–54
    Neo-Gramscianism in, 53–55
  soft law and, 52
  TRIPS and, 172–179
Human Rights Watch, 7
hunger. *See* global hunger
Hunt, Alan, 9, 58

ICCPR. *See* International Covenant on Civil and Political Rights
ICESCR. *See* International Covenant on Economic, Social and Cultural Rights
ICSID. *See* International Centre for the Settlement of Investment Disputes
IFIs. *See* international financial institutions
IGOs. *See* intergovernmental organisations
IMF. *See* International Monetary Fund
Immanent critique, 83–85
  social criticism as foundation of, 83–84
  of socioeconomic rights, 249
India, IPR protections in, 155–156
inequality, economic
  global increases in, 2–3
  neoliberalism and, 8–9
Institute Maritain International, 118–119
intellectual property rights (IPRs), 152–154
  access to medicine paradigm compared to, 166
  accumulation by dispossession and, 5
  under CAFTA, 185
  commodification of, 161
  expansion of, 33
  flexibility of, 158–159, 162
  as free trade, 159
  under FTAA, 185
  in FTAs, 152
  global regimes, for access to medicines, 154–158
    criticisms of, 157–158
    minimum standard settings under, 155–156
    R&D sector, expansion of, 154–155, 157–158
    TCC and, 156–157
  in Global South, 155–156
  in India, 155–156, 161
  innovation through, 158–161
  as natural rights, 158–162, 165
  ownership of, 158–166
    economic rationalism for, 161
    ethico-political framework for, 159–162
    individualism as factor in, 161
    Lockean influences on, 159–160

through patent protection, 155–156
personality theory and, 159–160
pharmaceutical industry and
  global drug gap and, 189
  in India, 155–156, 161
  parallel importing in, 162
as private rights, 159–160
second enclosure movement for, 153
under TRIPS
  accumulation by dispossession and, 5, 34
  flexibility of, 162
  minimum standards for, 155–156
  protections under, 152, 154–155
under TTIP, 185
under TTPA, 185
universalisation of, 81
in US, 156
WTO protections for, 152
Inter-American Vigilance for the Defence and Right to Water Network, 205
intergovernmental organisations (IGOs), 134–135
internal hegemony, 23
International Bill of Human Rights, 208–210
International Centre for the Settlement of Investment Disputes (ICSID), 33–34
International Covenant on Civil and Political Rights (ICCPR), 69–70
International Covenant on Economic, Social and Cultural Rights (ICESCR), 6–7, 68–73
  access to medicine and, 172–173
  negative feedback cycle, 70
  OP-ICESCR, 73–74
  right to food under, 110–115
  right to health and, 167
international financial institutions (IFIs), 32–34, 36–37
  socioeconomic rights and, 90–93
international law
  right to food under, 109–144
    Draft International Code of Conduct, 118–120

General Comment 12, by CESCR, 120–122
through international treaties, 111
legal basis for, 110–115
WFS and, 115–118, 122–126
socioeconomic rights and, 67–79
  standards clarification of, 68–74
  UN human rights framework for, 74–79
International Monetary Fund (IMF), 3
  CESCR critique of, 76–77
  corporate power as influence on, 4
  in global political society, 33–34
  global poor and, 1–2
  in neoliberal food regime, 97–98
  PRSPs and, 39–40
  SAPs, 65–66
International Relations (IR), hegemony and, 17, 31
international trade
  formulation of rules for, 4–5
  TTIP, 4–5
IPRs. *See* intellectual property rights
IR. *See* International Relations

Kant, Immanuel, 61
Keck, Margret, 45
Kell, Susan K., 154
Keynes, John Maynard, 61
Keynesian economics, 28–29
Klug, Heinz, 171–172
Kohl, Benjamin, 36–37
Kothari, Miloon, 218

La Vía Campesina, 105–106, 119–120, 140–142
  peasant rights and, 142–144
    Draft Declaration on the Rights of Peasants, 142–143
Landless Peasants Movement (MST), 149
Lang, Andrew, 75, 182–183
Langford, Malcolm, 228
Langley, Paul, 46–47
Larner, Wendy, 26
legal positivism, human rights and, 51–53
  disadvantages of, 52–53
  human agency and, 52

legalism, narrow, 88–90
Lenin, Vladimir, 20
Locke, John, 59–60
  on IPRs, 159–160

malnutrition. *See also* right to food
  global hunger and, 95–96
Mamani, Abel, 229
Manchester School, 27
Marx, Karl, 18, 61. *See also* historical materialism
  on structure-superstructure metaphor, 18
Marxism
  hegemony and, 17–18
  historical materialism and, 18–20
    structure-superstructure metaphor and, 18–20
  reductionist forms of, 18–19
Matthews, Duncan, 171
Mazibuko ruling, in South Africa, 236–238
McMichael, Philip, 96–97, 104–105
Medical Research and Development Treaty (MRDT), 189–191
Medicines Transparency Alliances (MeTA) Philippines, 186
Médicins Sans Frontières (MSF), 164
  NDG and, 189–190
Meire, Benjamin Mason, 168, 188
MERCUSOR. *See* Southern Common Market
MeTA Philippines. *See* Medicines Transparency Alliances Philippines
Miliband, Ralph, 84
minimal hegemony, 23, 42
Minkler, Lanse, 82
Mirosa, Oriol, 245
Monsave, Sofia, 141–142
Mont Pelerin Society, 28
moral rights, 52
Morales, Evo, 229
Mowbray, Jacqueline, 133
Moyn, Samuel, 8, 89

MRDT. *See* Medical Research and Development Treaty
MSF. *See* Médicins Sans Frontières
MST. *See* Landless Peasants Movement

NAFTA. *See* North American Free Trade Agreement
Nash, Katie, 59
Neglected Diseases Group (NDG), 189–190
Nemeth, Thomas, 24
Neo-Gramscianism, 30, 32
  human rights and, 50–60
    constructivist perspectives for, 56–58
    realist perspectives for, 56–58
    social constructivist approach to, 53–55
neoliberal food regime
  agricultural trade liberalisation, 98–101
  AoA and, 97–100
  criticisms of, 98–101
  development of, 96–97
  food price crisis and, 100
  in Global South, 96–101
  NAFTA and, 99–100
  political-institutional organisations, 97–98
  structural adjustment measures, 98–101
  TNCs in, 97, 100–101
neoliberal transnational historical bloc (NTHB), 36
neoliberalism
  CESCR critiques on, 75–79
  defined, 26–27
  disciplinary, 29–30
  economic inequality and, 8–9
  embedded liberalism and, 27–28
  free market forces and, 63–64
  global hegemony and, 31–40
    global civil society and, 35–40
    global political society and, 32–35
    HST, 31
    Neo-Gramscianism and, 32
  globalisation of, 25–42
    crises in, 48

historical development of, 29–30
TCC and, 30
through TNCs, 29
human rights and, 8–9
intellectual origins of, 27–29
limits of rights under, 62–64
minimal state under, 67
Mont Pelerin Society and, 28
Neo-Gramscianism and, 30, 32
political-institutional context of, 12–13
right to health and, 167–168
right to water and, 198–201, 210–212, 224–232
  criticisms of, 200–201
  TNCs and, 199
  water as commodity, 201–204, 206–207, 211
  water as 'commons', 204–207
  WWC and, 199–200
socioeconomic rights and, 61–67
  conflicts between, 62–67
state power and, 26–27
new constitutionalism, 32–34
new rights, advocacy of
global justice movements and, 7, 9–10
NGOs and, 7–8
UN organisations for, 6–7, 6–9, 37
NGOs. *See* non-government organisations
Nicholson, Paul, 144
non-government organisations (NGOs)
in global access to medicine movement, 170–171
in global civil society, 35–36
global justice movements and, 46
new rights advocacy by, 7–8
at WSF events, 7–8
North American Free Trade Agreement (NAFTA), 99–100
Nozick, Robert, 63
NTHB. *See* neoliberal transnational historical bloc

O'Connell, Paul, 86, 247
Olivera, Oscar, 214

Optional Protocol to the International Covenant on Economic, Social and Cultural Rights (OP-ICESCR), 73–74

Page, Ben, 244
Paine, Thomas, 61
parallel importing, in pharmaceutical industry, 162
passive revolution, 23, 105
patent protection, 155–156
peasant rights, 142–144
  Draft Declaration on the Rights of Peasants, 142–143
  MST, 149
Peck, Jamie, 38
personality theory, 159–160
pharmaceutical industry
  global drug gap and, 189
  in India, 155–156, 161
  parallel importing in, 162
Plant, Raymond, 63
Pogge, Thomas, 1
Polanyi, Karl, 5–6
political rights. *See* first generation political rights
political society, global
  BITs and, 35
  BTAs and, 35
  G8 countries, 34
  hegemony and, 20–22, 32–35
  IFIs and, 32–34
  IMF and, 33–34
  international trade agreements as component of, 34–35
  new constitutionalism and, 32–34
  World Bank and, 33–34
  WTO and, 33–34
Post Washington Consensus (PWC), 38–39, 47–48
  right to water and, 204
  on socioeconomic rights, 78–79
poverty, global increase in, 3
Poverty Reduction Strategy Papers (PRSPs), 39–40
praxis. *See* Marxism
Prison Notebooks (Gramsci), 17, 53
private rights, IPRs as, 159–160

privatisation of water. *See* water privatisation
pro-poor policies, for right to water, 203
PWC. *See* Post Washington Consensus

rational choice theory, 55
Rawls, John, 61
R&D. *See* research and development
RDS. *See* Revised Drug Strategy
redistributivist state, 83
Reidel, Eide, 218
research and development (R&D)
  food security through, 103–104
  for IPRs, 154–155, 157–158
Revised Drug Strategy (RDS), of WHO, 164
Ricardo, David, 102–103
Riedel, Eibe, 250–251
Rifkin, Jeremy, 216
right to food, 87. *See also* food security; food sovereignty; neoliberal food regime
  CESCR and, 111
  as counter-hegemonic strategy, 144–149
  in voluntary guidelines, 134
  extraterritorial obligations, 113
  food sovereignty and, 144–149
  global hunger and, 94–96
    reduction strategies for, 94
  under ICESCR, 110–115
  under international law, 109–144
    Draft International Code of Conduct, 118–120
    General Comment 12, by CESCR, 120–122
    through international treaties, 111
    legal basis for, 110–115
    WFS and, 115–118, 122–126
  limitations of, 142–144
  peasant rights
    Draft Declaration on the Rights of Peasants, 142–143
    MST, 149
  peasant rights and, 142–144
  social movements for, 127
  in UDHR, 110
  La Vía Campesina campaign for, 140–142, 144–149
  voluntary guidelines on, 126–137
    counter-hegemonic approaches, 134
    development strategy for, 128
    formulation of, 134–135
    policy promotion as part of, 131–132
    state obligation in, 131–132
    during World Food Crisis of 2008, 137–140
    CESCR statement on, 139–140
right to health, 87, 166–193
  global drug gap and, 189
  ICESCR and, 167
  IPRs and, alternatives to, 188–193
  neoliberalism and, 167–168
  TRIPS and, 171–172, 193–195
    Doha Declaration, 179–183
    in Global South, 183–188
    in post-Doha landscape, 183–188
    reform of, 179–183
  under UDHR, 166–167
right to water, 87, 208–216, 245–247
  access to water and, 196–198
  as anthropocentric, 241–242
  CESCR and, 77–78
  commercialisation of, 211
  commodification of, 211
  as counter-hegemonic strategy, 239–245
  through litigation, 240
  under Declaration of Cochabamba, 214–216, 242
  under Declaration of Marrakech, 212
  under Dublin Principles, 210–211, 225
  in England and Wales, 230
  foundations of, 196–198
  in Global South, 203
  Global Water Regime and, 246
  as human right, 208
  under neoliberal regimes, 198–201, 210–212, 224–232
    criticisms of, 200–201
    TNCs and, 199

INDEX 299

water as commodity, 201–204, 206–207, 211
water as 'commons', 204–207
  WWC and, 199–200
  oppositional framework for, 212–216
  pro-poor policies for, 203
  PWC and, 204
  in South Africa, 232–238
    through FBW policy, 232–239
  TCC and, 204
  TNI and, 244
  in 20th Century, 208–210
  under UDHR, 208–210
  UN and, 216–224
    General Comments by, 217–222
    historical development of, 217
    Human Rights Council, 223
    international developments with, 222–224
  water justice movement and, 9, 227–232
  water shortages and, 196–198
  World Bank and, 225–228
  WWC and, 199–200, 228–232
Right2Water movement, 247
rights. *See* human rights; new rights
Rights of Man, 59–60
Risse, Thomas, 257–258
Rodriguez-Garavito, Cesar A., 59
Rome Declaration on World Food Security, 115, 117–118
Rooney, Eoin, 251–252
Roosevelt, Franklin, 61
Rose, John, 143–144
Rothbard, Murray, 63
Rucket, Arne, 204
ruling class, hegemony by, 22–23

Santos, Boaventura de Sousa, 59
SAPRIN. *See* Structural Adjustment Participatory Review Initiative Network
SAPs. *See* structural adjustment programs
Schanbacher, William, 107–108
Schutter, Olivier de, 143, 147–148
second generation socioeconomic rights, 61

self-determination rights, 61
Shamir, Ronen, 46, 182–183
Sikkink, Kathryn, 45, 257–258
Smith, Adam, 27
social change, human rights and, 55–60
  constructivist perspective on, 55–59
  realist perspective on, 55–59
  TNHB and, 57
social constructivism, human rights and, 53–54
  neo-Gramscianism and, 53–55
socioeconomic rights. *See also* human rights; new rights
  boomerang model for, 257–258
  CESCR and, 6–7, 50, 67–74
  counter-hegemonic potential of, 79–93, 252–257
    as Immanent critique, 83–85
    normative obligations and, 81–83
    through participation in intergovernmental forums, 252–254
    universalism and, 80–81
  defined, 61–62
  as ESC rights, 61–62
  global justice movements and, 9–10
  globalisation of, 74–79
  under ICESCR, 6–7, 68–73
  IFIs and, 90–93
  Immanent critique of, 249
  under international law, development of, 67–79
    standards clarification, 68–74
    UN human rights framework for, 74–79
  international legal standards for, 248
  jurisprudence of, 254–255
  limitations of, 85–93
    economic liberalism and, 85–88
    human rights failures, 90–93
    narrow legalism, 88–90
  neo-Gramscian approach to, 10–11
  neoliberalism and, 61–67
    conflicts between, 62–67
  origins of, 61
  public discourse on, 248–251
  purpose of, 64
  PWC on, 78–79

socioeconomic rights. (cont.)
    SAPs and, 65–66
    second generation, 61
    subaltern counterpublics and, 256–257
    subsistence rights, 62
    as subsistence rights, 62
    TNCs and, 90–93
    under UDHR, 68
    World Bank and, 87–88
    WTO and, 66
soft law, 52
solidarity rights, 61
South Africa
    FBW policy in, 232–238
        CAWP and, 236
        historical development of, 232–234
        Mazibuko ruling and, 236–238
        problems with, 234–236, 239
    global access to medicine movement in, 169–170
    right to water in, 232–238
South Asian Peasants Coalition, 127
Southern Common Market (MERCUSOR), 34
The State and Revolution (Lenin), 20
state formation, 28
state power. *See also* hegemony
    defined, 20
    Gramsci on, 20–22
    neoliberalism and, 26–27
structural adjustment measures, in neoliberal food regime, 98–101
Structural Adjustment Participatory Review Initiative Network (SAPRIN), 65
structural adjustment programs (SAPs), 65–66, 75–76
subaltern counterpublics, 256–257
subsistence rights, 62

TAC. *See* Treatment Action Campaign
TANs. *See* transnational advocacy networks
TCC. *See* transnational capitalist class
Thatcher, Margaret, 36–37
TNCs. *See* transnational corporations

TNHB. *See* transnational neo-liberal hegemonic bloc
TNI. *See* Transnational Institute
TNLB. *See* transnational neo-liberal historical bloc
tragedy of the commons theory, 160
Transatlantic Trade and Investment Partnership (TTIP), 4–5, 185
transformismo process, 23, 40
    Rome Declaration on World Food Security, 118
transnational advocacy networks (TANs), 45
transnational capitalist class (TCC), 30, 49
    global justice movements and, 42–43, 47
    IPRs and, 156–157
    right to water and, 204
transnational corporations (TNCs), 29
    neoliberal food regime and, 97, 100–101
    right to water and, 199
    socioeconomic rights and, 90–93
Transnational Institute (TNI), 244
transnational neo-liberal hegemonic bloc (TNHB), 57
transnational neo-liberal historical bloc (TNLB), 47–48
Trans-Pacific Partnership Agreement (TTPA), 185
Treatment Action Campaign (TAC), 169–170
TRIMs. *See* Agreement on Trade-Related Investment Measures
TRIPS. *See* Agreement on Trade-Related Intellectual Property Rights
TTIP. *See* Transatlantic Trade and Investment Partnership
TTPA. *See* Trans-Pacific Partnership Agreement

UDHR. *See* Universal Declaration of Human Rights
UN. *See* United Nations
UN Conference on Trade and Development (UNCTD), 37

UNCTC. *See* United Nations Centre on Transnational Corporations
UNCTD. *See* UN Conference on Trade and Development
UNDP. *See* United Nations Development Programme
unemployment, global increases in, 3
United Nations (UN). *See also* human rights regimes
  in global civil society, 37
  new rights and, advocacy through, 6–9, 37
  right to water and, 216–224
    General Comments on, 217–222
    historical development of, 217
    Human Rights Council, 223
    international developments with, 222–224
United Nations Centre on Transnational Corporations (UNCTC), 37
United Nations Development Programme (UNDP), 2
United States (US)
  IPR standards in, 156
  PWC and, 38–39, 47–48
  on socioeconomic rights, 78–79
  Washington Consensus and, 3–4, 33–34, 204
Universal Declaration of Human Rights (UDHR), 64
  right to food in, 110
  right to health under, 166–167
  right to water under, 208–210
  socioeconomic rights under, 68
Universal Declaration on the Eradication of Hunger and Malnutrition, 116
universalism, 80–81
U.S. *See* United States

Von Hayek, Friedrich, 28

Wager, Hannu, 87
Wales. *See* England and Wales
WANAHR. *See* World Alliance for Nutrition and Human Rights

war of position, counter-hegemony and, 24–25, 44–45
Washington Consensus, 3–4, 33–34, 204
water democracy, 227
water justice movement, 9, 227–232
water privatisation, 200–201, 212–216
  monopolies and, 205
  water as commodity, 201–204, 206–207, 211
water shortages, 196–198
WBI. *See* World Bank Institute
Wealth of Nations (Smith), 27
WFS. *See* World Food Summit
WFS:fyl. *See* World Food Summit: Five Years Later
WHA. *See* World Health Assembly
Whelan, Daniel, 69
WHO. *See* World Health Organization
Williams, Patricia, 58
Woodiwiss, Alan, 60
World Alliance for Nutrition and Human Rights (WANAHR), 118–119
World Bank, 3
  CESCR critique of, 76–77
  corporate power as influence on, 4
  in global political society, 33–34
  global poor and, 1–2
  PRSPs and, 39–40
  right to water and, 225–228
  SAPs and, 65–66
  socioeconomic rights and, 87–88
World Bank Institute (WBI), 36–37
World Food Crisis of 2008 137–140
  CESCR statement on, 139–140
World Food Summit (WFS), in Rome (1996), 115–118
  Rome Declaration on World Food Security, 115, 117–118
World Food Summit: Five Years Later (WFS:fyl), 122–126
World Forum of Fisher Peoples, 127
World Health Assembly (WHA), 164, 191–193
World Health Organization (WHO), 6–7
  MRDT and, 189–191
  RDS of, 164
world hegemony, 56

world hunger. *See* global hunger
World Social Forum (WSF), 7–8, 45
World Trade Organisation (WTO)
    CESCR critique of, 76–77
    corporate power as influence on, 4
    in global political society, 33–34
    global poor and, 1–2
    IPR protections by, 152
    in neoliberal food regime, 97–98
    socioeconomic rights and, 66
    TRIPS defended by, 174–176

World Water Council (WWC), 199–200, 228–232
World Water Forum (WWF), 228–232
WSF. *See* World Social Forum
WTO. *See* World Trade Organisation
WWC. *See* World Water Council
WWF. *See* World Water Forum

Young, Katherine, 179

Zeigler, Jean, 126, 147, 218, 253

Lightning Source UK Ltd.
Milton Keynes UK
UKHW010606131218
333930UK00018B/645/P

9 781316 628249